STUDENT SOLUTIONS MANUAL

Richard N. Aufmann
Palomar College

Vernon C. Barker
Palomar College

Joanne S. Lockwood
Plymouth State University

INTRODUCTORY ALGEBRA: AN APPLIED APPROACH

SEVENTH EDITION

Aufmann/Barker/Lockwood

HOUGHTON MIFFLIN COMPANY BOSTON NEW YORK

Editor-in-Chief: Jack Shira
Senior Sponsoring Editor: Lynn Cox
Associate Editor: Melissa Parkin
Editorial Assistant: Noel Kamm
Editorial Assistant: Julia Keller
Assistant Manufacturing Coordinator: Karmen Chong
Senior Marketing Manager: Ben Rivera
Marketing Associate: Lisa Lawler

Printed in the U.S.A.

ISBN: 0-618-52028-7

3 4 5 6 7 8 9-CRS-09 08 07 06 05

Contents

Chapter 1: Prealgebra Review 1

Chapter 2: Variable Expressions 18

Chapter 3: Solving Equations 26

Chapter 4: Polynomials 60

Chapter 5: Factoring 71

Chapter 6: Rational Expressions 103

Chapter 7: Linear Equations in Two Variables 137

Chapter 8: Systems of Linear Equations 151

Chapter 9: Inequalities 175

Chapter 10: Radical Expressions 187

Chapter 11: Quadratic Equations 202

Final Exam 229

Student Solutions Manual

Chapter 1: Prealgebra Review

Prep Test

1. 127.16

2. 46,514

3. 4517

4. 11,396

5. 508

6. 24

7. 4

8. $3 \cdot 7$

9. $\frac{2}{5}$

10. d

Go Figure

Place 3 coins on each balance pan. If they balance, the fake coin is among the remaining 3 coins. If it does not balance, then the fake coin is one of the three coins on the side that goes up. From the three coins that contain the fake coin, place one coin on one side of the balance and a second coin on the other side of the balance. If they balance, the third coin is counterfeit. If they do not balance, then the side that goes up has the counterfeit coin.

Section 1.1

Objective A Exercises

1. The natural numbers are the positive integers and do not include 0 or the negative integers.

3. $8 > -6$

5. $-12 < 1$

7. $42 > 19$

9. $0 > -31$

11. $53 > -46$

13. False

15. True

17. False

19. True

21. True

23. $A = \{1, 2, 3, 4, 5, 6, 7, 8\}$

25. $A = \{1, 2, 3, 4, 5, 6, 7, 8\}$

27. $A = \{-6, -5, -4, -3, -2, -1\}$

29. $-7 < 2$
$0 < 2$
$2 = 2$
$5 > 2$
The element 5 is greater than 2.

31. $-23 < -8$
$-18 < -8$
$-8 = -8$
$0 > -8$
The elements -23 and -18 are less than -8.

33. $-35 < -10$
$-13 < -10$
$21 > -10$
$37 > -10$
The elements 21 and 37 are greater than -10.

35. $-52 < 0$
$-46 < 0$
$0 = 0$
$39 > 0$
$58 > 0$
The elements -52, -46, and 0 are less than or equal to 0.

37. $-23 < -17$
$-17 = -17$
$0 > -17$
$4 > -17$
$29 > -17$
The elements $-17, 0, 4$, and 29 are greater than or equal to -17.

39. $1 < 5$
$2 < 5$
$3 < 5$
$4 < 5$
$5 = 5$
$6 > 5$
$7 > 5$
$8 > 5$
$9 > 5$
The elements 5, 6, 7, 8, and 9 are greater than or equal to 5.

41. $-10 < -4$
$-9 < -4$
$-8 < -4$
$-7 < -4$
$-6 < -4$
$-5 < -4$
$-4 = -4$
$-3 > -4$
$-2 > -4$
$-1 > -4$
The elements -10, -9, -8, -7, -6, and -5 are less than -4.

Objective B Exercises

43. -4

45. 9

47. 36

49. 40

51. -39

53. 74

55. -82

57. -81

59. $|22| > |-19|$

61. $|-71| < |-92|$

63. $|12| < |-31|$

65. $|-28| < |43|$

67. **a.** $-(-11) = 11$
$-(-7) = 7$
$-(-3) = 3$
$-(1) = -1$
$-(5) = -5$

b. $|-11| = 11$
$|-7| = 7$
$|-3| = 3$
$|1| = 1$
$|5| = 5$

Applying the Concepts

69. Never true

Section 1.2

Objective A Exercises

1. Add the absolute values. The sign is the sign of the addends.

3. -11

5. -5

7. -83

9. -46

11. 0

13. -5

15. $-17 + (-3) + 29 = -20 + 29$
$= 9$

17. $-3 + (-8) + 12 = -11 + 12$
$= 1$

19. $13 + (-22) + 4 + (-5) = -9 + 4 + (-5)$
$= -5 + (-5)$
$= -10$

21. $-22 + 20 + 2 + (-18) = -2 + 2 + (-18)$
$= 0 + (-18)$
$= -18$

23. $-16 + (-17) + (-18) + 10 = -33 + (-18) + 10$
$= -51 + 10$
$= -41$

25. $26 + (-15) + (-11) + (-12) = 11 + (-11) + (-12)$
$= 0 + (-12)$
$= -12$

27. $-17 + (-18) + 45 + (-10) = -35 + 45 + (-10)$
$= 10 + (-10)$
$= 0$

29. $46 + (-17) + (-13) + (-50) = 29 + (-13) + (-50)$
$= 16 + (-50)$
$= -34$

31. $-14 + (-15) + (-11) + 40 = -29 + (-11) + 40$
$= -40 + 40$
$= 0$

33. $-23 + (-22) + (-21) + 5 = -45 + (-21) + 5$
$= -66 + 5$
$= -61$

35. $-42 + (-23) = -65$

37. $-31 + 16 = -15$

39. $-17 + (-23) + 43 + 19 = -40 + 43 + 19$
$= 3 + 19$
$= 22$

Objective B Exercises

41. *Minus* is the operation subtraction; *negative* indicates the opposite of a number.

43. $16 - 8 = 16 + (-8) = 8$

45. $7 - 14 = 7 + (-14) = -7$

47. $-7 - 2 = -7 + (-2) = -9$

49. $7 - (-2) = 7 + 2 = 9$

51. $-6 - (-3) = -6 + 3 = -3$

53. $6 - (-12) = 6 + 12 = 18$

55.
$$\begin{aligned}-4 - 3 - 2 &= -4 + (-3) + (-2)\\ &= -7 + (-2)\\ &= -9\end{aligned}$$

57.
$$\begin{aligned}12 - (-7) - 8 &= 12 + 7 + (-8)\\ &= 19 + (-8)\\ &= 11\end{aligned}$$

59.
$$\begin{aligned}-19 - (-19) - 18 &= -19 + 19 + (-18)\\ &= 0 + (-18)\\ &= -18\end{aligned}$$

61.
$$\begin{aligned}-17 - (-8) - (-9) &= -17 + 8 + 9\\ &= -9 + 9\\ &= 0\end{aligned}$$

63.
$$\begin{aligned}-30 - (-65) - 29 - 4 &= -30 + 65 + (-29) + (-4)\\ &= 35 + (-29) + (-4)\\ &= 6 + (-4)\\ &= 2\end{aligned}$$

65.
$$\begin{aligned}-16 - 47 - 63 - 12 &= -16 + (-47) + (-63) + (-12)\\ &= -63 + (-63) + (-12)\\ &= -126 + (-12)\\ &= -138\end{aligned}$$

67.
$$\begin{aligned}-47 - (-67) - 13 - 15 &= -47 + 67 + (-13) + (-15)\\ &= 20 + (-13) + (-15)\\ &= 7 + (-15)\\ &= -8\end{aligned}$$

69.
$$\begin{aligned}-19 - 17 - (-36) - 12 &= -19 + (-17) + 36 + (-12)\\ &= -36 + 36 + (-12)\\ &= 0 + (-12)\\ &= -12\end{aligned}$$

71.
$$\begin{aligned}21 - (-14) - 43 - 12 &= 21 + 14 + (-43) + (-12)\\ &= 35 + (-43) + (-12)\\ &= -8 + (-12)\\ &= -20\end{aligned}$$

73.
$$\begin{aligned}-21 - (-36) &= -21 + 36\\ &= 15\end{aligned}$$

75.
$$\begin{aligned}-27 - 12 &= -27 + (-12)\\ &= -39\end{aligned}$$

77.
$$\begin{aligned}-21 - (-37) &= -21 + 37\\ &= 16\end{aligned}$$

Objective C Exercises

79. Strategy To find the difference, subtract the temperature at which mercury boils (360°C) from the temperature at which mercury freezes (−39°C).

Solution
$$\begin{aligned}360 - (-39) &= 360 + 39\\ &= 399°C\end{aligned}$$
The difference in temperature is 399°C.

81. Strategy To find the difference, subtract the elevation of Death Valley (−86 m) from the elevation of Mt. Aconcagua (6960 m).

Solution
$$\begin{aligned}6960 - (-86) &= 6960 + 86\\ &= 7046\text{ m}\end{aligned}$$
The difference in elevation is 7046 m.

83. Strategy To find which continent has the greatest difference in elevation: Find the difference in elevation for each continent and then compare the differences.

Solution

Africa: $5895 - (-133) = 5895 + 133 = 6028$ m

Asia: $8850 - (-400) = 8850 + 400 = 9250$ m

Europe: $5634 - (-28) = 5634 + 28 = 5662$ m

America: $6960 - (-86) = 6960 + 86 = 7046$ m

The continent with the greatest difference between the highest and lowest elevation is Asia.

85. Strategy To find the difference, subtract the melting point of mercury (−39°C) from the boiling point of mercury (357°C).

Solution
$$\begin{aligned}357 - (-39) &= 357 + 39\\ &= 396°C\end{aligned}$$
The difference in temperature is 396°C.

87. Strategy To determine if Mt. Everest could fit in the Tonga Trench, compare the height of Mt. Everest (8850 m) with the absolute value of the depth of the Tonga Trench (−10,630 m).

Solution $8850 < (-10{,}630)$

Yes. Mt. Everest could fit in the Tonga Trench.

89. Strategy To find the difference, subtract the depth of the Mariana Trench (−11,520 m) from the height of Mt. Everest (8850 m).

Solution
$$\begin{aligned}&8850 - (-11{,}520)\\ &= 8850 + 11{,}520\\ &= 20{,}370\end{aligned}$$
The difference is 20,370 m.

91. Strategy To find the difference in the wind chill factor, subtract the wind chill factor when the wind is blowing at 25 mph (−51°F) from the wind chill factor when the wind is blowing at 15 mph (−45°F).

Solution $-45 - (-51) = 6°F$
The difference in wind chill factor is 6°F.

Applying the Concepts

93. No. For example, the difference between 10 and −8 is 18, which is greater than either 10 or −8.

Section 1.3

Objective A Exercises

1. Multiply the absolute values. The answer is positive.

3. 42

5. 60

7. −253

9. −114

11. $7(5)(-3) = 35(-3)$
$= -105$

13. $-3(-8)(-9) = 24(-9)$
$= -216$

15. $(-9)7(5) = -63(5)$
$= -315$

17. $(-3)7(-2)8 = -21(-2)8$
$= (42)8$
$= 336$

19. $7(9)(-11)4 = 63(-11)4$
$= (-693)4$
$= -2772$

21. $(-14)9(-11)0 = -126(-11)0$
$= (1386)0$
$= 0$

23. $(-14)(-25) = 350$

25. $16(-21) = -336$

27. $4(-8)(11) = -32(11)$
$= -352$

Objective B Exercises

29. −2

31. 8

33. −7

35. −12

37. −6

39. 11

41. −14

43. 15

45. −16

47. 0

49. −29

51. undefined

53. −11

55. undefined

57. $(-132) \div (11) = -12$

59. $-96 \div (-4) = 24$

61. $-196 \div (-7) = 28$

Objective C Exercises

63. Strategy To find the average daily high temperature:
- Add the six temperature readings.
- Divide by 6.

Solution

$$\begin{aligned} -23 + (-29) + (-21) + (-28) + (-28) + (-27) &= -52 + (-21) + (-28) + (-28) + (-27) \\ &= -73 + (-28) + (-28) + (-27) \\ &= -101 + (-28) + (-27) \\ &= -129 + (-27) \\ &= -156 \end{aligned}$$

$= -156 \div 6 = -26$

The average daily temperature was −26°F.

65. Strategy To find the five-day moving average, determine the average of the stock for days 1 through 5, 2 through 6, 3 through 7, and so on.

Solution

Day 1–5	Day 2–6	Day 3–7
−100	+20	+0
+20	+0	+30
+0	+30	−10
+30	−10	+100
−10	+100	+ 80
Sum = −60	Sum = +140	Sum = +200

$\text{Avg} = \frac{-60}{5} = -12$ $\text{Avg} = \frac{+140}{5} = +28$ $\text{Avg} = \frac{+200}{5} = +40$

Day 4–8	Day 5–9	Day 6–10
+30	−10	+100
−10	+100	+80
+100	+80	−60
+80	−60	+130
− 60	+130	+ 10
Sum = +140	Sum = +240	Sum = +260

$\text{Avg} = \frac{+140}{5} = +28$ $\text{Avg} = \frac{+240}{5} = +48$ $\text{Avg} = \frac{+260}{5} = +52$

The five-day moving average is −12, 28, 40, 28, 48, 52.

67. Strategy To find the score:
- Multiply the number of correct answers by 5.
- Multiply the number of incorrect numbers by −2.
- Add the results.

Solution $(20)(5) = 100$
$(13)(-2) = -26$
$100 + (-26) = 74$
The student's score was 74.

Applying the Concepts

69.

x	$-3x$
−6	$-3(-6) = 18$
−2	$-3(-2) = 6$
7	$-3(7) = -21$

The expression $-3x$ is greatest for $x = -6$.

71. The product of two negative integers is a positive number. Because −4 is a negative number, it must be multiplied by a negative integer for the product to be positive. Therefore, if $-4x$ equals a positive integer, then x must be a negative integer.

Section 1.4

Objective A Exercises

1. $6^2 = 6 \cdot 6 = 36$

3. $-7^2 = -(7 \cdot 7) = -49$

5. $(-3)^2 = (-3)(-3) = 9$

7. $(-3)^4 = (-3)(-3)(-3)(-3) = 81$

9. $-4^4 = -(4 \cdot 4 \cdot 4 \cdot 4) = -256$

11. $2 \cdot (-3)^2 = 2 \cdot (-3)(-3) = 2 \cdot 9 = 18$

13. $(-1)^9 \cdot 3^3 = -1 \cdot (3)(3)(3)$
$= -1 \cdot 27$
$= -27$

15. $(3)^3 \cdot 2^3 = (3)(3)(3) \cdot 2 \cdot 2 \cdot 2$
$= (27) \cdot 8$
$= 216$

17. $(-3) \cdot 2^2 = (-3) \cdot 2 \cdot 2 = -3 \cdot 4 = -12$

19. $(-2) \cdot (-2)^3 = -2 \cdot (-2)(-2)(-2) = 16$

21. $2^3 \cdot 3^3 \cdot (-4) = 2 \cdot 2 \cdot 2 \cdot 3 \cdot 3 \cdot 3 \cdot (-4)$
$= 8 \cdot 27 \cdot (-4)$
$= 216(-4) = -864$

23. $(-7) \cdot 4^2 \cdot 3^2 = (-7) \cdot 4 \cdot 4 \cdot 3 \cdot 3$
$= (-7) \cdot 16 \cdot 9$
$= -112 \cdot 9 = -1008$

25. $-3^2 \cdot (-3)^2 = -3 \cdot 3 \cdot (-3)(-3)$
$= -9 \cdot 9$
$= -81$

27. $8^2 \cdot (-3)^5 \cdot 5 = 8 \cdot 8 \cdot (-3)(-3)(-3)(-3)(-3) \cdot 5$
$= 64 \cdot (-243) \cdot 5$
$= -15{,}552 \cdot 5$
$= -77{,}760$

Objective B Exercises

29.
1. Perform operations within grouping symbols.
2. Simplify exponential expressions.
3. Perform multiplications and divisions from left to right.
4. Perform additions and subtractions from left to right.

31. $2^2 \cdot 3 - 3 = 4 \cdot 3 - 3$
$= 12 - 3$
$= 9$

33. $16 - 32 \div 2^3 = 16 - 32 \div 8$
$= 16 - 4$
$= 12$

35. $8 - (-3)^2 - (-2) = 8 - 9 - (-2)$
$= 8 - 9 + 2$
$= -1 + 2$
$= 1$

37. $16 - 16 \cdot 2 \div 4 = 16 - 32 \div 4$
$= 16 - 8$
$= 8$

39. $16 - 2 \cdot 4^2 = 16 - 2 \cdot 16$
$= 16 - 32$
$= -16$

41. $4 + 12 \div 3 \cdot 2 = 4 + 4 \cdot 2$
$= 4 + 8$
$= 12$

43. $14 - 2^2 - (4 - 7) = 14 - 2^2 - (-3)$
$= 14 - 4 - (-3)$
$= 10 + 3$
$= 13$

45. $-2^2 + 4[16 \div (3 - 5)] = -2^2 + 4[16 \div (-2)]$
$= -2^2 + 4[-8]$
$= -4 + 4[-8]$
$= -4 + [-32]$
$= -4 - 32$
$= -36$

47. $24 \div \dfrac{3^2}{8-5} - (-5) = 24 \div \dfrac{9}{3} - (-5)$
$= 24 \div 3 - (-5)$
$= 8 - (-5)$
$= 8 + 5$
$= 13$

49. $4[16-(7-1)]\div 10 = 4[16-6]\div 10$
$= 4[10]\div 10$
$= 40\div 10$
$= 4$

51. $18\div(9-2^3)+(-3) = 18\div(9-8)+(-3)$
$= 18\div(1)+(-3)$
$= 18+(-3)$
$= 18-3$
$= 15$

53. $4(-8)\div[2(7-3)^2] = 4(-8)\div[2(4)^2]$
$= 4(-8)\div[2(16)]$
$= 4(-8)\div[32]$
$= -32\div 32$
$= -1$

55. $16-4\cdot\dfrac{3^3-7}{2^3+2}-(-2)^2 = 16-4\cdot\dfrac{27-7}{8+2}-4$
$= 16-4\cdot\dfrac{20}{10}-4$
$= 16-4\cdot 2-4$
$= 16-8-4$
$= 8-4$
$= 4$

57. $-18\div(-6)\div(1-(-2)) = -18\div(-6)\div(1+2)$
$= -18\div(-6)\div 3$
$= 3\div 3$
$= 1$

59. $(8-3^2)^{10}+(2\cdot 3-7)^{11} = (8-9)^{10}+(6-7)^{11}$
$= (-1)^{10}+(-1)^{11}$
$= 1+(-1)$
$= 0$

61. $-6^2\cdot 3-2^2(1-5)^2 = -6^2\cdot 3-2^2(-4)^2$
$= -36\cdot 3-4\cdot 16$
$= -108-64$
$= -172$

Applying the Concepts

63. The Order of Operations Agreement was not followed in the given simplification of $6+2(4-9)$ because the addition $6+2$ was performed before the multiplication $2(-5)$. The expression should be simplified as follows: $6+2(4-9) = 6+2(-5)$ Perform operations inside parentheses $= -4$ Do the multiplication Do the addition.

Section 1.5

Objective A Exercises

1. 1, 2, 4

3. 1, 2, 3, 4, 6, 12

5. 1, 2, 4, 8

7. 1, 13

9. 1, 2, 4, 7, 8, 14, 28, 56

11. 1, 3, 5, 9, 15, 45

13. 1, 29

15. 1, 2, 4, 13, 26, 52

17. 1, 2, 41, 82

19. 1, 3, 19, 57

21. 1, 2, 3, 4, 6, 8, 12, 16, 24, 48

23. 1, 2, 5, 10, 25, 50

25. 1, 7, 11, 77

27. 1, 2, 4, 5, 10, 20, 25, 50, 100

29. 1, 5, 17, 85

Objective B Exercises

31. $2\cdot 7$

33. $2^3\cdot 3^2$

35. $2^3\cdot 3$

37. $2^2\cdot 3^2$

39. $2\cdot 13$

41. 7^2

43. Prime

45. $2\cdot 31$

47. Prime

49. $2\cdot 43$

51. $5\cdot 19$

53. $2\cdot 3\cdot 13$

55. $2^4\cdot 3^2$

57. $5^2\cdot 7$

59. $2^4\cdot 5^2$

Objective C Exercises

61. 24

63. 12

65. 36

67. 140

69. 36

71. 240

73. 720

75. 216

77. 360

79. 160

81. 24

83. 30

85. 72

87. 150

89. 108

91. 2

93. 1

95. 2

97. 6

99. 4

101. 10

103. 8

105. 7

107. 30

109. 1

111. 2

113. 6

115. 12

117. 26

119. 18

Applying the Concepts

121. Answers may vary. One possibility is 3, 5; 5, 7; 11, 13.

123. 2 is the only even prime number because all other even numbers have 2 as a factor.

125. The product of the two numbers.

127. Divide the product of the two numbers by the GCF of the numbers.

Section 1.6

Objective A Exercises

1. $\frac{1}{3}$

3. $-\frac{4}{11}$

5. $-\frac{2}{3}$

7. $\frac{3}{2}$

9. 0

11. $-\frac{3}{5}$

13. $-\frac{7}{5}$

15. -15

17. $\frac{1}{2}$

19. 0.8

21. $0.1\overline{6}$

23. $-0.\overline{3}$

25. $-0.\overline{2}$

27. $-0.58\overline{3}$

29. $0.91\overline{6}$

31. $0.3\overline{8}$

33. 0.5625

35. $-0.\overline{857142}$

Objective B Exercises

37. $\frac{13}{12}$

39. $\frac{35}{24}$

41. $\frac{1}{24}$

43. $-\frac{29}{26}$

45. $-\frac{19}{12}$

47. $\frac{25}{18}$

49. $-\frac{7}{48}$

51. $-\frac{17}{24}$

53. 8.022

55. -25.4

57. -9.964

59. -2.84

61. -6.961

63. 0

65. -108.677

67. $-\frac{1}{12}$

69. $-\frac{7}{18}$

71. -5.55

Objective C Exercises

73. $-\frac{2}{27}$

75. $-\frac{1}{8}$

77. $-\frac{37}{24}$

79. $-\frac{23}{18}$

81. $\frac{73}{60}$

83. $\frac{29}{16}$

85. -1

87. $\frac{143}{72}$

89. -2.801

91. 2.088

93. 21.16

95. 0

97. 14.452

99. -17.606

101. 4.001

103. $-\frac{29}{24}$

105. $-\frac{1}{12}$

107. $-\frac{7}{18}$

109. Strategy To list the coins from thinnest to thickest, order thickness measures given in the table from smallest to largest.

Solution Dime: 1.35 mm
Penny: 1.55 mm
Quarter: 1.75 mm
Nickel: 1.95 mm
From thinnest to thickest, the coins are dime, penny, quarter, nickel.

111. Strategy To find the sum of the weights of the four coins, add the weights given in the table.

Solution $2.5 + 5 + 2.268 + 5.67 = 15.438$
The sum of the weights of the four coins is 15.438 g.

113. Strategy To find the largest difference in diameter between two coins, subtract the smallest diameter length (17.91) from the largest diameter length (24.26).

Solution $24.26 - 17.91 = 6.35$
The largest difference in diameter between two coins is 6.35 mm.

115. Strategy To find the additional broth needed, subtract $\frac{2}{3}$ c from $\frac{3}{4}$ c.

Solution $\frac{3}{4} - \frac{2}{3} = \frac{9}{12} - \frac{8}{12} = \frac{1}{12}$

$\frac{1}{12}$ c of additional broth is needed.

117. Strategy To determine the years that Apple Computer had negative earnings, read the years associated with negative values on the graph.

Solution According to the graph, earnings were negative in 1996, 1997, and 2001. Apple Computer had negative earnings per share in 1996, 1997, and 2001.

119. Strategy To find the decrease in earnings per share between 1996 and 1997, subtract the earnings per share for 1997 (−$8.29) from the earnings per share in 1996 (−$6.59).

Solution $-6.59 - (-8.29) = 1.70$
Between 1996 and 1997, the decrease in earnings per share was $1.70.

121. Strategy To find the number of barrels of oil per day consumed by the five countries, add the five numbers given that represent the consumption in millions of barrels.

Solution $19.7 + 5.3 + 5.0 + 2.7 + 2.2 = 34.9$
The five countries consume 34.9 million barrels of oil per day.

123. Strategy To find the difference between the numbers of millions of barrels of oil consumed and the number imported, find the difference between the sum of the five numbers representing consumption and the sum of the five numbers representing what is imported.

Solution Consumed:
$19.7 + 5.3 + 5.0 + 2.7 + 2.2 = 34.9$
Imported:
$11.1 + 5.6 + 1.6 + 2.6 + 2.5 = 23.4$
$34.9 - 23.4 = 11.5$
The difference between the number of barrels of oil consumed per day and the number imported is 11.5 million barrels.

125. The LCM of 2, 6, and 8 is 24

$$\frac{1}{2} - \frac{5}{6} + \frac{7}{8} = \frac{12}{24} - \frac{20}{24} + \frac{21}{24} = \frac{12 - 20 + 21}{24} = \frac{13}{24}$$

127. The LCM of 4, 3, and 8 is 24.

$$-\frac{3}{4} - \left(-\frac{2}{3}\right) + \left(-\frac{1}{8}\right) = -\frac{18}{24} - \left(-\frac{16}{24}\right) + \left(-\frac{3}{24}\right) = \frac{-18 - (-16) - 3}{24} = \frac{-18 + 16 - 3}{24} = -\frac{5}{24}$$

129. The LCM of 5, 4, and 3 is 60.

$$\frac{4}{5} - \left(\frac{1}{4} - \left(-\frac{2}{3}\right)\right) = \frac{48}{60} - \left(\frac{15}{60} - \left(-\frac{40}{60}\right)\right) = \frac{48}{60} - \left(\frac{15 + 40}{60}\right) = \frac{48}{60} - \frac{55}{60} = \frac{48 - 55}{60} = -\frac{7}{60}$$

131. $\frac{1}{4}-\frac{1}{3}=\frac{3}{12}-\frac{4}{12}=-\frac{1}{12}=-0.08\overline{3}$

$0.25-0.33=-0.08$

So, $\frac{1}{4}-\frac{1}{3}<0.25-0.33$

133. $0.15-\frac{3}{4}=0.15-0.75=-0.6$

$0.75-\frac{27}{20}=0.75-1.35=-0.6$

So, $0.15-\frac{3}{4}=0.75-\frac{27}{20}$

135. $\frac{1}{9}$

137. $\frac{5}{12}$

139. 2, 5

141. larger

Section 1.7

Objective A Exercises

1. $\frac{2}{3}\cdot\frac{5}{7}=\frac{2\cdot 5}{3\cdot 7}=\frac{10}{21}$

3. $\frac{5}{8}\cdot\left(-\frac{3}{10}\right)=-\frac{5}{8}\cdot\frac{3}{10}=-\frac{\overset{1}{\cancel{5}}\cdot 3}{2\cdot 2\cdot 2\cdot 2\cdot \underset{1}{\cancel{5}}}=-\frac{3}{16}$

5. $\frac{5}{12}\left(-\frac{3}{10}\right)=-\frac{5}{12}\cdot\frac{3}{10}=-\frac{\overset{1}{\cancel{5}}\cdot\overset{1}{\cancel{3}}}{2\cdot 2\cdot\underset{1}{\cancel{3}}\cdot 2\cdot\underset{1}{\cancel{5}}}=-\frac{1}{8}$

7. $\frac{6}{13}\left(-\frac{26}{27}\right)=-\frac{6}{13}\cdot\frac{26}{27}=-\frac{6\cdot 26}{13\cdot 27}=-\frac{2\cdot\overset{1}{\cancel{3}}\cdot 2\cdot\overset{1}{\cancel{13}}}{\underset{1}{\cancel{13}}\cdot\underset{1}{\cancel{3}}\cdot 3\cdot 3}=-\frac{4}{9}$

9. $\left(-\frac{3}{5}\right)\left(-\frac{3}{10}\right)=\frac{3}{5}\cdot\frac{3}{10}=\frac{3\cdot 3}{5\cdot 10}=\frac{9}{50}$

11. $\left(-\frac{3}{4}\right)^2=\left(-\frac{3}{4}\right)\left(-\frac{3}{4}\right)=\frac{3}{4}\cdot\frac{3}{4}=\frac{3\cdot 3}{4\cdot 4}=\frac{9}{16}$

13. $\left(-\frac{3}{4}\right)\frac{5}{6}\left(-\frac{2}{9}\right)=\frac{3}{4}\cdot\frac{5}{6}\cdot\frac{2}{9}=\frac{\overset{1}{\cancel{3}}\cdot 5\cdot\overset{1}{\cancel{2}}}{\underset{1}{\cancel{2}}\cdot 2\cdot 2\cdot\underset{1}{\cancel{3}}\cdot 3\cdot 3}=\frac{5}{36}$

15. $\left(-\frac{3}{8}\right)\left(-\frac{5}{12}\right)\left(\frac{3}{10}\right)=\frac{3}{8}\cdot\frac{5}{12}\cdot\frac{3}{10}=\frac{3\cdot 5\cdot 3}{8\cdot 12\cdot 10}=\frac{\overset{1}{\cancel{3}}\cdot\overset{1}{\cancel{5}}\cdot 3}{2\cdot 2\cdot 2\cdot 2\cdot 2\cdot\underset{1}{\cancel{3}}\cdot 2\cdot\underset{1}{\cancel{5}}}=\frac{3}{64}$

17. $\left(-\frac{15}{2}\right)\left(-\frac{4}{3}\right)\left(-\frac{7}{10}\right)=-\frac{15}{2}\cdot\frac{4}{3}\cdot\frac{7}{10}=-\frac{15\cdot 4\cdot 7}{2\cdot 3\cdot 10}=-\frac{\overset{1}{\cancel{3}}\cdot\overset{1}{\cancel{5}}\cdot\overset{1}{\cancel{2}}\cdot\overset{1}{\cancel{2}}\cdot 7}{\underset{1}{\cancel{2}}\cdot\underset{1}{\cancel{3}}\cdot\underset{1}{\cancel{2}}\cdot\underset{1}{\cancel{5}}}=-\frac{7}{1}=-7$

19. $\left(\frac{8}{9}\right)\left(-\frac{11}{12}\right)\left(\frac{3}{4}\right)=-\frac{8}{9}\cdot\frac{11}{12}\cdot\frac{3}{4}=-\frac{\overset{1}{\cancel{2}}\cdot\overset{1}{\cancel{2}}\cdot\overset{1}{\cancel{2}}\cdot 11\cdot\overset{1}{\cancel{3}}}{3\cdot 3\cdot\underset{1}{\cancel{2}}\cdot\underset{1}{\cancel{2}}\cdot\underset{1}{\cancel{3}}\cdot\underset{1}{\cancel{2}}\cdot 2}=-\frac{11}{18}$

21. The product is negative: $0.46(-3.9)=-1.794$

23. The product is positive: $(-8.23)(-0.09) = 0.7407$

25. The product is negative: $-0.48(0.85) = -0.408$

27. The product is negative: $-6.5(0.0341) = -0.22165$

29. The product is positive: $(-8.004)(-3.4) = 27.2136$

31. The product is negative: $0.089(-1.098) = -0.097722$

33. $\left(-\frac{3}{4}\right)\left(\frac{4}{5}\right) = -\frac{3 \cdot 4}{4 \cdot 5} = -\frac{12}{20} = -\frac{3}{5}$

35. The product is positive: $(-0.23)(-4.5) = 1.035$

Objective B Exercises

37. $\frac{3}{8} \div \left(-\frac{9}{10}\right) = -\left(\frac{3}{8} \cdot \frac{10}{9}\right) = -\frac{\overset{1}{\cancel{3}} \cdot \overset{1}{\cancel{2}} \cdot 5}{\underset{1}{\cancel{2}} \cdot 2 \cdot 2 \cdot \underset{1}{\cancel{3}} \cdot 3} = -\frac{5}{12}$

39. $\left(-\frac{8}{9}\right) \div \left(-\frac{4}{5}\right) = \frac{8}{9} \cdot \frac{5}{4} = \frac{\overset{1}{\cancel{2}} \cdot \overset{1}{\cancel{2}} \cdot 2 \cdot 5}{3 \cdot 3 \cdot \underset{1}{\cancel{2}} \cdot \underset{1}{\cancel{2}}} = \frac{10}{9}$

41. $\left(-\frac{11}{12}\right) \div \left(-\frac{7}{6}\right) = \frac{11}{12} \cdot \frac{6}{7} = \frac{11 \cdot \overset{1}{\cancel{2}} \cdot \overset{1}{\cancel{3}}}{\underset{1}{\cancel{2}} \cdot 2 \cdot \underset{1}{\cancel{3}} \cdot 7} = \frac{11}{14}$

43. $\left(-\frac{6}{11}\right) \div 6 = -\frac{6}{11} \cdot \frac{1}{6} = -\frac{6}{66} = -\frac{1}{11}$

45. $\left(-\frac{11}{12}\right) \div \frac{5}{3} = -\frac{11}{12} \cdot \frac{3}{5} = -\frac{33}{60} = -\frac{11}{20}$

47. $\left(-\frac{5}{8}\right) \div \frac{15}{16} = -\frac{5}{8} \cdot \frac{16}{15} = -\frac{\overset{1}{\cancel{5}} \cdot 2 \cdot \overset{1}{\cancel{8}}}{\underset{1}{\cancel{8}} \cdot 3 \cdot \underset{1}{\cancel{5}}} = -\frac{2}{3}$

49. $\left(-\frac{7}{9}\right) \div \left(-\frac{5}{18}\right) = \frac{7}{9} \cdot \frac{18}{5} = \frac{7 \cdot 2 \cdot \overset{1}{\cancel{9}}}{\underset{1}{\cancel{9}} \cdot 5} = \frac{14}{5}$

51. $\frac{1}{2} \div \left(-\frac{1}{4}\right) = -\frac{1}{2} \cdot \frac{4}{1} = -\frac{2}{1} = -2$

53. The quotient is negative: $25.61 \div (-5.2) = -4.925$

55. The quotient is positive: $(-0.2205) \div (-0.21) = 1.05$

57. The quotient is negative: $-0.0647 \div 0.75 \approx -0.09$

59. The quotient is positive: $-2.45 \div (-21.44) \approx 0.11$

61. The quotient is positive: $(-75.469) \div (-77.8) \approx 0.970$

63. The quotient is positive: $-0.1142 \div (-17.2) \approx 0.007$

65. The quotient is positive: $(-0.3045) \div (-0.203) = 1.5$

67. The quotient is negative: $-0.00552 \div 1.2 = -0.0046$

69. $$\begin{aligned} \frac{2}{3} - \frac{1}{4}\left(-\frac{2}{5}\right) &= \frac{2}{3} - \left(-\frac{1}{4} \cdot \frac{2}{5}\right) \\ &= \frac{2}{3} - \left(-\frac{1}{10}\right) \\ &= \frac{2}{3} + \frac{1}{10} \\ &= \frac{20}{30} + \frac{3}{30} \\ &= \frac{23}{30} \end{aligned}$$

71. $$\begin{aligned} \frac{3}{4}\left(\frac{1}{2}\right)^2 - \frac{5}{16} &= \frac{3}{4} \cdot \frac{1}{4} - \frac{5}{16} \\ &= \frac{3}{16} - \frac{5}{16} \\ &= \frac{3-5}{16} \\ &= -\frac{2}{16} \\ &= -\frac{1}{8} \end{aligned}$$

73. $$\begin{aligned} \frac{7}{12} - \left(\frac{2}{3}\right)^2 + \left(-\frac{3}{4}\right) &= \frac{7}{12} - \frac{4}{9} + \left(-\frac{3}{4}\right) \\ &= \frac{21}{36} - \frac{16}{36} + \left(-\frac{3}{4}\right) \\ &= \frac{5}{36} + \left(-\frac{3}{4}\right) \\ &= \frac{5}{36} + \left(-\frac{27}{36}\right) \\ &= \frac{5-27}{36} \\ &= -\frac{22}{36} \\ &= -\frac{11}{18} \end{aligned}$$

75. $$\begin{aligned} \left(\frac{2}{3}\right)\left(\frac{3}{4}\right) - \left(\frac{4}{5}\right)\left(\frac{5}{8}\right) &= \frac{\overset{1}{\cancel{2}} \cdot \overset{1}{\cancel{3}}}{\underset{1}{\cancel{3}} \cdot \underset{1}{\cancel{2}} \cdot 2} - \left(\frac{4}{5}\right)\left(\frac{5}{8}\right) \\ &= \frac{1}{2} - \frac{4 \cdot 5}{5 \cdot 8} \\ &= \frac{1}{2} - \frac{1}{2} \\ &= 0 \end{aligned}$$

77. $-\frac{3}{4}\div\frac{5}{8}-\frac{4}{5}=-\frac{3}{4}\cdot\frac{8}{5}-\frac{4}{5}$
$=-\frac{24}{20}-\frac{4}{5}$
$=-\frac{6}{5}-\frac{4}{5}$
$=-\frac{10}{5}$
$=-2$

79. $\left(\frac{2}{3}\right)^3-\left(\frac{2}{3}\right)^2=\frac{8}{27}-\frac{4}{9}$
$=\frac{8}{27}-\frac{12}{27}$
$=\frac{8-12}{27}$
$=-\frac{4}{27}$

81. $1.2-2.3^2=1.2-5.29$
$=-4.09$

83. $0.03\cdot 0.2^2-0.5^3=0.03\cdot 0.04-0.125$
$=0.0012-0.125$
$=-0.1238$

85. $\frac{3.8-5.2}{-0.35}-\left(\frac{1.2}{0.6}\right)^2=\frac{-1.4}{-0.35}-2^2$
$=\frac{-1.4}{-0.35}-4$
$=4-4$
$=0$

Objective C Exercises

87. Replace the % symbol by $\frac{1}{100}$. Then multiply.

89. $75\%=75\left(\frac{1}{100}\right)=\frac{75}{100}=\frac{3}{4}$

$75\%=75(0.01)=0.75$

91. $64\%=64\left(\frac{1}{100}\right)=\frac{64}{100}=\frac{16}{25}$

$64\%=64(0.01)=0.64$

93. $175\%=175\left(\frac{1}{100}\right)=\frac{175}{100}=\frac{7}{4}$

$175\%=175(0.01)=1.75$

95. $19\%=19\left(\frac{1}{100}\right)=\frac{19}{100}$

$19\%=19(0.01)=0.19$

97. $5\%=5\left(\frac{1}{100}\right)=\frac{5}{100}=\frac{1}{20}$

$5\%=5(0.01)=0.05$

99. $11\frac{1}{9}\%=11\frac{1}{9}\left(\frac{1}{100}\right)=\frac{100}{9}\left(\frac{1}{100}\right)=\frac{1}{9}$

101. $12\frac{1}{2}\%=12\frac{1}{2}\left(\frac{1}{100}\right)=\frac{25}{2}\left(\frac{1}{100}\right)=\frac{1}{8}$

103. $66\frac{2}{3}\%=66\frac{2}{3}\left(\frac{1}{100}\right)=\frac{200}{3}\left(\frac{1}{100}\right)=\frac{2}{3}$

105. $\frac{1}{2}\%=\frac{1}{2}\left(\frac{1}{100}\right)=\frac{1}{200}$

107. $83\frac{1}{3}\%=83\frac{1}{3}\left(\frac{1}{100}\right)=\frac{250}{3}\left(\frac{1}{100}\right)=\frac{5}{6}$

109. $7.3\%=7.3(0.01)=0.073$

111. $15.8\%=15.8(0.01)=0.158$

113. $0.3\%=0.3(0.01)=0.003$

115. $9.9\%=9.9(0.01)=0.099$

117. $121.2\%=121.2(0.01)=1.212$

119. $0.15=0.15(100\%)=15\%$

121. $0.05=0.05(100\%)=5\%$

123. $0.175=0.175(100\%)=17.5\%$

125. $1.15=1.15(100\%)=115\%$

127. $0.008=0.008(100\%)=0.8\%$

129. $\frac{27}{50}=\frac{27}{50}(100\%)=\frac{2700}{50}\%=54\%$

131. $\frac{1}{3}=\frac{1}{3}(100\%)=\frac{100}{3}\%=33\frac{1}{3}\%$

133. $\frac{5}{11}=\frac{5}{11}(100\%)=\frac{500}{11}\%=45\frac{5}{11}\%$

135. $\frac{7}{8}=\frac{7}{8}(100\%)=\frac{700}{8}\%=87.5\%$

137. $1\frac{2}{3}=1\frac{2}{3}(100\%)=\frac{5}{3}(100\%)=\frac{500}{3}\%=166\frac{2}{3}\%$

139. Strategy To find how far from the left side and how far from the bottom the center of the frame is, divide the width $\left(18\frac{1}{4}\right)$ by 2 and divide the height $\left(24\frac{1}{2}\right)$ by 2.

Solution $18\frac{1}{4}\div 2=9\frac{1}{8}$

$24\frac{1}{2}\div 2=12\frac{1}{4}$

The center of the frame is $9\frac{1}{8}$ in. from the left side of the frame and $12\frac{1}{4}$ in. from the bottom of the frame.

141. Strategy To find how far from the left side of the board the cut should be made, subtract of the width of the saw blade cut from the length of the board and then divide the result by 2.

Solution $\left(36\frac{5}{8}-\frac{1}{8}\right)\div 2=\left(36\frac{1}{2}\right)\div 2$
$=18\frac{1}{4}$

The cut should be made $18\frac{1}{4}$ in. from the left side of the board.

143. Strategy To find the height of the staircase:
- Add the height of a riser (8) and the height of a foot plate $\left(\frac{3}{4}\right)$.
- Multiply by the number of stairs (10).

Solution $8+\frac{3}{4}=8\frac{3}{4}$

$10\left(8\frac{3}{4}\right)=87\frac{1}{2}$

The staircase is $87\frac{1}{2}$ in. high.

145. $\frac{1}{2}\left(\frac{5}{11}-\frac{4}{11}\right)=\frac{1}{2}\left(\frac{1}{11}\right)=\frac{1}{22}$

147. 1st jump $2-\frac{1}{2}\cdot 2=1$

2nd jump $1-\frac{1}{2}\cdot 1=\frac{1}{2}$

3rd jump $\frac{1}{2}-\frac{1}{2}\cdot\frac{1}{2}=\frac{1}{4}$

4th jump $\frac{1}{4}-\frac{1}{2}\cdot\frac{1}{4}=\frac{1}{8}$

5th jump $\frac{1}{8}-\frac{1}{2}\cdot\frac{1}{8}=\frac{1}{16}$

The frog is $\frac{1}{16}$ ft from the wall.

149. The value of the new fraction is less than the value of the original fraction. For example, if 3 is added to both the numerator and the denominator, the resulting fraction is $-\frac{2+3}{5+3}=-\frac{5}{8}$.
We can compare the fractions by rewriting $-\frac{2}{5}$ and $-\frac{5}{8}$ with a denominator of 40: $\frac{2}{5}=-\frac{16}{40}$ and $-\frac{5}{8}=-\frac{25}{40}$; $-\frac{25}{40}<-\frac{16}{40}$.
Therefore, the value of the new fraction, $-\frac{5}{8}$, is less than the value of the original fraction, $-\frac{2}{5}$.

Section 1.8

Objective A Exercises

1. 90°

3. To find the complement of a 62° angle, subtract 62° from 90°.
90° – 62° = 28°
28° is the complement of 62°.

5. To find the supplement of a 48° angle, subtract 48° from 180°.
180° – 48° = 132°
132° is the supplement of 48°.

7. To find the complement of a 7° angle, subtract 7° from 90°.
90° – 7° = 83°
83° is the complement of 7°.

9. To find the supplement of a 89° angle, subtract 89° from 180°.
180° – 89° = 91°
91° is the supplement of 89°.

11. The $m\angle AOC$ is the difference between the measure of a straight angle (180°) and $m\angle COB$ (48°).
$m\angle AOC = 180° - 48° = 132°$.

13. The $m\angle x$ is the complement of a 39° angle.
90° – 39° = 51°.

15. $m\angle AOB$ is the sum of the measures of the two angles.
$m\angle AOB = 32° + 45° = 77°$.

17. $m\angle AOC$ is the difference between $m\angle AOB$ and $m\angle COB$.
$m\angle AOC = 138° - 59° = 79°$.

19. $m\angle A$ is the difference between 360° and 68°.
$m\angle A = 360° - 68° = 292°$.

Objective B Exercises

21. Perimeter = side 1 + side 2 + side 3
= 2.51 cm + 4.08 cm + 3.12 cm
= 9.71 cm

23. Perimeter = 2 · Length + 2 · Width
= 2(4 ft 8 in.) + 2(2 ft 5 in.)
= 8 ft 16 in. + 4 ft 10 in.
= 9 ft 4 in. + 4 ft 10 in.
= 13 ft 14 in.
= 13 ft + 12 in. + 2 in.
= 13 ft + 1 ft + 2 in.
= 14 ft 2 in.

25. Perimeter = 4 · side
= 4 · 13 in.
= 52 in.

27. $\text{Circumference} = 2 \cdot \pi \cdot \text{radius}$
$\approx 2(3.14) \cdot 21 \text{ cm}$
$= 131.88 \text{ cm}$

29. $\text{Circumference} = 2 \cdot \pi \cdot \text{radius}$
$= \pi \cdot \text{diameter}$
$\approx 3.14(1.2 \text{ m})$
$= 3.768 \text{ m}$

31. Strategy To find the cost of the wood framing:
- Find the perimeter of the rectangle.
- Multiply the perimeter by the cost per foot (4.81).

Solution $\text{Perimeter} = 2 \cdot \text{Length} + 2 \cdot \text{Width}$
$= 2(5 \text{ ft}) + 2(3 \text{ ft})$
$= 10 \text{ ft} + 6 \text{ ft}$
$= 16 \text{ ft}$
$16(4.81) = 76.96$
The wood framing would cost $76.96.

33. Strategy To find the cost of the binding:
- Find the circumference of the ring.
- Multiply the circumference by the cost per foot (1.05).

Solution $\text{Circumference} = 2 \cdot \pi \cdot \text{radius}$
$\approx 2(3.14)(3 \text{ ft})$
$= 18.84 \text{ ft}$
$18.84(1.05) = 19.782$
The cost of the binding is $19.78.

35. $\text{Area} = \text{length} \cdot \text{width}$
$= 8 \text{ ft} \cdot 4 \text{ ft}$
$= 32 \text{ ft}^2$

37. $\text{Area} = \text{base} \cdot \text{height}$
$= 27 \text{ cm} \cdot 14 \text{ cm}$
$= 378 \text{ cm}^2$

39. $\text{Area} = \pi(\text{radius})^2$
$\approx 3.14(4 \text{ in.})^2$
$= 50.24 \text{ in}^2$

41. $\text{Area} = \text{side} \cdot \text{side}$
$= 4.1 \text{ m} \cdot 4.1 \text{ m}$
$= 16.81 \text{ m}^2$

43. $\text{Area} = \frac{1}{2} \cdot \text{base} \cdot \text{height}$
$= \frac{1}{2}(15 \text{ cm})(7 \text{ cm})$
$= 52.5 \text{ cm}^2$

45. $\text{Area} = \pi(\text{radius})^2$
$\approx 3.14(8.5 \text{ in.})^2 \cdot \text{radius} = \frac{1}{2}\text{diameter}$
$= 226.865 \text{ in}^2$

47. Strategy To find how many gallons of water should be used:
- Find the area of the rectangular area.
- Multiply the area by the number of gallons needed for each square foot (0.1).

Solution $\text{Area} = \text{length} \cdot \text{width}$
$= 42 \text{ ft} \cdot 33 \text{ ft}$
$= 1386 \text{ ft}^2$
$1386(0.1) = 138.6$
138.6 gal of water should be used.

49. Strategy To find the cost to build the design:
- Find the area of the circular design in in^2.
- Convert the number of square inches to square feet. ($144 \text{ in}^2 = 1 \text{ ft}^2$)
- Multiply the number of square feet by the cost per square foot ($35).

Solution $\text{Area} = \pi(\text{radius})^2$
$\approx (3.14)(15 \text{ in.})^2$
$= 706.5 \text{ in}^2$
$706.5 \text{ in}^2 \cdot \frac{1 \text{ ft}^2}{144 \text{ in}^2} = 4.90625 \text{ ft}^2$
$4.90625(35) \approx 172$
The cost to build the design is $172.

51. Strategy To find the cost to plaster the walls of the rectangular room:
- Find the total area of the four walls.
- Subtract the number of square feet not plastered.
- Multiply the result by the cost per foot ($2.56).

Solution Area of each of the two larger walls: $(18 \text{ ft}) \cdot (8 \text{ ft}) = 144 \text{ ft}^2$
Area of each of the two smaller walls: $(14 \text{ ft}) \cdot (8 \text{ ft}) = 112 \text{ ft}^2$
Total area: $144 \text{ ft}^2 + 144 \text{ ft}^2 + 112 \text{ ft}^2 + 112 \text{ ft}^2 = 512 \text{ ft}^2$
$512 \text{ ft}^2 - 125 \text{ ft}^2 = 387 \text{ ft}^2$ to be plastered
$387(2.56) = 990.72$

The cost to plaster the room is $990.72.

53. The perimeter is the sum of two sides of 70 m each and the circumference of a circle with radius 20 m.

$$2(70\text{ m}) + 2\pi \cdot 20\text{ m} \approx 140\text{ m} + 40(3.14)\text{ m} = 265.6\text{ m}$$

perimeter: 265.6 m

The area is the sum of the area of a 10 m by 40 m rectangle and the area of a circle with radius 20 m.

$$\begin{aligned} 70\text{ m} \cdot 40\text{ m} + \pi(20\text{ m})^2 &\approx 2800\text{ m}^2 + (3.14)(20\text{ m}^2) \\ &= 2800\text{ m}^2 + 1256\text{ m}^2 \\ &= 4056\text{ m}^2 \end{aligned}$$

area: 4056 m^2

55. The outside perimeter of the shaded portion is $2 \cdot 90\text{ m} + 2 \cdot 80\text{ m} = 340\text{ m}$.

perimeter: 340 m

The area of the shaded portion is the difference between the area of the larger rectangle and the area of the smaller rectangle.

$$\begin{aligned} 90\text{ m} \cdot 80\text{ m} = 80\text{ m} \cdot 70\text{ m} &= 7200\text{ m}^2 - 5600\text{ m}^2 \\ &= 1600\text{ m}^2 \end{aligned}$$

area: 1600 m^2

57. **a.**
$$\begin{aligned} \text{Area} &= \frac{1}{2} \cdot \text{height}(\text{base } 1 + \text{base } 2) \\ &= \frac{1}{2} \cdot 6\text{ in.}\ (5\text{ in.} + 8\text{ in.}) \\ &= \frac{1}{2} \cdot 6\text{ in.}\ (13\text{ in.}) \\ &= 3\text{ in.}\ (13\text{ in.}) \\ &= 39\text{ in}^2 \end{aligned}$$

b.
$$\begin{aligned} \text{Area} &= \frac{1}{2} \cdot \text{height}(\text{base } 1 + \text{base } 2) \\ &= \frac{1}{2} \cdot 6\text{ in.}\ (13\text{ in.} + 16\text{ in.}) \\ &= \frac{1}{2} \cdot 6\text{ in.}\ (29\text{ in.}) \\ &= 3\text{ in.}\ (29\text{ in.}) \\ &= 87\text{ in}^2 \end{aligned}$$

59. Select one side of the triangle as the base. Draw a perpendicular line segment from the vertex opposite the base to the base. If necessary, extend the base so that the height meets the base.

Chapter 1 Review Exercises

1. $-13 + 7 = -6$

2.
$$\begin{array}{r} 0.28 \\ 25\overline{)7.00} \\ \underline{-50} \\ 200 \\ \underline{-200} \end{array}$$
$$\frac{7}{25} = 0.28$$

3. $-5^2 = -(5)(5) = -25$

4.
$$\begin{aligned} 5 - 2^2 + 9 &= 5 - 4 + 9 \\ &= 1 + 9 \\ &= 10 \end{aligned}$$

5. The measure of $\angle AOB$ is the difference between the measure of $\angle AOC$ and the measure of $\angle BOC$

$m\angle AOB = m\angle AOC - m\angle BOC = 82° - 45° = 37°$

6. $6.2\% = 6.2(0.01) = 0.062$

7. $(-6)(7) = -42$

8.
$$\begin{aligned} \frac{1}{3} - \frac{1}{6} + \frac{5}{12} &= \frac{4}{12} - \frac{2}{12} + \frac{5}{12} \\ &= \frac{4 - 2 + 5}{12} \\ &= \frac{7}{12} \end{aligned}$$

9. $90° - 56° = 34°$. The complement of 56° is 34°.

10. $x < -1$ is true for -4

11. The factors of 56 are 1, 2, 4, 7, 8, 14, 28, 56.

12. $5.17 - 6.238 = -1.068$

13. $\frac{5}{8} = \frac{5}{8}(100\%) = \frac{500}{8}\% = 62.5\%$

14.
$$\begin{array}{r} 0.133 \\ 15\overline{)2.000} \\ \underline{-15} \\ 50 \\ \underline{-45} \\ 50 \\ \underline{-45} \\ 5 \end{array}$$
$$\frac{2}{15} = 0.1\overline{3}$$

15. $9 - 13 = -4$

16. $\frac{4}{15} - \frac{2}{5} = \frac{4}{15} - \frac{6}{15} = \frac{4 - 6}{15} = -\frac{2}{15}$

17. 4

18.
$$\begin{aligned} \text{Area} &= \frac{1}{2} \cdot \text{base} \cdot \text{height} \\ &= \frac{1}{2} \cdot 4\text{ cm} \cdot 9\text{ cm} \\ &= 18\text{ cm}^2 \end{aligned}$$

19. $-100 \div 5 = -20$

20. $79\frac{1}{2}\% = 79\frac{1}{2}\left(\frac{1}{100}\right) = \frac{159}{2}\left(\frac{1}{100}\right) = \frac{159}{200}$

21. The prime factorization of 280 is $2 \cdot 2 \cdot 2 \cdot 5 \cdot 7$.

22.
$$\begin{aligned} -3^2 + 4[18 + (12 - 20)] &= -3^2 + 4[18 + (-8)] \\ &= -9 + 4[10] \\ &= -9 + 40 \\ &= 31 \end{aligned}$$

23. $-3+(-12)+6+(-4) = -15+6+(-4)$
$= -9+(-4)$
$= -13$

24. $\frac{4}{5}+\left(-\frac{3}{8}\right) = \frac{32}{40}+\left(-\frac{15}{40}\right)$
$= \frac{32+(-15)}{40}$
$= \frac{17}{40}$

25. $\frac{19}{35} = \frac{19}{35}(100\%) = \frac{1900}{35}\% = 54\frac{2}{7}\%$

26. Area $= \pi(\text{radius})^2$
$\approx 3.14(3\text{ m})^2 \cdot$ radius $= \frac{1}{2}$diameter
$= 28.26\text{ m}^2$

27. $4.32(-1.07) = -4.6224$

28. $-|-5| = -(5) = -5$

29. $16-(-3)-18 = 16+3-18$
$= 19-18$
$= 1$

30. $\frac{-18}{35} \times \frac{28}{27} = \frac{-18 \times 28}{35 \times 27}$
$= \frac{-2 \cdot \overset{1}{\cancel{3}} \cdot \overset{1}{\cancel{3}} \cdot 2 \cdot 2 \cdot \overset{1}{\cancel{7}}}{5 \cdot \underset{1}{\cancel{7}} \cdot \underset{1}{\cancel{3}} \cdot \underset{1}{\cancel{3}} \cdot 3}$
$= -\frac{8}{15}$

31. $180° - 28° = 152°$
The supplement of 28° is 152°.

32. Perimeter $= 2 \cdot$ Length $+ 2 \cdot$ Width
$= 2(12\text{ in.}) + 2(10\text{ in.})$
$= 24\text{ in.} + 20\text{ in.}$
$= 44\text{ in.}$

33. $-|6| < |-10|$

34. $\frac{5^2+11}{2^2+5} \div (2^3-2^2) = \frac{25+11}{4+5} \div (8-4)$
$= \frac{36}{9} \div 4$
$= 4 \div 4$
$= 1$

35. **Strategy** To find the score:
- Multiply the number of correct answers by 6.
- Multiply the number of incorrect answers by –4.
- Multiply the number of blank responses by –2.
- Add the results.

Solution
$21(6) = 126$
$5(-4) = -20$
$4(-2) = -8$
$126+(-20)+(-8) = 98$
The student's score was 98.

36. **Strategy** To find the percent:
- Add each number to find the total number of PDAs shipped.
- Divide the number of PDAs shipped by Palm by the total number of PDAs shipped and convert the decimal to a percent.

Solution $259.7+552+240+195$
$+131+67.5 = 1445.2$
$552 \div 1445.2 \approx 0.382$
38.2% of the PDAs were shipped by Palm.

37. **Strategy** To find the difference, subtract the freezing point of mercury (–39°C) from the boiling point of mercury (357°C).

Solution $357° - (-39°) = 357° + 39°$
$= 396°\text{C}$
The difference between the boiling and freezing point of mercury is 396°C.

38. **Strategy** To find the total cost of replacing the bed with sod:
- Find the area of the rectangular flower bed in square feet.
- Multiply the number of square feet by the cost per square foot ($2.51).

Solution Area = length · width
$= 12\text{ ft} \cdot 8\text{ ft}$
$= 96\text{ ft}^2$
$96(2.51) = 240.96$
The cost to replace with sod is $240.96.

Chapter 1 Test

1. $-561 \div (-33) = 17$

2. $\frac{5}{6} = \frac{5}{6}(100\%)$
$= \frac{500}{6}\%$
$= 83\frac{1}{3}\%$

3. $90° - 28° = 62°$
The complement of 28° is 62°.

4. $6.02(-0.89) = -5.3578$

5. $16-30 = 16+(-30) = -14$

6. $37\frac{1}{2}\% = 37\frac{1}{2}\left(\frac{1}{100}\right)$
$= \frac{75}{2}\left(\frac{1}{100}\right)$
$= \frac{75}{200}$
$= \frac{3}{8}$

7. $-\frac{5}{6} - \left(-\frac{7}{8}\right) = -\frac{5}{6} + \frac{7}{8}$
$= -\frac{20}{24} + \frac{21}{24}$
$= \frac{-20+21}{24}$
$= \frac{1}{24}$

8. $\frac{-10+2}{2+(-4)} \div 2 + 6 = \frac{-8}{-2} \div 2 + 6$
$= 4 \div 2 + 6$
$= 2 + 6$
$= 8$

9. $-5 \times (-6) \times 3 = 30 \times 3$
$= 90$

10. Circumference
$= 2 \cdot \pi \cdot \text{radius}$
$\approx 2(3.14)(13.5 \text{ in.}) \cdot \text{radius} = \frac{1}{2}\text{diameter}$
$= 84.78$ in.

11. $(-3^3)(2)^2 = -27(2)(2)$
$= -54(2)$
$= -108$

12. Area = base · height
$= 10 \text{ cm} \cdot 9 \text{ cm}$
$= 90 \text{ cm}^2$

13. $-2 > -40$

14. $-\frac{3}{4} + \frac{2}{5} = -\frac{15}{20} + \frac{8}{20}$
$= \frac{-15+8}{20}$
$= -\frac{7}{20}$

15. -4

16. $45\% = 45\left(\frac{1}{100}\right) = \frac{45}{100} = \frac{9}{20}$
$45\% = 45(0.01) = 0.45$

17. $-22 + 14 - 8 = -8 - 8$
$= -16$

18. $-4 \cdot 12 = -48$

19. The prime factorization of 990 is $2 \cdot 3 \cdot 3 \cdot 5 \cdot 11$.

20. $16 \div 2[8 - 3(4-2)] + 1 = 16 \div 2[8 - 3(4-2)] + 1$
$= 16 \div 2[8 - 3(2)] + 1$
$= 16 \div 2[8 - 6] + 1$
$= 16 \div 2[2] + 1$
$= 8[2] + 1$
$= 16 + 1$
$= 17$

21. $16 - (-30) - 42 = 16 + 30 - 42$
$= 46 - 42$
$= 4$

22. $\frac{5}{12} \div \frac{-5}{6} = \frac{5}{12} \cdot \frac{-6}{5} = \frac{\overset{1}{\cancel{5}} \cdot \overset{1}{\cancel{2}} \cdot \overset{1}{\cancel{3}}}{\underset{1}{\cancel{2}} \cdot 2 \cdot \underset{1}{\cancel{3}} \cdot \underset{1}{\cancel{5}}} = -\frac{1}{2}$

23. 47°

24. $3^2 - 4 + 20 \div 5 = 9 - 4 + 4$
$= 9$

25.
$$\begin{array}{r} 0.77 \\ 9\overline{)7.00} \\ \underline{-63} \\ 70 \\ \underline{-63} \\ 7 \end{array}$$
$\frac{7}{9} = 0.\overline{7}$

26. a. **Strategy** To find the annual loss, multiply the quarterly loss by 4.

Solution $-114.4 \times 4 = -457.6$
The annual earnings would be $-\$457.6$ million.

b. **Strategy** To find the average monthly loss divide the quarterly loss by 3.

Solution $-0.2 \div 3 = -0.0\overline{6} \approx -0.067$
The monthly earnings would be $-\$.067$ million, or $-\$67$ thousand.

27. **Strategy** To find the cost of the new fencing:
- Find the perimeter in feet.
- Multiply the number of feet by the cost per foot ($6.52).

Solution Perimeter = 2 · Length + 2 · Width
$= 2 \cdot 200 \text{ ft} + 2 \cdot 150 \text{ ft}$
$= 400 \text{ ft} + 300 \text{ ft}$
$= 700 \text{ ft}$
$700(6.52) = 4564$
The cost of the new fencing is $4564.

Chapter 2: Variable Expressions

Prep Test

1. $-12-(-15)=-12+15=3$

2. $-36\div(-9)=4$

3. $-\frac{3}{4}+\frac{5}{6}=-\frac{3}{4}\cdot\frac{3}{3}+\frac{5}{6}\cdot\frac{2}{2}$
$=-\frac{9}{12}+\frac{10}{12}$
$=\frac{1}{12}$

4. $-\frac{4}{9}$

5. $-\frac{3}{4}\div\left(-\frac{5}{2}\right)=-\frac{3}{4}\cdot\left(-\frac{2}{5}\right)$
$=\frac{3}{10}$

6. $-2^4=-(2\cdot2\cdot2\cdot2)$
$=-16$

7. $\left(\frac{2}{3}\right)^3=\frac{2}{3}\cdot\frac{2}{3}\cdot\frac{2}{3}=\frac{8}{27}$

8. $3\cdot4^2=3\cdot16=48$

9. $7-2\cdot3=7-6=1$

10. $5-7(3-2^2)=5-7(3-4)$
$=5-7(-1)$
$=5+7$
$=12$

Go Figure

Strategy To find the sum:

- Find the other two fractions.
- Find the difference between $\frac{1}{2}$ and $\frac{1}{4}$.
- Divide the difference by 3.
- Add the difference to $\frac{1}{4}$ to get the 3rd fraction.
- Subtract the difference from $\frac{1}{2}$ to get the 4th fraction.
- Add the four fractions.

Solution $\frac{1}{2}-\frac{1}{4}=\frac{2}{4}-\frac{1}{4}=\frac{1}{4}$

$\frac{1}{4}\div3=\frac{1}{4}\cdot\frac{1}{3}=\frac{1}{12}$

$\frac{1}{4}+\frac{1}{12}=\frac{3}{12}+\frac{1}{12}=\frac{4}{12}=\frac{1}{3}$

$\frac{1}{2}-\frac{1}{12}=\frac{6}{12}-\frac{1}{12}=\frac{5}{12}$

The four fractions are $\frac{1}{4}$, $\frac{1}{3}$, $\frac{5}{12}$, and $\frac{1}{2}$.

$\frac{1}{4}+\frac{1}{3}+\frac{5}{12}+\frac{1}{2}=\frac{3}{12}+\frac{4}{12}+\frac{5}{12}+\frac{6}{12}$
$=\frac{18}{12}$
$=\frac{3}{2}$

Section 2.1

Objective A Exercises

1. $2x^2$, $5x$, $\underline{-8}$

3. $-a^4$, $\underline{6}$

5. $7x^2\underline{y}$, $6\underline{xy^2}$

7. coefficient of x^2: 1
coefficient of $-9x$: -9

9. coefficient of n^3: 1
coefficient of $-4n^2$: -4
coefficient of $-n$: -1

11. The phrase "evaluate the variable expression" means to replace each variable in the expression with its value and then to simplify the resulting numerical expression.

13. $a-2c$
$2-2(-4)=2-(-8)$
$=10$

15. $2c^2$
$2(-4)^2=2(16)$
$=32$

17. $3b-3c$
$3(3)-3(-4)=9-(-12)$
$=21$

19. $-3c+4$
$-3(-4)+4=12+4$
$=16$

21. $6b \div (-a)$

$$6(3) \div (-2) = 18 \div (-2) = -9$$

23. $b^2 - 4ac$

$$(3)^2 - 4(2)(-4) = 9 - 4(2)(-4) = 9 - (-32) = 9 + 32 = 41$$

25. $b^2 - c^2$

$$3^2 - (-4)^2 = 9 - 16 = -7$$

27. $a^2 + b^2$

$$2^2 + 3^2 = 4 + 9 = 13$$

29. $(b - a)^2 + 4c$

$$(3 - 2)^2 + 4(-4) = 1^2 + 4(-4) = 1 + (-16) = -15$$

31. $\frac{5ab}{6} - 3cb$

$$\frac{5(2)(3)}{6} - 3(-4)(3) = \frac{30}{6} - (-36) = 5 - (-36) = 41$$

33. $\frac{b + c}{d}$

$$\frac{4 + (-1)}{3} = \frac{3}{3} = 1$$

35. $\frac{2d + b}{-a}$

$$\frac{2(3) + 4}{-(-2)} = \frac{6 + 4}{2} = \frac{10}{2} = 5$$

37. $\frac{b - d}{c - a}$

$$\frac{4 - 3}{-1 - (-2)} = \frac{1}{1} = 1$$

39. $(b + d)^2 - 4a$

$$(4 + 3)^2 - 4(-2) = 7^2 - 4(-2) = 49 - (-8) = 57$$

41. $(d - a)^2 \div 5$

$$[3 - (-2)]^2 \div 5 = 5^2 \div 5 = 25 \div 5 = 5$$

43. $\frac{b - 2a}{bc^2 - d}$

$$\frac{4 - 2(-2)}{4(-1)^2 - 3} = \frac{4 - (-4)}{4(1) - 3} = \frac{8}{4 - 3} = \frac{8}{1} = 8$$

45. $\frac{1}{3}d^2 - \frac{3}{8}b^2$

$$\frac{1}{3}(3)^2 - \frac{3}{8}(4)^2 = \frac{1}{3}(9) - \frac{3}{8}(16) = 3 - 6 = -3$$

47. $\frac{-4bc}{2a - b}$

$$\frac{-4(4)(-1)}{2(-2) - 4} = \frac{16}{-4 - 4} = \frac{16}{-8} = -2$$

49. $-\frac{2}{3}d - \frac{1}{5}(bd - ac)$

$$-\frac{2}{3}(3) - \frac{1}{5}[4(3) - (-2)(-1)] = -\frac{2}{3}(3) - \frac{1}{5}[12 - 2] = -\frac{2}{3}(3) - \frac{1}{5}(10) = -2 - 2 = -4$$

51. $z = a^2 - 2a$ when $a = -3$

$$z = (-3)^2 - 2(-3) = 9 + 6 = 15$$

$$z^2 = (15)^2 = 225$$

53. $c = a^2 + b^2$ when $a = 2$ and $b = -2$

$$c = (2)^2 + (-2)^2 = 4 + 4 = 8$$

$$c^2 - 4 = (8)^2 - 4 = 64 - 4 = 60$$

Applying the Concepts

55. $z^x = (-2)^2 = 4$

57. $y^{(x^2)} = 3^{2^2}$
$= 3^4$
$= 81$

59. $n^x > x^n$ if $x \geq n+1$

Section 2.2

Objective A Exercises

1. Like terms are variable terms with the same variable part. Constant terms are also like terms. Examples of like terms are $4x$ and $-9x$. Examples of terms that are not like terms are $4x^2$ and $-9x$. The terms 4 and 9 are also like terms; 4 and $4x$ are not.

3. $6x + 8x = 14x$

5. $9a - 4a = 5a$

7. $4y - 10y = -6y$

9. $7 - 3b = 7 - 3b$

11. $-12a + 17a = 5a$

13. $5ab - 7ab = -2ab$

15. $-12xy + 17xy = 5xy$

17. $-3ab + 3ab = 0$

19. $-\frac{1}{2}x - \frac{1}{3}x = -\frac{5}{6}x$

21. $2.3x + 4.2x = 6.5x$

23. $x - 0.55x = 0.45x$

25. $5a - 3a + 5a = 7a$

27. $-5x^2 - 12x^2 + 3x^2 = -14x^2$

29. $\frac{3}{4}x - \frac{1}{3}x - \frac{7}{8}x = -\frac{11}{24}x$

31. $7x - 3y + 10x = 17x - 3y$

33. $3a + (-7b) - 5a + b = -2a - 6b$

35. $3x + (-8y) - 10x + 4x = -3x - 8y$

37. $x^2 - 7x + (-5x^2) + 5x = -4x^2 - 2x$

Objective B Exercises

39. $12x$

41. $-21a$

43. $6y$

45. $8x$

47. $-6a$

49. $12b$

51. $-15x^2$

53. x^2

55. a

57. x

59. n

61. x

63. y

65. $3x$

67. $-2x$

69. $-8a^2$

71. $8y$

73. $4y$

75. $-2x$

77. $6a$

Objective C Exercises

79. $-(x+2) = -x - 2$

81. $2(4x - 3) = 8x - 6$

83. $-2(a+7) = -2a - 14$

85. $-3(2y - 8) = -6y + 24$

87. $(5 - 3b)7 = 35 - 21b$

89. $\frac{1}{3}(6 - 15y) = 2 - 5y$

91. $3(5x^2 + 2x) = 15x^2 + 6x$

93. $-2(-y + 9) = 2y - 18$

95. $(-3x - 6)5 = -15x - 30$

97. $2(-3x^2 - 14) = -6x^2 - 28$

99. $-3(2y^2 - 7) = -6y^2 + 21$

101. $3(x^2 - y^2) = 3x^2 - 3y^2$

103. $-\frac{2}{3}(6x - 18y) = -4x + 12y$

105. $-(6a^2 - 7b^2) = -6a^2 + 7b^2$

107. $4(x^2 - 3x + 5) = 4x^2 - 12x + 20$

109. $\frac{3}{4}(2x - 6y + 8) = \frac{3}{2}x - \frac{9}{2}y + 6$

111. $4(-3a^2 - 5a + 7) = -12a^2 - 20a + 28$

113. $-3(-4x^2 + 3x - 4) = 12x^2 - 9x + 12$

115. $5(2x^2 - 4xy - y^2) = 10x^2 - 20xy - 5y^2$

117. $-(8b^2 - 6b + 9) = -8b^2 + 6b - 9$

Objective D Exercises

119. $6a - (5a + 7) = 6a - 5a - 7$
$= a - 7$

121. $10 - (11x - 3) = 10 - 11x + 3$
$= -11x + 13$

123. $8 - (12 + 4y) = 8 - 12 - 4y$
$= -4y - 4$

125. $2(x - 4) - 4(x + 2) = 2x - 8 - 4x - 8$
$= -2x - 16$

127. $6(2y - 7) - (3 - 2y) = 12y - 42 - 3 + 2y$
$= 14y - 45$

129. $2(a + 2b) - (a - 3b) = 2a + 4b - a + 3b$
$= a + 7b$

131. $2[x + 2(x + 7)] = 2[x + 2x + 14]$
$= 2[3x + 14]$
$= 6x + 28$

133. $-5[2x + 3(5 - x)] = -5[2x + 15 - 3x]$
$= -5[-x + 15]$
$= 5x - 75$

135. $-2[3x - (5x - 2)] = -2[3x - 5x + 2]$
$= -2[-2x + 2]$
$= 4x - 4$

137. $-7x + 3[x - (3 - 2x)] = -7x + 3[x - 3 + 2x]$
$= -7x + 3[3x - 3]$
$= -7x + 9x - 9$
$= 2x - 9$

Applying the Concepts

139. a. False. For example, $8 \div 2 \neq 2 \div 8$.

b. False. For example,
$(12 \div 4) \div 2 = 3 \div 2 = 1.5$
$12 \div (4 \div 2) = 12 \div 2 = 6$.

c. False. For example,
$(9 - 2) - 3 = 7 - 3 = 4$
$9 - (2 - 3) = 9 - (-1) = 10$.

d. False. For example, $10 - 4 \neq 4 - 10$.

141. No. 0 does not have a multiplicative inverse.

143. Examples of two operations that occur in everyday experience that are not commutative are:
(1) unlocking the car door and starting the car, and;
(2) taking a shower and drying oneself off.

Section 2.3

Objective A Exercises

1. the sum of 8 and y
$8 + y$

3. t increased by 10
$t + 10$

5. z added to 14
$z + 14$

7. 20 less than the square of x
$x^2 - 20$

9. the sum of three-fourths of n and 12
$\frac{3}{4}n + 12$

11. 8 increased by the quotient of n and 4
$8 + \frac{n}{4}$

13. the product of 3 and the total of y and 7
$3(y + 7)$

15. the product of t and the sum of t and 16
$t(t + 16)$

17. 15 more than one-half of the square of x
$\frac{1}{2}x^2 + 15$

19. the total of five times the cube of n and the square of n
$5n^3 + n^2$

21. r decreased by the quotient of r and 3
$r - \frac{r}{3}$

23. the difference between the square of x and the total of x and 17
$x^2 - (x + 17)$

25. the product of 9 and the total of z and 4
$9(z + 4)$

Objective B Exercises

27. the unknown number x
$12 - x$

29. the unknown number x
$\frac{2}{3}x$

31. the unknown number x
twice the number: $2x$
$\frac{2x}{9}$

33. the unknown number x
the product of eleven and the number: $11x$
$11x - 8$

35. the unknown number x
the total of the number and two: $x + 2$
$(x + 2) - 9 = x + 2 - 9$
$= x - 7$

37. the unknown number x
the total of five and the number: $5+x$
$\frac{7}{5+x}$

39. the unknown number x
the sum of the number and three: $x+3$
one-half the sum of the number and three: $(1/2)(x+3)$
$$5+\frac{1}{2}(x+3)=5+\frac{1}{2}x+\frac{3}{2}$$
$$=\frac{1}{2}x+\frac{13}{2}$$

41. the unknown number x
twice the number: $2x$
the difference between twice the number and four: $2x-4$
$$(2x-4)+x=2x-4+x$$
$$=3x-4$$

43. the unknown number x
five less than the number: $x-5$
$(x-5)7=7x-35$

45. the unknown number x
twice the number: $2x$
five more than twice the number: $2x+5$
$\frac{2x+5}{x}$

47. the unknown number x
three times the number: $3x$
the difference between three times the number and eight: $3x-8$
$$x-(3x-8)=x-3x+8$$
$$=-2x+8$$

49. the unknown number x
the product of three and the number: $3x$
$x+3x=4x$

51. the unknown number x
the sum of the number and six: $x+6$
$(x+6)+5=x+11$

53. the unknown number x
the sum of the number and ten: $x+10$
$$x-(x+10)=x-x-10$$
$$=-10$$

55. the unknown number x
one-sixth of the number: $\frac{1}{6}x$
four-ninths of the number: $\frac{4}{9}x$
$$\frac{1}{6}x+\frac{4}{9}x=\frac{3}{18}x+\frac{8}{18}x$$
$$=\frac{11}{18}x$$

57. the unknown number x
the number divided by three: $\frac{x}{3}$
$$\frac{x}{3}+x=\frac{x}{3}+\frac{3x}{3}$$
$$=\frac{4x}{3}$$

Objective C Exercises

59. number of emails: A
number of spam emails: $\frac{1}{2}A$

61. length of first piece of rope: S
length of second piece of rope: $12-S$

63. distance traveled by the faster car: x
distance traveled by the slower car: $200-x$

65. number of bones in your body: N
number of bones in your foot: $\frac{1}{4}N$

67. salary needed in San Francisco: S
salary needed in Daytona Beach: $\frac{1}{2}S$

Applying the Concepts

69. number of oxygen atoms: x
number of hydrogen atoms: $2x$

71. Students should express the idea that variables are used to represent unknown quantities or quantities that can change according to different circumstances.

Chapter 2 Review Exercises

1. $$3(x^2-8x-7)=3(x^2)+3(-8x)+3(-7)$$
$$=3x^2-24x-21$$

2. $$7x+4x=(7+4)x$$
$$=11x$$

3. $$6a-4b+2a=6a+2a-4b$$
$$=(6+2)a-4b$$
$$=8a-4b$$

4. $$(-50n)\left(\frac{1}{10}\right)=(-50)\left(\frac{1}{10}\right)n$$
$$=-5n$$

5. $$(5c-4a)^2-b=[5(1)-4(-1)]^2-2$$
$$=[5+4]^2-2$$
$$=9^2-2$$
$$=81-2$$
$$=79$$

6. $$5(2x-7)=5(2x)+5(-7)$$
$$=10x-35$$

7. $$2(6y^2+4y-5)=2(6y^2)+2(4y)+2(-5)$$
$$=12y^2+8y-10$$

8. $\frac{1}{4}(-24)a = \frac{1}{4}(-24)a$
$= -6a$

9. $-6(7x^2) = -6(7)x^2$
$= -42x^2$

10. $-9(7+4x) = -9(7) - 9(4x)$
$= -63 - 36x$

11. $12y - 17y = (12-17)y$
$= -5y$

12. $\frac{2bc}{a+7} = \frac{2(-5)(4)}{3+7}$
$= \frac{-40}{10}$
$= -4$

13. $7 - 2(3x+4) = 7 - 2(3x) - 2(4)$
$= 7 - 6x - 8$
$= -6x - 1$

14. $6 + 2[2 - 5(4a-3)] = 6 + 2(2 - 20a + 15)$
$= 6 + 2(-20a + 17)$
$= 6 - 40a + 34$
$= -40a + 40$

15. $6(8y-3) - 8(3y-6) = 6(8y) + 6(-3) - 8(3y) - 8(-6)$
$= 48y - 18 - 24y + 48$
$= 48y - 24y + 48 - 18$
$= 24y + 30$

16. $5c + (-2d) - 3d - (-4c) = 5c - 2d - 3d + 4c$
$= 5c + 4c - 2d - 3d$
$= (5+4)c + (-2-3)d$
$= 9c - 5d$

17. $5(4x) = 5(4)x$
$= 20x$

18. $-4(2x-9) + 5(3x+2)$
$= -4(2x) - 4(-9) + 5(3x) + 5(2)$
$= -8x + 36 + 15x + 10$
$= 15x - 8x + 36 + 10$
$= 7x + 46$

19. $(b-a)^2 + c = [3-(-2)]^2 + 4$
$= [3+2]^2 + 4$
$= 5^2 + 4$
$= 25 + 4$
$= 29$

20. $-9r + 2s - 6s + 12s = -9r + (2 - 6 + 12)s$
$= -9r + 8s$

21. $(2x-y)^2 + (2x+y)^2 = (2(-2) - (-3))^2$
$+ (2(-2) + (-3))^2$
$= (-4+3)^2 + (-4-3)^2$
$= (-1)^2 + (-7)^2$
$= 1 + 49$
$= 50$

22. $b^2 - 4ac = (-4)^2 - 4(1)(-3)$
$= 16 + 12$
$= 28$

23. $4x - 3x^2 + 2x - x^2 = -3x^2 - x^2 + 4x + 2x$
$= (-3-1)x^2 + (4+2)x$
$= -4x^2 + 6x$

24. $5[2 - 3(6x-1)] = 5[2 - 18x + 3]$
$= 5[5 - 18x]$
$= 25 - 90x$
$= -90x + 25$

25. $0.4x + 0.6(250 - x) = 0.4x + 0.6(250) - 0.6(x)$
$= 0.4x + 150 - 0.6x$
$= -0.2x + 150$

26. $\frac{2}{3}x - \frac{3}{4}x = \left(\frac{2}{3} - \frac{3}{4}\right)x$
$= \left(\frac{8}{12} - \frac{9}{12}\right)x$
$= -\frac{1}{12}x$

27. $(7a^2 - 2a + 3)4$
$= 7a^2(4) - 2a(4) + 3(4)$
$= 28a^2 - 8a + 12$

28. $18 - (4x - 2) = 18 - 4x + 2$
$= 18 + 2 - 4x$
$= 20 - 4x$
$= -4x + 20$

29. $a^2 - b^2 = (3)^2 - (4)^2$
$= 9 - 16$
$= -7$

30. $-3(-12y) = -3(-12)y$
$= 36y$

31. $\frac{2}{3}(x + 10)$

32. $4x$

33. $x - 6$

34. the unknown number x
$x + 2x = 3x$

35. the unknown number x
$2x - \frac{1}{2}x = \frac{4}{2}x - \frac{1}{2}x$
$= \frac{3}{2}x$

36. the unknown number x
$3x + 5(x-1) = 3x + 5x - 5$
$= 8x - 5$

37. number of American League cards: A
number of National League cards: $5A$

38. number of ten-dollar bills: T
number of five-dollar bills: $35 - T$

39. number of calories in an apple: a
number of calories in a candy bar: $2a + 8$

40. width of Parthenon: w
length of Parthenon: $1.6w$

41. kneeling height: h
standing height: $1.3h$

Chapter 2 Test

1. $\begin{aligned} 3x - 5x + 7x &= (3 - 5 + 7)x \\ &= 5x \end{aligned}$

2. $\begin{aligned} -3(2x^2 - 7y^2) &= -3(2x^2) - 3(-7y^2) \\ &= -6x^2 + 21y^2 \end{aligned}$

3. $\begin{aligned} 2x - 3(x - 2) &= 2x - 3(x) - 3(-2) \\ &= 2x - 3x + 6 \\ &= -x + 6 \end{aligned}$

4. $\begin{aligned} 2x + 3[4 - (3x - 7)] &= 2x + 3[4 - 3x + 7] \\ &= 2x + 3[11 - 3x] \\ &= 2x + 33 - 9x \\ &= 2x - 9x + 33 \\ &= -7x + 33 \end{aligned}$

5. $\begin{aligned} 3x - 7y - 12x &= 3x - 12x - 7y \\ &= -9x - 7y \end{aligned}$

6. $\begin{aligned} b^2 - 3(a)(b) &= (-2)^2 - 3(3)(-2) \\ &= 4 - 3(-6) \\ &= 4 + 18 \\ &= 22 \end{aligned}$

7. $\begin{aligned} \frac{1}{5}(10x) &= \frac{1}{5}(10)x \\ &= 2x \end{aligned}$

8. $\begin{aligned} 5(2x + 4) - 3(x - 6) &= 5(2x) + 5(4) - 3(x) - 3(-6) \\ &= 10x + 20 - 3x + 18 \\ &= 10x - 3x + 20 + 18 \\ &= 7x + 38 \end{aligned}$

9. $\begin{aligned} -5(2x^2 - 3x + 6) &= 5(2x^2) - 5(-3x) - 5(6) \\ &= -10x^2 + 15x - 30 \end{aligned}$

10. $\begin{aligned} 3x + (-12y) - 5x - (-7y) &= 3x - 12y - 5x + 7y \\ &= 3x - 5x - 12y + 7y \\ &= -2x - 5y \end{aligned}$

11. $\begin{aligned} \frac{-2ab}{2b - a} &= \frac{-2(-4)(6)}{2(6) - (-4)} \\ &= \frac{8(6)}{12 + 4} \\ &= \frac{48}{16} = 3 \end{aligned}$

12. $\begin{aligned} (12x)\frac{1}{4} &= 12\left(\frac{1}{4}\right)x \\ &= 3x \end{aligned}$

13. $\begin{aligned} -7y^2 + 6y^2 - (-2y^2) &= -7y^2 + 6y^2 + 2y^2 \\ &= (-7 + 6 + 2)y^2 \\ &= y^2 \end{aligned}$

14. $\begin{aligned} -2(2x - 4) &= -2(2x) - 2(-4) \\ &= -4x + 8 \end{aligned}$

15. $\begin{aligned} \frac{2}{3}(-15a) &= \frac{2}{3}(-15)a \\ &= -10a \end{aligned}$

16. $\begin{aligned} -2[x - 2(x - y)] + 5y &= -2[x - 2x + 2y] + 5y \\ &= -2[-x + 2y] + 5y \\ &= 2x - 4y + 5y \\ &= 2x + y \end{aligned}$

17. $\begin{aligned} (-3)(-12)y &= (-3)(-12)y \\ &= 36y \end{aligned}$

18. $\begin{aligned} 5(3 - 7b) &= 5(3) + 5(-7b) \\ &= 15 - 35b \end{aligned}$

19. $a^2 - b^2$

20. $\begin{aligned} 10(x - 3) &= 10(x) + 10(-3) \\ &= 10x - 30 \end{aligned}$

21. $x + 2x^2$

22. $\frac{6}{x} - 3$

23. $b - 7b$

24. speed of return throw: s
speed of fastball: $2s$

25. length of shorter piece: x
length of longer piece: $4x - 3$

Cumulative Review Exercises

1. $\begin{aligned} -4 + 7 + (-10) &= -4 - 10 + 7 \\ &= -14 + 7 \\ &= -7 \end{aligned}$

2. $\begin{aligned} -16 - (-25) - 4 &= -16 + 25 - 4 \\ &= -16 - 4 + 25 \\ &= -20 + 25 \\ &= 5 \end{aligned}$

3. $\begin{aligned} (-2)(3)(-4) &= (-2)(-12) \\ &= 24 \end{aligned}$

4. $(-60) \div 12 = -5$

5. $90^\circ - 37^\circ = 53^\circ$
The complement of 37° is 53°.

6. $\frac{7}{12}-\frac{11}{16}-\left(-\frac{1}{3}\right)=\frac{7}{12}-\frac{11}{16}+\frac{1}{3}$
$=\frac{28}{48}-\frac{33}{48}+\frac{16}{48}$
$=\frac{28+16-33}{48}$
$=\frac{44-33}{48}$
$=\frac{11}{48}$

7. $-\frac{5}{12}\div\frac{5}{2}=-\frac{5}{12}\cdot\frac{2}{5}$
$=-\frac{\overset{1}{\cancel{5}}\cdot\overset{1}{\cancel{2}}}{\underset{1}{\cancel{2}}\cdot 2\cdot 3\cdot\underset{1}{\cancel{5}}}$
$=-\frac{1}{6}$

8. $\left(\frac{\overset{-1}{\cancel{-9}}}{\underset{2}{\cancel{16}}}\right)\left(\frac{\overset{1}{\cancel{8}}}{\underset{\underset{1}{\cancel{3}}}{\cancel{27}}}\right)\left(\frac{\overset{-1}{\cancel{-3}}}{2}\right)=\frac{1}{4}$

9. $\frac{3}{4}=\frac{3}{4}(100\%)=\frac{300}{4}\%$
$=75\%$

10. $-2^5\div(3-5)^2-(-3)=-32\div(-2)^2-(-3)$
$=-32\div 4+3$
$=-8+3$
$=-5$

11. $\left(\frac{-3}{4}\right)^2\div\left(\frac{3}{8}-\frac{11}{12}\right)=\frac{9}{16}\div\left(\frac{9}{24}-\frac{22}{24}\right)$
$=\frac{9}{16}\div\left(\frac{-13}{24}\right)$
$=\frac{9}{16}\cdot\left(-\frac{24}{13}\right)$
$=-\frac{27}{26}$

12. $a^2-3b=(2)^2-3(-4)$
$=4+12$
$=16$

13. $-2x^2-(-3x^2)+4x^2=-2x^2+3x^2+4x^2$
$=-2x^2+7x^2$
$=5x^2$

14. $5a-10b-12a=5a-12a-10b$
$=-7a-10b$

15. Area $=\pi(\text{radius})^2$
$\approx 3.14(7\text{ cm})^2$
$=153.86\text{ cm}^2$

16. Perimeter $=4\cdot$ side
$=4(24\text{ ft})$
$=96\text{ ft}$

17. $3(8-2x)=3(8)+3(-2x)$
$=24-6x$

18. $-2(-3y+9)=-2(-3y)-2(9)$
$=6y-18$

19. $37\frac{1}{2}\%=37\frac{1}{2}\left(\frac{1}{100}\right)=\frac{75}{2}\left(\frac{1}{100}\right)=\frac{75}{200}=\frac{3}{8}$

20. $1.05\%=1.05(0.01)=0.0105$

21. $-4(2x^2-3y^2)=-4(2x^2)-4(-3y^2)$
$=-8x^2+12y^2$

22. $-3(3y^2-3y-7)=-3(3y^2)-3(-3y)-3(-7)$
$=-9y^2+9y+21$

23. $-3x-2(2x-7)=-3x-2(2x)-2(-7)$
$=-3x-4x+14$
$=-7x+14$

24. $4(3x-2)-7(x+5)=4(3x)+4(-2)-7(x)-7(5)$
$=12x-8-7x-35$
$=12x-7x-8-35$
$=5x-43$

25. $2x+3[x-2(4-2x)]=2x+3[x-8+4x]$
$=2x+3[5x-8]$
$=2x+3(5x)+3(-8)$
$=2x+15x-24$
$=17x-24$

26. $3[2x-3(x-2y)]+3y=3[2x-3x+6y]+3y$
$=3[-x+6y]+3y$
$=3(-x)+3(6y)+3y$
$=-3x+18y+3y$
$=-3x+21y$

27. $\frac{1}{2}b+b$

28. $\frac{10}{y-2}$

29. $8-\frac{x}{12}$

30. $x+(x+2)=x+x+2$
$=2x+2$

31. Area $=$ side $\cdot$ side
$=60\text{ ft}\cdot 60\text{ ft}$
$=3600\text{ ft}^2$

32. speed of dial-up connection: s
speed of DSL connection: $10s$

Chapter 3: Solving Equations

Prep Test

1. $\frac{9}{100} = 0.09$

2. 75%

3. $3x^2 - 4x - 1$
$$\begin{aligned} 3(-4)^2 - 4(-4) - 1 &= 3 \cdot 16 + 16 - 1 \\ &= 48 + 16 - 1 \\ &= 63 \end{aligned}$$

4. $$\begin{aligned} R - 0.35R &= (1 - 0.35)R \\ &= 0.65R \end{aligned}$$

5. $$\begin{aligned} \frac{1}{2}x + \frac{2}{3}x &= \left(\frac{1}{2} + \frac{2}{3}\right)x \\ &= \left(\frac{3}{6} + \frac{4}{6}\right)x \\ &= \frac{7}{6}x \end{aligned}$$

6. $$\begin{aligned} 6x - 3(6 - x) &= 6x - 18 + 3x \\ &= 9x - 18 \end{aligned}$$

7. $$\begin{aligned} 0.22(3x + 6) + x &= 0.66x + 1.32 + x \\ &= 1.66x + 1.32 \end{aligned}$$

8. $5 - 2n$

9. speed of the old card: s
speed of the new card: $5s$

10. length of the longer piece: x
length of the shorter piece: $5 - x$

Go Figure

Strategy With the donut sitting on a table, slice the donut parallel to the table. Now cut the halved donut in quarters.

Section 3.1

Objective A Exercises

1. An equation contains an equals sign; an expression does not contain an equals sign.

3. $$\begin{array}{c|c} 2x & 8 \\ \hline 2(4) & 8 \end{array}$$
$8 = 8$
Yes, 4 is a solution.

5. $$\begin{array}{r|l} 2b - 1 & 3 \\ \hline 2(-1) - 1 & 3 \end{array}$$
$-3 \neq 3$
No, -1 is not a solution.

7. $$\begin{array}{r|l} 4 - 2m & 3 \\ \hline 4 - 2(1) & 3 \\ 4 - 2 & 3 \end{array}$$
$2 \neq 3$
No, 1 is not a solution.

9. $$\begin{array}{r|l} 2x + 5 & 3x \\ \hline 2(5) + 5 & 3(5) \\ 10 + 5 & 15 \end{array}$$
$15 = 15$
Yes, 5 is a solution.

11. $$\begin{array}{r|l} 3a + 2 & 2 - a \\ \hline 3(-2) + 2 & 2 - (-2) \\ -6 + 2 & 2 + 2 \end{array}$$
$-4 \neq 4$
No, 2 is not a solution.

13. $$\begin{array}{r|l} 2x^2 - 1 & 4x - 1 \\ \hline 2(2)^2 - 1 & 4(2) - 1 \\ 2(4) - 1 & 8 - 1 \\ 8 - 1 & 7 \end{array}$$
$7 = 7$
Yes, 2 is a solution.

15. $$\begin{array}{r|l} x(x + 1) & x^2 + 5 \\ \hline 4(4 + 1) & 4^2 + 5 \\ 4(5) & 16 + 5 \end{array}$$
$20 \neq 21$
No, 4 is not a solution.

17. $$\begin{array}{r|l} 8t + 1 & -1 \\ \hline 8\left(-\frac{1}{4}\right) + 1 & -1 \\ -2 + 1 & -1 \end{array}$$
$-1 = -1$
Yes, $-\left(\frac{1}{4}\right)$ is a solution.

19. $$\begin{array}{r|l} 5m + 1 & 10m - 3 \\ \hline 5\left(\frac{2}{5}\right) + 1 & 10\left(\frac{2}{5}\right) - 3 \\ 2 + 1 & 4 - 3 \end{array}$$
$3 \neq 1$
No, $\frac{2}{5}$ is not a solution.

Objective B Exercises

21. Zero can be a solution of an equation. For instance, the solution of $x + 9 = 9$ is 0.

23. $$\begin{aligned} x + 5 &= 7 \\ x + 5 + (-5) &= 7 + (-5) \\ x &= 2 \end{aligned}$$
The solution is 2.

25. $$\begin{aligned} b - 4 &= 11 \\ b - 4 + 4 &= 11 + 4 \\ b &= 15 \end{aligned}$$
The solution is 15.

27.
$$\begin{aligned} 2+a &= 8 \\ 2+(-2)+a &= 8+(-2) \\ a &= 6 \end{aligned}$$
The solution is 6.

29.
$$\begin{aligned} n-5 &= -2 \\ n-5+5 &= -2+5 \\ n &= 3 \end{aligned}$$
The solution is 3.

31.
$$\begin{aligned} b+7 &= 7 \\ b+7+(-7) &= 7+(-7) \\ b &= 0 \end{aligned}$$
The solution is 0.

33.
$$\begin{aligned} z+9 &= 2 \\ z+9+(-9) &= 2+(-9) \\ z &= -7 \end{aligned}$$
The solution is -7.

35.
$$\begin{aligned} 10+m &= 3 \\ 10+(-10)+m &= 3+(-10) \\ m &= -7 \end{aligned}$$
The solution is -7.

37.
$$\begin{aligned} 9+x &= -3 \\ 9+(-9)+x &= -3+(-9) \\ x &= -12 \end{aligned}$$
The solution is -12.

39.
$$\begin{aligned} 2 &= x+7 \\ 2+(-7) &= x+7+(-7) \\ -5 &= x \end{aligned}$$
The solution is -5.

41.
$$\begin{aligned} 4 &= m-11 \\ 4+11 &= m-11+11 \\ 15 &= m \end{aligned}$$
The solution is 15.

43.
$$\begin{aligned} 12 &= 3+w \\ 12+(-3) &= 3+(-3)+w \\ 9 &= w \end{aligned}$$
The solution is 9.

45.
$$\begin{aligned} 4 &= -10+b \\ 4+10 &= -10+10+b \\ 14 &= b \end{aligned}$$
The solution is 14.

47.
$$\begin{aligned} m+\frac{2}{3} &= -\frac{1}{3} \\ m+\frac{2}{3}+\left(-\frac{2}{3}\right) &= -\frac{1}{3}+\left(-\frac{2}{3}\right) \\ m &= -1 \end{aligned}$$
The solution is -1.

49.
$$\begin{aligned} x-\frac{1}{2} &= \frac{1}{2} \\ x-\frac{1}{2}+\frac{1}{2} &= \frac{1}{2}+\frac{1}{2} \\ x &= 1 \end{aligned}$$
The solution is 1.

51.
$$\begin{aligned} \frac{5}{8}+y &= \frac{1}{8} \\ \frac{5}{8}+\left(-\frac{5}{8}\right)+y &= \frac{1}{8}+\left(-\frac{5}{8}\right) \\ y &= -\frac{4}{8} = -\frac{1}{2} \end{aligned}$$
The solution is $-\frac{1}{2}$.

53.
$$\begin{aligned} m+\frac{1}{2} &= -\frac{1}{4} \\ m+\frac{1}{2}+\left(-\frac{1}{2}\right) &= -\frac{1}{4}+\left(-\frac{1}{2}\right) \\ m &= -\frac{1}{4}+\left(-\frac{2}{4}\right) \\ m &= -\frac{3}{4} \end{aligned}$$
The solution is $-\frac{3}{4}$.

55.
$$\begin{aligned} x+\frac{2}{3} &= \frac{3}{4} \\ x+\frac{2}{3}+\left(-\frac{2}{3}\right) &= \frac{3}{4}+\left(-\frac{2}{3}\right) \\ x &= \frac{9}{12}+\left(-\frac{8}{12}\right) \\ x &= \frac{1}{12} \end{aligned}$$
The solution is $\frac{1}{12}$.

57.
$$\begin{aligned} -\frac{5}{6} &= x-\frac{1}{4} \\ -\frac{5}{6}+\frac{1}{4} &= x-\frac{1}{4}+\frac{1}{4} \\ -\frac{10}{12}+\frac{3}{12} &= x \\ -\frac{7}{12} &= x \end{aligned}$$
The solution is $-\frac{7}{12}$.

59.
$$\begin{aligned} d+1.3619 &= 2.0148 \\ d+1.3619+(-1.3619) &= 2.0148+(-1.3619) \\ d &= 0.6529 \end{aligned}$$
The solution is 0.6529.

61.
$$\begin{aligned} -0.813+x &= -1.096 \\ -0.813+0.813+x &= -1.096+0.813 \\ x &= -0.283 \end{aligned}$$
The solution is -0.283.

63.
$$\begin{aligned} 6.149 &= -3.108 + z \\ 6.149 + 3.108 &= -3.108 + 3.108 + z \\ 9.257 &= z \end{aligned}$$
The solution is 9.257.

Objective C Exercises

65. x is greater than 0 because a negative number times a positive number is a negative number.

67.
$$\begin{aligned} 5x &= -15 \\ \frac{1}{5}(5x) &= \frac{1}{5}(-15) \\ x &= -3 \end{aligned}$$
The solution is -3.

69.
$$\begin{aligned} 3b &= 0 \\ \frac{1}{3}(3b) &= \frac{1}{3}(0) \\ b &= 0 \end{aligned}$$
The solution is 0.

71.
$$\begin{aligned} -3x &= 6 \\ -\frac{1}{3}(-3x) &= -\frac{1}{3}(6) \\ x &= -2 \end{aligned}$$
The solution is -2.

73.
$$\begin{aligned} -3x &= -27 \\ -\frac{1}{3}(-3x) &= -\frac{1}{3}(-27) \\ x &= 9 \end{aligned}$$
The solution is 9.

75.
$$\begin{aligned} 20 &= \frac{1}{4}c \\ 4(20) &= 4\left(\frac{1}{4}c\right) \\ 80 &= c \end{aligned}$$
The solution is 80.

77.
$$\begin{aligned} 0 &= -5x \\ -\frac{1}{5}(0) &= -\frac{1}{5}(-5x) \\ 0 &= x \end{aligned}$$
The solution is 0.

79.
$$\begin{aligned} 49 &= -7t \\ -\frac{1}{7}(49) &= -\frac{1}{7}(-7t) \\ -7 &= t \end{aligned}$$
The solution is -7.

81.
$$\begin{aligned} \frac{x}{4} &= 3 \\ 4\left(\frac{1}{4}x\right) &= 4(3) \\ x &= 12 \end{aligned}$$
The solution is 12.

83.
$$\begin{aligned} -\frac{b}{3} &= 6 \\ -3\left(-\frac{1}{3}b\right) &= -3(6) \\ b &= -18 \end{aligned}$$
The solution is -18.

85.
$$\begin{aligned} \frac{2}{5}x &= 6 \\ \frac{5}{2}\left(\frac{2}{5}x\right) &= \frac{5}{2}(6) \\ x &= 15 \end{aligned}$$
The solution is 15.

87.
$$\begin{aligned} -\frac{3}{5}m &= 12 \\ -\frac{5}{3}\left(-\frac{3}{5}m\right) &= -\frac{5}{3}(12) \\ m &= -20 \end{aligned}$$
The solution is -20.

89.
$$\begin{aligned} \frac{5x}{6} &= 0 \\ \frac{6}{5}\left(\frac{5}{6}x\right) &= \frac{6}{5}(0) \\ x &= 0 \end{aligned}$$
The solution is 0.

91.
$$\begin{aligned} \frac{3x}{4} &= 2 \\ \frac{4}{3}\left(\frac{3x}{4}\right) &= \frac{4}{3}(2) \\ x &= \frac{8}{3} \end{aligned}$$
The solution is $\frac{8}{3}$.

93.
$$\begin{aligned} \frac{2}{9} &= \frac{2}{3}y \\ \frac{3}{2}\left(\frac{2}{9}\right) &= \frac{3}{2}\left(\frac{2}{3}y\right) \\ \frac{1}{3} &= y \end{aligned}$$
The solution is $\frac{1}{3}$.

95.
$$\begin{aligned} \frac{1}{5}x &= -\frac{1}{10} \\ 5\left(\frac{1}{5}x\right) &= 5\left(-\frac{1}{10}\right) \\ x &= -\frac{1}{2} \end{aligned}$$
The solution is $-\frac{1}{2}$.

97.
$$\begin{aligned} -1 &= \frac{2n}{3} \\ \frac{3}{2}(-1) &= \frac{3}{2}\left(\frac{2n}{3}\right) \\ -\frac{3}{2} &= n \end{aligned}$$
The solution is $-\frac{3}{2}$.

99.
$$\begin{aligned} -\frac{2}{5}m &= -\frac{6}{7} \\ -\frac{5}{2}\left(-\frac{2}{5}m\right) &= -\frac{5}{2}\left(-\frac{6}{7}\right) \\ m &= \frac{15}{7} \end{aligned}$$
The solution is $\frac{15}{7}$.

101.
$$\begin{aligned} 3n + 2n &= 20 \\ 5n &= 20 \\ \frac{1}{5}(5n) &= \frac{1}{5}(20) \\ n &= 4 \end{aligned}$$
The solution is 4.

103.
$$\begin{aligned} 10y - 3y &= 21 \\ 7y &= 21 \\ \frac{1}{7}(7y) &= \frac{1}{7}(21) \\ y &= 3 \end{aligned}$$
The solution is 3.

105.
$$\begin{aligned} \frac{x}{1.46} &= 3.25 \\ 1.46\left(\frac{1}{1.46}x\right) &= 1.46(3.25) \\ x &= 4.745 \end{aligned}$$
The solution is 4.745.

107.
$$\begin{aligned} 3.47a &= 7.1482 \\ \frac{1}{3.47}(3.47a) &= \frac{1}{3.47}(7.1482) \\ a &= 2.06 \end{aligned}$$
The solution is 2.06.

109.
$$\begin{aligned} -3.7x &= 7.881 \\ -\frac{1}{3.7}(-3.7x) &= -\frac{1}{3.7}(7.881) \\ x &= -2.13 \end{aligned}$$
The solution is -2.13.

Objective D Exercises

111. $0.40(80) = 0.80(40)$.

113. $0.35(80) = A$
$28 = A$
35% of 80 is 28.

115. $0.012(60) = A$
$0.72 = A$
1.2% of 60 is 0.72.

117. $(1.25)B = 80$
$\frac{1}{1.25}(1.25)B = \frac{1}{1.25}(80)$
$B = 64$
The number is 64.

119. $P(50) = 12$
$P(50)\left(\frac{1}{50}\right) = 12\left(\frac{1}{50}\right)$
$P = 0.24$
The percent is 24%.

121. $0.18(40) = A$
$7.20 = A$
18% of 40 is 7.2.

123. $(0.12)B = 48$
$\frac{1}{0.12}(0.12)B = \frac{1}{0.12}(48)$
$B = 400$
The number is 400.

125. $\frac{1}{3}(27) = A \quad \left(33\frac{1}{3}\% = \frac{1}{3}\right)$
$9 = A$
$33\frac{1}{3}\%$ of 27 is 9.

127. $P(12) = 3$
$P(12)\left(\frac{1}{12}\right) = 3\left(\frac{1}{12}\right)$
$P = \frac{1}{4}$
$P = 0.25$
The percent is 25%.

129. $12 = P(6)$
$12\left(\frac{1}{6}\right) = P(6)\left(\frac{1}{6}\right)$
$2 = P$
12 is 200% of 6.

131. $(0.0525)B = 21$
$\frac{1}{0.0525}(0.0525)B = \frac{1}{0.0525}(21)$
$B = 400$
The number is 400.

133. $(0.154)(50) = A$
$7.7 = A$
15.4% of 50 is 7.7.

135. $1 = (0.005)B$
$\frac{1}{0.005}(1) = \frac{1}{0.005}(0.005)B$
$200 = B$
The number is 200.

137. $(0.0075)B = 3$
$\frac{1}{0.0075}(0.0075)B = \frac{1}{0.0075}(3)$
$B = 400$
The number is 400.

139. $(2.50)(12) = A$
$30 = A$
250% of 12 is 30.

141. **Strategy** To find the amount of oxygen, solve the basic percent equation, using $P = 21\%$, or 0.21, and $B = 21{,}600$. The amount is unknown.

Solution
$P \times B = A$
$0.21 \times 21{,}600 = A$
$4536 = A$
There are 4536 L of oxygen in the room.

143. **Strategy** To find the median income the next year:
- Use the basic percent equation to find the amount of decrease. Use $P = 0.011$ and $B = 42{,}900$. The amount is unknown.
- Subtract the amount of decrease from \$42,900.

Solution
$P \cdot B = A$
$0.011(42{,}900) = A$
$472 \approx A$
$\$42{,}900 - \$472 = \$42{,}428$
The median income was \$42,428.

145. **Strategy** To find the percent, use the basic percent equation. $B = 290$, $A = 138.9$, P is unknown.

Solution
$P \cdot B = A$
$P(290) = 138.9$
$P = \frac{138.9}{290}$
$P \approx 0.479$
47.9% of the U.S. population watched Super Bowl XXXVIII.

147. You need to know the number of people three years old and older in the U.S.

149. Strategy To find how much money Andrea should invest, solve $I = Prt$ for P using $I = 300$, $r = 0.08$, and $t = 2$.

Solution

$$I = Prt$$
$$300 = P(0.08)2$$
$$300 = 0.16P$$
$$\frac{300}{0.16} = \frac{0.16P}{0.16}$$
$$1875 = P$$

Andrea must invest $1875.

151. Strategy To find whether Americo or Octavia will earn the greater amount of interest:

- Find the interest earned by Americo by solving $I = Prt$ using $P = 2500$, $r = 0.08$, and $t = 1$.
- Find the interest earned by Octavia by solving $I = Prt$ using $P = 3000$, $r = 0.07$, and $t = 1$.
- Compare the interest amounts to determine who earned the larger amount.

Solution Americo: $I = Prt$, $I = 2500(0.08)1$, $I = 200$ Octavia: $I = Prt$, $I = 3000(0.07)1$, $I = 210$

$$210 > 200$$

Octavia will earn the greater amount of interest.

153. Strategy To find how much was invested at 8%:

- Solve $I = Prt$ for I using $P = 2000$, $r = 0.06$, and $t = 1$.
- Solve $I = Prt$ for P using the amount of interest found in the first step for I, $r = 0.08$, and $t = 1$.

Solution

$$I = Prt$$
$$I = 2000(0.06) \cdot 1$$
$$I = 120$$

The interest on $2000 at 6% is $120.

$$I = Prt$$
$$120 = P(0.08) \cdot 1$$
$$\frac{120}{0.08} = \frac{0.08P}{0.08}$$
$$1500 = P$$

$1500 was invested at 8%.

155. Strategy To find the number of grams of platinum in the necklace, solve $Q = Ar$ for Q using $A = 12$ and $r = 15\% = 0.15$.

Solution

$$Q = Ar$$
$$Q = 12(0.15)$$
$$Q = 1.8$$

There are 1.8g of platinum in the necklace.

157. Strategy To find the number of pounds of wool in the carpet, solve $Q = Ar$ for Q using $A = 175$ and $r = 0.75$.

Solution

$$Q = Ar$$
$$Q = 175(0.75)$$
$$Q = 131.25$$

There are 131.25 lb of wool in the carpet.

159. Strategy To find the percent concentration of sugar, solve $Q = Ar$ for r using $A = 1000$ and $Q = 500$.

Solution

$$Q = Ar$$
$$500 = 1000r$$
$$\frac{500}{1000} = \frac{1000r}{1000}$$
$$0.5 = r$$

The percent concentration of sugar is 50%.

161. Strategy To find the percent concentration of the mixture:
- Solve $Q = Ar$ for Q using $A = 100$ and $r = 9\%$.
- Using this value for Q, solve for r using $A = 150$.

Solution
$$Q = Ar$$
$$Q = 100(0.09)$$
$$Q = 9$$
$$Q = Ar$$
$$9 = 150r$$
$$\frac{9}{150} = \frac{150r}{150}$$
$$0.06 = r$$
The percent concentration of the resulting mixture is 6%.

163. Strategy To find the number of miles traveled, solve $d = rt$ for d using $r = 9$ mph and $t = \frac{20}{60} = \frac{1}{3}$ h.

Solution
$$d = rt$$
$$d = 9 \cdot \frac{1}{3}$$
$$d = 3$$
The runner will travel 3 mi.

165. Strategy To find the number of miles per hour, solve $d = rt$ for r using $d = 27$ mi and $t = \frac{45}{60} = \frac{3}{4}$ h.

Solution
$$d = rt$$
$$27 = r \cdot \frac{3}{4}$$
$$\frac{4}{3} \cdot 27 = r \cdot \frac{3}{4} \cdot \frac{4}{3}$$
$$36 = r$$
Marcella's average rate of speed is 36 mph.

167. Strategy To find the number of hours to walk the course:
- Find the rate to run the course by solving $d = rt$ for r using $d = 30$ km and $t = 2$ h.
- Decrease the rate by 3 km/h to find his walking rate.
- Solve $d = rt$ for t using $d = 30$ km and r equal to his walking rate.

Solution
$$d = rt$$
$$30 = r \cdot 2$$
$$\frac{30}{2} = \frac{r \cdot 2}{2}$$
$$15 = r \quad \text{His running rate}$$
$$15 - 3 = 12 \quad \text{His walking rate}$$
$$d = rt$$
$$30 = 12t$$
$$\frac{30}{12} = \frac{12t}{12}$$
$$2.5 = t$$
It would take Palmer 2.5 h to walk the course.

169. Strategy The distance is 8 mi. Therefore $d = 8$. The joggers are running toward each other, one at 5 mph and one at 7 mph. The rate is the sum of the two rates, or 12 mph. So, $r = 12$. To find the time, solve $d = rt$ for t. Convert the answer to minutes.

Solution
$$d = rt$$
$$8 = 12t$$
$$\frac{8}{12} = \frac{12t}{12}$$
$$\frac{2}{3} = t$$
$$\frac{2}{3}\text{h} = \frac{2}{3} \cdot 60 \text{ min} = 40 \text{ min}$$
The two joggers will meet 40 min after they start.

171. Strategy The distance is 4 miles. So, $d = 4$. The canoe is traveling against a 2 mph current. In calm water they can paddle at 10 mph. The rate is 10 mph − 2 mph = 8 mph. So, $r = 8$. Solve $d = rt$ for t.

Solution
$$d = rt$$
$$4 = 8t$$
$$\frac{4}{8} = \frac{8t}{8}$$
$$\frac{1}{2} = t$$
It will take them 0.5 h.

173.
$$3x + 2x + 4x = 360°$$
$$9x = 360°$$
$$\frac{9x}{9} = \frac{360°}{9}$$
$$x = 40°$$

175.
$$6x + 3x = 90°$$
$$9x = 90°$$
$$\frac{9x}{9} = \frac{90°}{9}$$
$$x = 10°$$

Applying the Concepts

177. a. Answers will vary.

$$x+7=9$$
$$x=2$$

b. Answers will vary.

$$3x=-3$$
$$x=-1$$

179. Students should paraphrase the definitions of the properties given in the text.

The Addition Property of Equations states that the same number can be added to each side of an equation without changing the solution of the equation.

The Multiplication Property of Equations states that each side of an equation can be multiplied by the same nonzero number without changing the solution of the equation.

Section 3.2

Objective A Exercises

1.
$$\begin{aligned} 3x+1 &= 10 \\ 3x+1+(-1) &= 10+(-1) \\ 3x &= 9 \\ \frac{1}{3}(3x) &= \frac{1}{3}\cdot 9 \\ x &= 3 \end{aligned}$$
The solution is 3.

3.
$$\begin{aligned} 2a-5 &= 7 \\ 2a-5+5 &= 7+5 \\ 2a &= 12 \\ \frac{1}{2}\cdot 2a &= \frac{1}{2}\cdot 12 \\ a &= 6 \end{aligned}$$
The solution is 6.

5.
$$\begin{aligned} 5 &= 4x+9 \\ 5+(-9) &= 4x+9+(-9) \\ -4 &= 4x \\ \frac{1}{4}(-4) &= \frac{1}{4}(4x) \\ -1 &= x \end{aligned}$$
The solution is −1.

7.
$$\begin{aligned} 2x-5 &= -11 \\ 2x-5+5 &= -11+5 \\ 2x &= -6 \\ \frac{1}{2}(2x) &= \frac{1}{2}(-6) \\ x &= -3 \end{aligned}$$
The solution is −3.

9.
$$\begin{aligned} 4-3w &= -2 \\ 4+(-4)-3w &= -2+(-4) \\ -3w &= -6 \\ -\frac{1}{3}(-3w) &= -\frac{1}{3}(-6) \\ w &= 2 \end{aligned}$$
The solution is 2.

11.
$$\begin{aligned} 8-3t &= 2 \\ 8+(-8)-3t &= 2+(-8) \\ -3t &= -6 \\ -\frac{1}{3}(-3t) &= -\frac{1}{3}(-6) \\ t &= 2 \end{aligned}$$
The solution is 2.

13.
$$\begin{aligned} 4a-20 &= 0 \\ 4a-20+20 &= 0+20 \\ 4a &= 20 \\ \frac{1}{4}(4a) &= \frac{1}{4}\cdot 20 \\ a &= 5 \end{aligned}$$
The solution is 5.

15.
$$\begin{aligned} 6+2b &= 0 \\ 6+(-6)+2b &= 0+(-6) \\ 2b &= -6 \\ \frac{1}{2}\cdot 2b &= \frac{1}{2}(-6) \\ b &= -3 \end{aligned}$$
The solution is −3.

17.
$$\begin{aligned} -2x+5 &= -7 \\ -2x+5+(-5) &= -7+(-5) \\ -2x &= -12 \\ -\frac{1}{2}(-2x) &= -\frac{1}{2}(-12) \\ x &= 6 \end{aligned}$$
The solution is 6.

19.
$$\begin{aligned} -12x+30 &= -6 \\ -12x+30+(-30) &= -6+(-30) \\ -12x &= -36 \\ -\frac{1}{12}(-12x) &= -\frac{1}{12}(-36) \\ x &= 3 \end{aligned}$$
The solution is 3.

21.
$$\begin{aligned} 2 &= 7-5a \\ 2+(-7) &= 7+(-7)-5a \\ -5 &= -5a \\ -\frac{1}{5}(-5) &= -\frac{1}{5}(-5a) \\ 1 &= a \end{aligned}$$
The solution is 1.

23.
$$\begin{aligned} -35 &= -6b + 1 \\ -35 + (-1) &= -6b + 1 + (-1) \\ -36 &= -6b \\ -\frac{1}{6}(-36) &= -\frac{1}{6}(-6b) \\ 6 &= b \end{aligned}$$
The solution is 6.

25.
$$\begin{aligned} -3m - 21 &= 0 \\ -3m - 21 + 21 &= 0 + 21 \\ -3m &= 21 \\ -\frac{1}{3}(-3m) &= -\frac{1}{3}(21) \\ m &= -7 \end{aligned}$$
The solution is -7.

27.
$$\begin{aligned} -4y + 15 &= 15 \\ -4y + 15 + (-15) &= 15 + (-15) \\ -4y &= 0 \\ -\frac{1}{4}(-4y) &= -\frac{1}{4}(0) \\ y &= 0 \end{aligned}$$
The solution is 0.

29.
$$\begin{aligned} 9 - 4x &= 6 \\ 9 + (-9) - 4x &= 6 + (-9) \\ -4x &= -3 \\ -\frac{1}{4}(-4x) &= -\frac{1}{4}(-3) \\ x &= \frac{3}{4} \end{aligned}$$
The solution is $\frac{3}{4}$.

31.
$$\begin{aligned} 9x - 4 &= 0 \\ 9x - 4 + 4 &= 0 + 4 \\ 9x &= 4 \\ \frac{1}{9}(9x) &= \frac{1}{9}(4) \\ x &= \frac{4}{9} \end{aligned}$$
The solution is $\frac{4}{9}$.

33.
$$\begin{aligned} 1 - 3x &= 0 \\ 1 + (-1) - 3x &= 0 + (-1) \\ -3x &= -1 \\ -\frac{1}{3}(-3x) &= -\frac{1}{3}(-1) \\ x &= \frac{1}{3} \end{aligned}$$
The solution is $\frac{1}{3}$.

35.
$$\begin{aligned} 12w + 11 &= 5 \\ 12w + 11 + (-11) &= 5 + (-11) \\ 12w &= -6 \\ \frac{1}{12}(12w) &= \frac{1}{12}(-6) \\ w &= -\frac{6}{12} = -\frac{1}{2} \end{aligned}$$
The solution is $-\frac{1}{2}$.

37.
$$\begin{aligned} 8b - 3 &= -9 \\ 8b - 3 + 3 &= -9 + 3 \\ 8b &= -6 \\ \frac{1}{8}(8b) &= \frac{1}{8}(-6) \\ b &= -\frac{6}{8} = -\frac{3}{4} \end{aligned}$$
The solution is $-\frac{3}{4}$.

39.
$$\begin{aligned} 7 - 9a &= 4 \\ 7 + (-7) - 9a &= 4 + (-7) \\ -9a &= -3 \\ -\frac{1}{9}(-9a) &= -\frac{1}{9}(-3) \\ a &= \frac{3}{9} = \frac{1}{3} \end{aligned}$$
The solution is $\frac{1}{3}$.

41.
$$\begin{aligned} 10 &= -18x + 7 \\ 10 + (-7) &= -18x + 7 + (-7) \\ 3 &= -18x \\ -\frac{1}{18}(3) &= -\frac{1}{18}(-18x) \\ -\frac{3}{18} &= x \\ -\frac{1}{6} &= x \end{aligned}$$
The solution is $-\frac{1}{6}$.

43.
$$\begin{aligned} 4a + \frac{3}{4} &= \frac{19}{4} \\ 4a + \frac{3}{4} + \left(-\frac{3}{4}\right) &= \frac{19}{4} + \left(-\frac{3}{4}\right) \\ 4a &= \frac{16}{4} \\ 4a &= 4 \\ \frac{1}{4}(4a) &= \frac{1}{4}(4) \\ a &= 1 \end{aligned}$$
The solution is 1.

45.
$$3x - \frac{5}{6} = \frac{13}{6}$$
$$3x - \frac{5}{6} + \frac{5}{6} = \frac{13}{6} + \frac{5}{6}$$
$$3x = \frac{18}{6}$$
$$3x = 3$$
$$\frac{1}{3}(3x) = \frac{1}{3}(3)$$
$$x = 1$$
The solution is 1.

47.
$$9x + \frac{4}{5} = \frac{4}{5}$$
$$9x + \frac{4}{5} + \left(-\frac{4}{5}\right) = \frac{4}{5} + \left(-\frac{4}{5}\right)$$
$$9x = 0$$
$$\frac{1}{9}(9x) = \frac{1}{9}(0)$$
$$x = 0$$
The solution is 0.

49.
$$8 = 10x - 5$$
$$8 + 5 = 10x - 5 + 5$$
$$13 = 10x$$
$$\frac{1}{10}(13) = \frac{1}{10}(10x)$$
$$\frac{13}{10} = x$$
The solution is $\frac{13}{10}$.

51.
$$7 = 9 - 5a$$
$$7 + (-9) = 9 + (-9) - 5a$$
$$-2 = -5a$$
$$-\frac{1}{5}(-2) = -\frac{1}{5}(-5a)$$
$$\frac{2}{5} = a$$
The solution is $\frac{2}{5}$.

53.
$$12x + 19 = 3$$
$$12x + 19 + (-19) = 3 + (-19)$$
$$12x = -16$$
$$\frac{1}{12}(12x) = \frac{1}{12}(-16)$$
$$x = -\frac{16}{12} = -\frac{4}{3}$$
The solution is $-\frac{4}{3}$.

55.
$$-4x + 3 = 9$$
$$-4x + 3 + (-3) = 9 + (-3)$$
$$-4x = 6$$
$$-\frac{1}{4}(-4x) = -\frac{1}{4}(6)$$
$$x = -\frac{6}{4} = -\frac{3}{2}$$
The solution is $-\frac{3}{2}$.

57.
$$\frac{1}{3}m - 1 = 5$$
$$\frac{1}{3}m - 1 + 1 = 5 + 1$$
$$\frac{1}{3}m = 6$$
$$3\left(\frac{1}{3}m\right) = 3 \cdot 6$$
$$m = 18$$
The solution is 18.

59.
$$\frac{3}{4}n + 7 = 13$$
$$\frac{3}{4}n + 7 + (-7) = 13 + (-7)$$
$$\frac{3}{4}n = 6$$
$$\frac{4}{3}\left(\frac{3}{4}n\right) = \frac{4}{3}(6)$$
$$n = 8$$
The solution is 8.

61.
$$-\frac{3}{8}b + 4 = 10$$
$$-\frac{3}{8}b + 4 + (-4) = 10 + (-4)$$
$$-\frac{3}{8}b = 6$$
$$-\frac{8}{3}\left(-\frac{3}{8}b\right) = -\frac{8}{3}(6)$$
$$b = -16$$
The solution is -16.

63.
$$\frac{y}{5} - 2 = 3$$
$$\frac{y}{5} - 2 + 2 = 3 + 2$$
$$\frac{y}{5} = 5$$
$$5\left(\frac{1}{5}y\right) = 5 \cdot 5$$
$$y = 25$$
The solution is 25.

65.
$$\frac{2}{3}x - \frac{5}{6} = -\frac{1}{3}$$
$$6\left(\frac{2}{3}x - \frac{5}{6}\right) = 6\left(-\frac{1}{3}\right)$$
$$4x - 5 = -2$$
$$4x - 5 + 5 = -2 + 5$$
$$4x = 3$$
$$\frac{4x}{4} = \frac{3}{4}$$
$$x = \frac{3}{4}$$
The solution is $\frac{3}{4}$.

67.
$$\frac{1}{2} - \frac{2}{3}x = \frac{1}{4}$$
$$12\left(\frac{1}{2} - \frac{2}{3}x\right) = 12\left(\frac{1}{4}\right)$$
$$6 - 8x = 3$$
$$6 - 8x - 6 = 3 - 6$$
$$-8x = -3$$
$$\frac{-8x}{-8} = \frac{-3}{-8}$$
$$x = \frac{3}{8}$$
The solution is $\frac{3}{8}$.

69.
$$\frac{3}{2} = \frac{5}{6} + \frac{3x}{8}$$
$$\frac{3}{2} - \frac{5}{6} = \frac{5}{6} + \frac{3x}{8} - \frac{5}{6}$$
$$\frac{2}{3} = \frac{3x}{8}$$
$$\frac{8}{3}\left(\frac{2}{3}\right) = \frac{8}{3}\left(\frac{3x}{8}\right)$$
$$\frac{16}{9} = x$$
The solution is $\frac{16}{9}$.

71.
$$\frac{11}{27} = \frac{4}{9} - \frac{2x}{3}$$
$$\frac{11}{27} - \frac{4}{9} = \frac{4}{9} - \frac{2x}{3} - \frac{4}{9}$$
$$-\frac{1}{27} = -\frac{2x}{3}$$
$$-\frac{3}{2}\left(-\frac{1}{27}\right) = -\frac{3}{2}\left(\frac{-2x}{3}\right)$$
$$\frac{1}{18} = x$$
The solution is $\frac{1}{18}$.

73.
$$\begin{aligned} 7 &= \frac{2x}{5} + 4 \\ 7 + (-4) &= \frac{2x}{5} + 4 + (-4) \\ 3 &= \frac{2x}{5} \\ \frac{5}{2}(3) &= \frac{5}{2}\left(\frac{2}{5}x\right) \\ \frac{15}{2} &= x \end{aligned}$$
The solution is $\frac{15}{2}$.

75.
$$\begin{aligned} 7 - \frac{5}{9}y &= 9 \\ 7 + (-7) - \frac{5}{9}y &= 9 + (-7) \\ -\frac{5}{9}y &= 2 \\ -\frac{9}{5}\left(-\frac{5}{9}y\right) &= -\frac{9}{5}(2) \\ y &= -\frac{18}{5} \end{aligned}$$
The solution is $-\frac{18}{5}$.

77.
$$\begin{aligned} 5y + 9 + 2y &= 23 \\ 7y + 9 &= 23 \\ 7y + 9 + (-9) &= 23 + (-9) \\ 7y &= 14 \\ \frac{1}{7}(7y) &= \frac{1}{7}(14) \\ y &= 2 \end{aligned}$$
The solution is 2.

79.
$$\begin{aligned} 11z - 3 - 7z &= 9 \\ 4z - 3 &= 9 \\ 4z - 3 + 3 &= 9 + 3 \\ 4z &= 12 \\ \frac{1}{4}(4z) &= \frac{1}{4}(12) \\ z &= 3 \end{aligned}$$
The solution is 3.

81.
$$\begin{aligned} 3x + 4y &= 13 \\ \text{when } y &= -2: \\ 3x + 4(-2) &= 13 \\ 3x - 8 &= 13 \\ 3x &= 21 \\ \frac{1}{3} \cdot 3x &= \frac{1}{3} \cdot 21 \\ x &= 7 \end{aligned}$$

83.
$$\begin{aligned} -4x + 3y &= 9 \\ \text{when } x &= 0: \\ -4 \cdot 0 + 3y &= 9 \\ -0 + 3y &= 9 \\ 3y &= 9 \\ \frac{1}{3} \cdot 3y &= \frac{1}{3} \cdot 9 \\ y &= 3 \end{aligned}$$

85.
$$\begin{aligned} 2x - 3 &= 7 \\ 2x &= 7 + 3 \\ \frac{1}{2}(2x) &= \frac{1}{2}(10) \\ x &= 5 \end{aligned} \qquad \begin{aligned} 3x + 4 &= 3(5) + 4 \\ &= 15 + 4 \\ &= 19 \end{aligned}$$
The solution is 19.

87.
$$\begin{aligned} 4 - 5x &= -1 \\ -5x &= -1 - 4 \\ -5x &= -5 \\ -\frac{1}{5}(-5x) &= -\frac{1}{5}(-5) \\ x &= 1 \end{aligned} \qquad \begin{aligned} x^2 - 3x + 1 &= (1)^2 - 3(1) + 1 \\ &= 1 - 3 + 1 \\ &= 2 - 3 \\ &= -1 \end{aligned}$$
The solution is -1.

89.
$$\begin{aligned} 5x + 3 - 2x &= 12 \\ 3x + 3 &= 12 \\ 3x &= 12 - 3 \\ 3x &= 9 \\ \frac{1}{3}(3x) &= \frac{1}{3}(9) \\ x &= 3 \end{aligned} \qquad \begin{aligned} 4 - 5x &= 4 - 5(3) \\ &= 4 - 15 \\ &= -11 \end{aligned}$$
The solution is -11.

Objective B Exercises

91. Strategy Given: $S = \$156.80$
$C = \$98$

Unknown: r

Solution

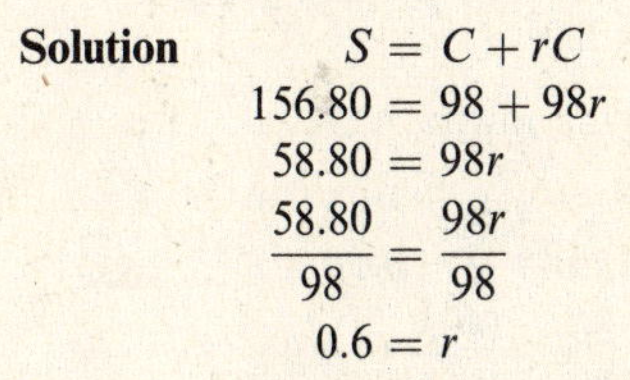

$$\begin{aligned} S &= C + rC \\ 156.80 &= 98 + 98r \\ 58.80 &= 98r \\ \frac{58.80}{98} &= \frac{98r}{98} \\ 0.6 &= r \end{aligned}$$
The markup rate is 60%.

93. Strategy Given: $S = \$82.60$
$r = 40\% = 0.4$

Unknown: C

Solution
$$\begin{aligned} S &= C + rC \\ 82.60 &= C + 0.4C \\ 82.60 &= 1.4C \\ \frac{82.60}{1.4} &= \frac{1.4C}{1.4} \\ 59 &= C \end{aligned}$$
The cost of the basketball is $59.

95. Strategy Given: $S = \$520$
$C = \$360$

Unknown: r

Solution
$$S = C + rC$$
$$520 = 360 + 360r$$
$$160 = 360r$$
$$\frac{160}{360} = \frac{360r}{360}$$
$$0.4444 \approx r$$
The markup rate is 44.4%.

97. Strategy Given: $S = \$11.90$
$r = 40\% = 0.4$

Unknown: C

Solution
$$S = C + rC$$
$$11.90 = C + .4C$$
$$11.90 = 1.4C$$
$$\frac{11.90}{1.4} = \frac{1.4C}{1.4}$$
$$8.5 = C$$
The cost of the CD is \$8.50.

99. Strategy Given: $S = \$995$
$R = \$1295$

Unknown: r

Solution
$$S = R - rR$$
$$995 = 1295 - 1295r$$
$$-300 = -1295r$$
$$\frac{-300}{-1295} = \frac{-1295r}{-1295}$$
$$0.2316 \approx r$$
The discount rate is 23.2%.

101. Strategy Given: $S = \$180$
$r = 40\% = 0.4$

Unknown: R

Solution
$$S = R - rR$$
$$180 = R - 0.4R$$
$$180 = 0.6R$$
$$\frac{180}{0.6} = \frac{0.6R}{0.6}$$
$$300 = R$$
The regular price of the tool set was \$300.

103. Strategy Given: $S = \$201.50$
$R = \$325$

Unknown: r

Solution
$$S = R - rR$$
$$201.50 = 325 - 325r$$
$$-123.50 = -325r$$
$$\frac{-123.50}{-325} = \frac{-325r}{-325}$$
$$0.38 = r$$
The markdown rate is 38%.

105. Strategy Given: $S = \$165$
$r = 40\% = 0.4$

Unknown: R

Solution
$$S = R - rR$$
$$165 = R - 0.4R$$
$$165 = 0.6R$$
$$\frac{165}{0.6} = \frac{0.6R}{0.6}$$
$$275 = R$$
The regular price of the telescope is \$275.

107.
$$S = 16t^2 + vt$$
$$80 = 16(2)^2 + v(2)$$
$$80 = 16(4) + 2v$$
$$80 = 64 + 2v$$
$$80 - 64 = 2v$$
$$16 = 2v$$
$$\frac{1}{2}(16) = \frac{1}{2}(2v)$$
$$8 = v$$
The initial velocity is 8 ft/s.

109.
$$V = C - 6000t$$
$$38{,}000 = 50{,}000 - 6000t$$
$$-12{,}000 = -6000t$$
$$-\frac{1}{6000}(-12{,}000) = -\frac{1}{6000}(-6000t)$$
$$2 = t$$
The depreciated value will be \$38,000 after 2 years.

111.
$$H = 1.2L + 27.8$$
$$66 = 1.2L + 27.8$$
$$66 - 27.8 = 1.2L$$
$$38.2 = 1.2L$$
$$\frac{38.2}{1.2} = L$$
$$31.8 \approx L$$
The approximate length is 31.8 in. to the nearest tenth.

113.

$$C = \frac{1}{4}D - 45$$
$$-3 = \frac{1}{4}D - 45$$
$$-3 + 45 = \frac{1}{4}D$$
$$42 = \frac{1}{4}D$$
$$4(42) = 4\left(\frac{1}{4}D\right)$$
$$168 = D$$

The distance the car will skid is 168 ft.

115.

$$N = \frac{2.51P_1P_2}{d^2}$$
$$1{,}100{,}000 = \frac{2.51(48{,}000)P_1}{75^2}$$
$$1{,}100{,}000 = \frac{2.51(48{,}000)P_1}{5625}$$
$$\frac{1{,}100{,}000 \times 5625}{2.51 \times 48{,}000} = P_1$$
$$\frac{6187500}{120.48} = P_1$$
$$51{,}357.072 \approx P_1$$

The estimated population is 51,000 people to the nearest thousand.

117. $2x + 1 = a = 3x - 2$

Solve $2x + 1 = 3x - 2$ for x and then find the value of a.

$$2x + 1 - 1 = 3x - 2 - 1$$
$$2x = 3x - 3$$
$$2x - 3x = 3x - 3x - 3$$
$$-x = -3$$
$$x = 3$$

Evaluate $2x + 1$ or $3x - 2$ when $x = 3$ to find a.

$2 \cdot 3 + 1 = 7 = a$ or $3 \cdot 3 - 2 = 7 = a$

The solution is $a = 7$.

Applying the Concepts

119. Strategy First find the reduced price, P.

Given: $S = \$180$

$r = 15\% = 0.15$

Unknown: P

Use the equation $S = P - rP$.

Then, find the regular price.

Given: P

$r = \frac{1}{3}$

Unknown: R

Use the equation $P = R - rR$.

Solution

$$S = P - rP$$
$$180 = P - 0.15P$$
$$180 = 0.85P$$
$$211.7647 \approx P$$
$$P = R - rR$$
$$211.7647 = R - \frac{1}{3}R$$
$$211.7647 = \frac{2}{3}R$$
$$317.65 \approx R$$

The regular price is \$317.65.

121.

$$A = \frac{1}{2}bh$$
$$40 = \frac{1}{2} \cdot 10(x + 2)$$
$$40 = 5(x + 2)$$
$$40 = 5x + 10$$
$$40 - 10 = 5x + 10 - 10$$
$$30 = 5x$$
$$\frac{30}{5} = \frac{5x}{5}$$
$$6 = x$$
$$x = 6\text{ m}$$

123. $x \div 15 = 25$ remainder 10

$$\frac{x}{15} = 25\frac{10}{15}$$
$$\frac{x}{15} = \frac{385}{15}$$
$$15\left(\frac{x}{15}\right) = 15\left(\frac{385}{15}\right)$$
$$x = 385$$

The solution is 385.

125. The solution of this equation is 0. Therefore, dividing each side of the equation by x means that each side was divided by 0. However, division by 0 is not allowed.

127. Students should note that in order to answer the questions, we need to know the distance from New York to Los Angeles, and we need to know the speed of the airplane.

Section 3.3

Objective A Exercises

1.

$$8x + 5 = 4x + 13$$
$$8x + (-4x) + 5 = 4x + (-4x) + 13$$
$$4x + 5 = 13$$
$$4x + 5 + (-5) = 13 + (-5)$$
$$4x = 8$$
$$\frac{1}{4}(4x) = \frac{1}{4}(8)$$
$$x = 2$$

The solution is 2.

3.
$$\begin{aligned} 5x - 4 &= 2x + 5 \\ 5x + (-2x) - 4 &= 2x + (-2x) + 5 \\ 3x - 4 &= 5 \\ 3x - 4 + 4 &= 5 + 4 \\ 3x &= 9 \\ \frac{1}{3}(3x) &= \frac{1}{3}(9) \\ x &= 3 \end{aligned}$$
The solution is 3.

5.
$$\begin{aligned} 15x - 2 &= 4x - 13 \\ 15x + (-4x) - 2 &= 4x + (-4x) - 13 \\ 11x - 2 &= -13 \\ 11x - 2 + 2 &= -13 + 2 \\ 11x &= -11 \\ \frac{1}{11}(11x) &= \frac{1}{11}(-11) \\ x &= -1 \end{aligned}$$
The solution is -1.

7.
$$\begin{aligned} 3x + 1 &= 11 - 2x \\ 3x + 2x + 1 &= 11 - 2x + 2x \\ 5x + 1 &= 11 \\ 5x + 1 + (-1) &= 11 + (-1) \\ 5x &= 10 \\ \frac{1}{5} \cdot 5x &= \frac{1}{5} \cdot 10 \\ x &= 2 \end{aligned}$$
The solution is 2.

9.
$$\begin{aligned} 2x - 3 &= -11 - 2x \\ 2x + 2x - 3 &= -11 - 2x + 2x \\ 4x - 3 &= -11 \\ 4x - 3 + 3 &= -11 + 3 \\ 4x &= -8 \\ \frac{1}{4}(4x) &= \frac{1}{4}(-8) \\ x &= -2 \end{aligned}$$
The solution is -2.

11.
$$\begin{aligned} 2b + 3 &= 5b + 12 \\ 2b + (-5b) + 3 &= 5b + (-5b) + 12 \\ -3b + 3 &= 12 \\ -3b + 3 + (-3) &= 12 + (-3) \\ -3b &= 9 \\ -\frac{1}{3}(-3b) &= -\frac{1}{3}(9) \\ b &= -3 \end{aligned}$$
The solution is -3.

13.
$$\begin{aligned} 4y - 8 &= y - 8 \\ 4y + (-y) - 8 &= y + (-y) - 8 \\ 3y - 8 &= -8 \\ 3y - 8 + 8 &= -8 + 8 \\ 3y &= 0 \\ \frac{1}{3}(3y) &= \frac{1}{3}(0) \\ y &= 0 \end{aligned}$$
The solution is 0.

15.
$$\begin{aligned} 6 - 5x &= 8 - 3x \\ 6 - 5x + 3x &= 8 - 3x + 3x \\ 6 - 2x &= 8 \\ 6 + (-6) - 2x &= 8 + (-6) \\ -2x &= 2 \\ -\frac{1}{2}(-2x) &= -\frac{1}{2}(2) \\ x &= -1 \end{aligned}$$
The solution is -1.

17.
$$\begin{aligned} 5 + 7x &= 11 + 9x \\ 5 + 7x + (-9x) &= 11 + 9x + (-9x) \\ 5 - 2x &= 11 \\ 5 + (-5) - 2x &= 11 + (-5) \\ -2x &= 6 \\ -\frac{1}{2}(-2x) &= -\frac{1}{2}(6) \\ x &= -3 \end{aligned}$$
The solution is -3.

19.
$$\begin{aligned} 2x - 4 &= 6x \\ 2x + (-2x) - 4 &= 6x + (-2x) \\ -4 &= 4x \\ \frac{1}{4}(-4) &= \frac{1}{4}(4x) \\ -1 &= x \end{aligned}$$
The solution is -1.

21.
$$\begin{aligned} 8m &= 3m + 20 \\ 8m + (-3m) &= 3m + (-3m) + 20 \\ 5m &= 20 \\ \frac{1}{5} \cdot 5m &= \frac{1}{5} \cdot 20 \\ m &= 4 \end{aligned}$$
The solution is 4.

23.
$$\begin{aligned} 8b + 5 &= 5b + 7 \\ 8b + (-5b) + 5 &= 5b + (-5b) + 7 \\ 3b + 5 &= 7 \\ 3b + 5 + (-5) &= 7 + (-5) \\ 3b &= 2 \\ \frac{1}{3} \cdot 3b &= \frac{1}{3} \cdot 2 \\ b &= \frac{2}{3} \end{aligned}$$
The solution is $\frac{2}{3}$.

25.
$$\begin{aligned} 7x - 8 &= x - 3 \\ 7x + (-x) - 8 &= x + (-x) - 3 \\ 6x - 8 &= -3 \\ 6x - 8 + 8 &= -3 + 8 \\ 6x &= 5 \\ \frac{1}{6} \cdot 6x &= \frac{1}{6} \cdot 5 \\ x &= \frac{5}{6} \end{aligned}$$
The solution is $\frac{5}{6}$.

27.
$$\begin{aligned} 2m - 1 &= -6m + 5 \\ 2m + 6m - 1 &= -6m + 6m + 5 \\ 8m - 1 &= 5 \\ 8m - 1 + 1 &= 5 + 1 \\ 8m &= 6 \\ \frac{1}{8} \cdot 8m &= \frac{1}{8} \cdot 6 \\ m &= \frac{3}{4} \end{aligned}$$
The solution is $\frac{3}{4}$.

29.
$$\begin{aligned} 7x + 3 &= 5x - 7 \\ 7x - 5x &= -7 - 3 \\ 2x &= -10 \\ \frac{1}{2}(2x) &= \frac{1}{2}(-10) \\ x &= -5 \end{aligned} \qquad \begin{aligned} 3x - 2 &= 3(-5) - 2 \\ &= -15 - 2 \\ &= -17 \end{aligned}$$
The answer is -17.

31.
$$\begin{aligned} 1 - 5c &= 4 - 4c \\ -5c + 4c &= 4 - 1 \\ -c &= 3 \\ c &= -3 \end{aligned} \qquad \begin{aligned} 3c^2 - 4c + 2 &= 3(-3)^2 - 4(-3) + 2 \\ &= 3(9) + 12 + 2 \\ &= 27 + 12 + 2 \\ &= 41 \end{aligned}$$
The answer is 41.

33.
$$\begin{aligned} 3z + 1 &= 1 - 5z \\ 3z + 5z &= 1 - 1 \\ 8z &= 0 \\ z &= 0 \end{aligned} \qquad \begin{aligned} 3z^2 - 7z + 8 &= 3(0)^2 - 7(0) + 8 \\ &= 3(0) - 0 + 8 \\ &= 0 - 0 + 8 \\ &= 8 \end{aligned}$$
The answer is 8.

Objective B Exercises

35.
$$\begin{aligned} 6y + 2(2y + 3) &= 16 \\ 6y + 4y + 6 &= 16 \\ 10y + 6 &= 16 \\ 10y + 6 + (-6) &= 16 + (-6) \\ 10y &= 10 \\ \frac{1}{10} \cdot 10y &= \frac{1}{10} \cdot 10 \\ y &= 1 \end{aligned}$$
The solution is 1.

37.
$$\begin{aligned} 12x - 2(4x - 6) &= 28 \\ 12x - 8x + 12 &= 28 \\ 4x + 12 &= 28 \\ 4x + 12 + (-12) &= 28 + (-12) \\ 4x &= 16 \\ \frac{1}{4} \cdot 4x &= \frac{1}{4} \cdot 16 \\ x &= 4 \end{aligned}$$
The solution is 4.

39.
$$\begin{aligned} 9m - 4(2m - 3) &= 11 \\ 9m - 8m + 12 &= 11 \\ m + 12 &= 11 \\ m + 12 + (-12) &= 11 + (-12) \\ m &= -1 \end{aligned}$$
The solution is -1.

41.
$$\begin{aligned} 4(1 - 3x) + 7x &= 9 \\ 4 - 12x + 7x &= 9 \\ 4 - 5x &= 9 \\ 4 + (-4) - 5x &= 9 + (-4) \\ -5x &= 5 \\ -\frac{1}{5}(-5x) &= -\frac{1}{5}(5) \\ x &= -1 \end{aligned}$$
The solution is -1.

43.
$$\begin{aligned} 0.22(x + 6) &= 0.2x + 1.8 \\ 0.22x + 1.32 &= 0.2x + 1.8 \\ 0.22x - 0.2x + 1.32 &= 0.2x - 0.2x + 1.8 \\ 0.02x + 1.32 &= 1.8 \\ 0.02x + 1.32 - 1.32 &= 1.8 - 1.32 \\ 0.02x &= 0.48 \\ \frac{0.02x}{0.02} &= \frac{0.48}{0.02} \\ x &= 24 \end{aligned}$$
The solution is 24.

45.
$$\begin{aligned} 0.3x + 0.3(x + 10) &= 300 \\ 0.3x + 0.3x + 3 &= 300 \\ 0.6x + 3 &= 300 \\ 0.6x + 3 - 3 &= 300 - 3 \\ 0.6x &= 297 \\ \frac{0.6x}{0.6} &= \frac{297}{0.6} \\ x &= 495 \end{aligned}$$
The solution is 495.

47.
$$\begin{aligned} 5 - (9 - 6x) &= 2x - 2 \\ 5 - 9 + 6x &= 2x - 2 \\ -4 + 6x &= 2x - 2 \\ -4 + 6x + (-2x) &= 2x + (-2x) - 2 \\ -4 + 4x &= -2 \\ -4 + 4 + 4x &= -2 + 4 \\ 4x &= 2 \\ \frac{1}{4} \cdot 4x &= \frac{1}{4} \cdot 2 \\ x &= \frac{1}{2} \end{aligned}$$
The solution is $\frac{1}{2}$.

49.
$$\begin{aligned} 3[2-4(y-1)] &= 3(2y+8) \\ 3[2-4y+4] &= 6y+24 \\ 3[6-4y] &= 6y+24 \\ 18-12y &= 6y+24 \\ 18-12y+(-6y) &= 6y+(-6y)+24 \\ 18-18y &= 24 \\ 18+(-18)-18y &= 24+(-18) \\ -18y &= 6 \\ -\frac{1}{18}(-18y) &= -\frac{1}{18}(6) \\ y &= -\frac{1}{3} \end{aligned}$$
The solution is $-\frac{1}{3}$.

51.
$$\begin{aligned} 3a+2[2+3(a-1)] &= 2(3a+4) \\ 3a+2[2+3a-3] &= 6a+8 \\ 3a+2[-1+3a] &= 6a+8 \\ 3a-2+6a &= 6a+8 \\ 9a-2 &= 6a+8 \\ 9a+(-6a)-2 &= 6a+(-6a)+8 \\ 3a-2 &= 8 \\ 3a-2+2 &= 8+2 \\ 3a &= 10 \\ \frac{1}{3}(3a) &= \frac{1}{3}(10) \\ a &= \frac{10}{3} \end{aligned}$$
The solution is $a = \frac{10}{3}$.

53.
$$\begin{aligned} -2[4-(3b+2)] &= 5-2(3b+6) \\ -2[4-3b-2] &= 5-6b-12 \\ -2[2-3b] &= -7-6b \\ -4+6b &= -7-6b \\ -4+6b+6b &= -7-6b+6b \\ -4+12b &= -7 \\ -4+4+12b &= -7+4 \\ 12b &= -3 \\ \frac{1}{12}(12b) &= \frac{1}{12}(-3) \\ b &= -\frac{1}{4} \end{aligned}$$
The solution is $-\frac{1}{4}$.

55.
$$\begin{aligned} 4-3a &= 7-2(2a+5) \\ 4-3a &= 7-4a-10 \\ 4-3a &= -3-4a \\ -3a+4a &= -3-4 \\ a &= -7 \end{aligned} \qquad \begin{aligned} a^2+7a &= (-7)^2+(7)(-7) \\ &= 49-49 \\ &= 0 \end{aligned}$$
The answer is 0.

57.
$$\begin{aligned} 2z-5 &= 3(4z+5) \\ 2z-5 &= 12z+15 \\ 2z-12z &= 15+5 \\ -10z &= 20 \\ -\frac{1}{10}(-10z) &= -\frac{1}{10}(20) \\ z &= -2 \end{aligned} \qquad \begin{aligned} \frac{z^2}{z-2} &= \frac{(-2)^2}{-2-2} \\ &= \frac{4}{-4} = -1 \end{aligned}$$
The answer is -1.

Objective C Exercises

59. Strategy To find the force when the system balances, replace the variables F_1, x, and d in the lever system equation by the given values and solve for F_2.

Solution
$$\begin{aligned} F_1 \cdot x &= F_2(d-x) \\ 100 \cdot 2 &= F_2 \cdot (10-2) \\ 100 \cdot 2 &= F_2 \cdot 8 \\ 200 &= 8F_2 \\ \frac{1}{8}(200) &= \frac{1}{8}(8F_2) \\ 25 &= F_2 \end{aligned}$$
A force of 25 lb must be applied to the other end.

61. Strategy To find the location of the fulcrum when the system balances, replace the variables F_1, F_2, and d in the lever system equation and solve for x.

Solution
$$\begin{aligned} F_1x &= F_2(d-x) \\ 180x &= 120(15-x) \\ 180x &= 1800-120x \\ 180x+120x &= 1800 \\ 300x &= 1800 \\ \frac{300x}{300} &= \frac{1800}{300} \\ x &= 6 \end{aligned}$$
The fulcrum is 6 ft from the 180-lb person.

63. Strategy To find the location of the fulcrum when the system balances, replace the variables F_1, F_2, and d in the lever system equation by the given values and solve for x.

Solution
$$\begin{aligned} F_1 \cdot x &= F_2(d-x) \\ 128x &= 160(18-x) \\ 128x &= 2880-160x \\ 128x+160x &= 2880 \\ 288x &= 2880 \\ \frac{1}{288}(288x) &= \frac{1}{288}(2880) \\ x &= 10 \end{aligned}$$
The fulcrum is 10 ft from the 128-lb acrobat.

65. Strategy To find the minimum force when the system balances, replace the variables, F_1, d, and x in the lever system equation and solve for F_2.

Solution

$$\begin{aligned} F_1x &= F_2(d-x) \\ 150(1.5) &= F_2(8-1.5) \\ 225 &= 6.5F_2 \\ \frac{225}{6.5} &= \frac{6.5F_2}{6.5} \\ 34.62 &\approx F_2 \end{aligned}$$

The minimum force to move the rock is 34.6 lb.

67. Strategy To find the break-even point, replace the variables P, C, and F in the break-even equation and solve for x.

Solution

$$\begin{aligned} Px &= Cx+F \\ 325x &= 175x+39{,}000 \\ 325x-175x &= 39{,}000 \\ 150x &= 39{,}000 \\ \frac{150x}{150} &= \frac{39{,}000}{150} \\ x &= 260 \end{aligned}$$

The break-even point is 260 barbecues.

69. Strategy To find the break-even point, replace the variables P, C, and F in the break-even equation and solve for x.

Solution

$$\begin{aligned} Px &= Cx+F \\ 49x &= 12x+19{,}240 \\ 49x-12x &= 19{,}240 \\ 37x &= 19{,}240 \\ \frac{37x}{37} &= \frac{19{,}240}{37} \\ x &= 520 \end{aligned}$$

The break-even point is 520 desk lamps.

71. Strategy To find the break-even point, replace the variables P, C and F in the cost equation by the given values and solve for x.

Solution

$$\begin{aligned} Px &= Cx+F \\ 7.00x &= 1.50x+16500 \\ 7.00x-1.50x &= 16500 \\ 5.50x &= 16500 \\ x &= \frac{16500}{5.50} \\ x &= 3000 \end{aligned}$$

The break-even point is 3000 softball bats.

Applying the Concepts

73.
$$\begin{aligned} 3(2x-1)-(6x-4) &= -9 \\ 6x-3-6x+4 &= -9 \\ 1 &= -9 \end{aligned}$$

No solution.

75.
$$\begin{aligned} \frac{1}{5}(25-10a)+4 &= \frac{1}{3}(12a-15)+14 \\ 5-2a+4 &= 4a-5+14 \\ 9-2a &= 4a+9 \\ 9+(-9)-2a &= 4a+9+(-9) \\ -2a &= 4a \\ -2a+(-4a) &= 4a+(-4a) \\ -6a &= 0 \\ -\frac{1}{6}(-6a) &= -\frac{1}{6}(0) \\ a &= 0 \end{aligned}$$

The solution is 0.

77. Students should explain that the solution of $2x+3=3$ is the (real) number zero. However, there is no solution of $x=x+1$ because there is no real number that is equal to itself plus 1.

Section 3.4

Objective A Exercises

1. The unknown number: x

[The difference between a number and fifteen] is [seven]

$$\begin{aligned} x-15 &= 7 \\ x-15+15 &= 7+15 \\ x &= 22 \end{aligned}$$

The number is 22.

3. The unknown number: x

[The product of seven and a number] is [negative twenty-one]

$$\begin{aligned} 7x &= -21 \\ \frac{1}{7}(7x) &= \frac{1}{7}(-21) \\ x &= -3 \end{aligned}$$

The number is -3.

5. The unknown number: x

[The difference between nine and a number] is [seven]

$$\begin{aligned} 9-x &= 7 \\ 9-9-x &= 7-9 \\ -x &= -2 \\ x &= 2 \end{aligned}$$

The number is 2.

7. The unknown number: x

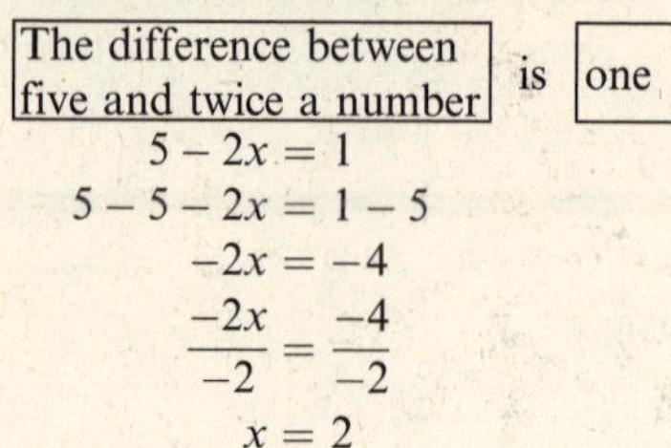

$$
\begin{aligned}
5 - 2x &= 1 \\
5 - 5 - 2x &= 1 - 5 \\
-2x &= -4 \\
\frac{-2x}{-2} &= \frac{-4}{-2} \\
x &= 2
\end{aligned}
$$

The number is 2.

9. The unknown number: x

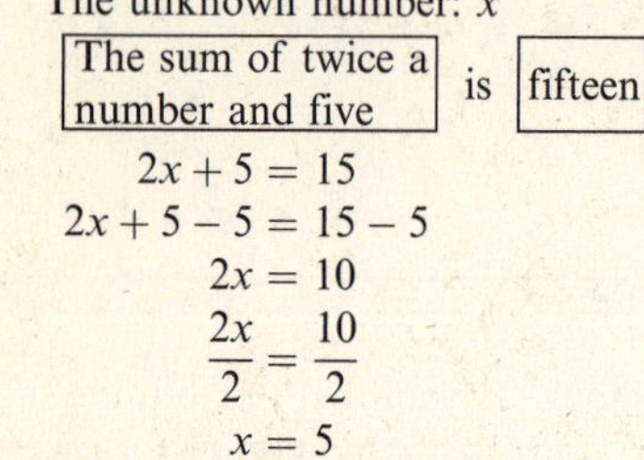

$$
\begin{aligned}
2x + 5 &= 15 \\
2x + 5 - 5 &= 15 - 5 \\
2x &= 10 \\
\frac{2x}{2} &= \frac{10}{2} \\
x &= 5
\end{aligned}
$$

The number is 5.

11. The unknown number: x

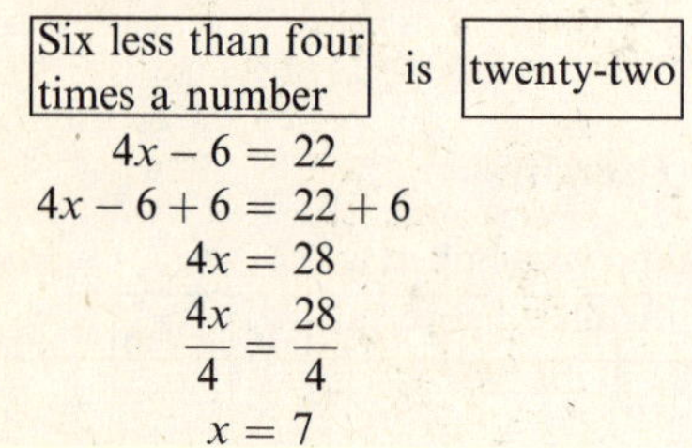

$$
\begin{aligned}
4x - 6 &= 22 \\
4x - 6 + 6 &= 22 + 6 \\
4x &= 28 \\
\frac{4x}{4} &= \frac{28}{4} \\
x &= 7
\end{aligned}
$$

The number is 7.

13. The unknown number: x

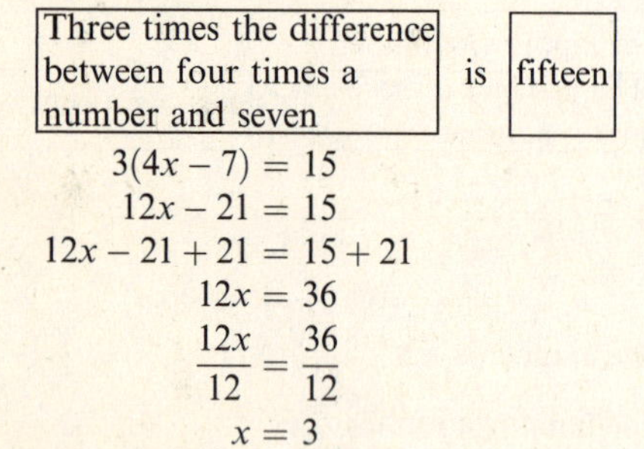

$$
\begin{aligned}
3(4x - 7) &= 15 \\
12x - 21 &= 15 \\
12x - 21 + 21 &= 15 + 21 \\
12x &= 36 \\
\frac{12x}{12} &= \frac{36}{12} \\
x &= 3
\end{aligned}
$$

The number is 3.

15. The smaller number: x
The larger number: $20 - x$

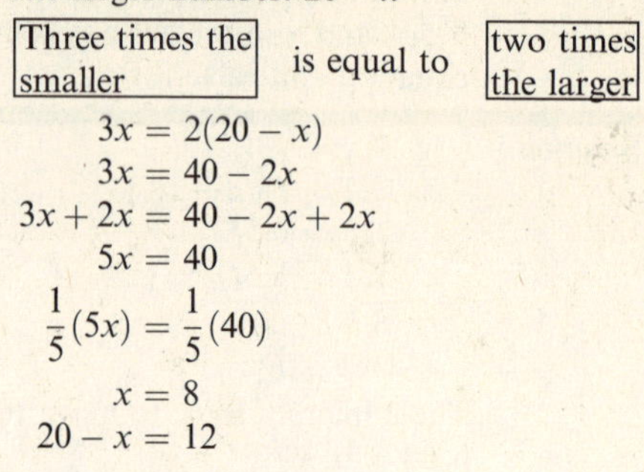

$$
\begin{aligned}
3x &= 2(20 - x) \\
3x &= 40 - 2x \\
3x + 2x &= 40 - 2x + 2x \\
5x &= 40 \\
\frac{1}{5}(5x) &= \frac{1}{5}(40) \\
x &= 8 \\
20 - x &= 12
\end{aligned}
$$

The smaller number is 8.
The larger number is 12.

17. The smaller number: x
The larger number: $14 - x$

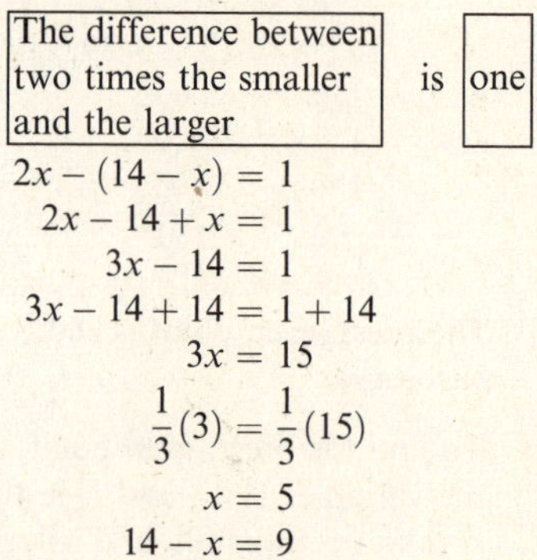

$$
\begin{aligned}
2x - (14 - x) &= 1 \\
2x - 14 + x &= 1 \\
3x - 14 &= 1 \\
3x - 14 + 14 &= 1 + 14 \\
3x &= 15 \\
\frac{1}{3}(3) &= \frac{1}{3}(15) \\
x &= 5 \\
14 - x &= 9
\end{aligned}
$$

The smaller number is 5.
The larger number is 9.

19. Strategy
- First odd integer: n
  Second odd integer: $n + 2$
  Third odd integer: $n + 4$
- The sum of the three integers is 51.

Solution

$$
\begin{aligned}
n + (n + 2) + (n + 4) &= 51 \\
3n + 6 &= 51 \\
3n &= 45 \\
n &= 15 \\
n + 2 &= 15 + 2 = 17 \\
n + 4 &= 15 + 4 = 19
\end{aligned}
$$

The three integers are 15, 17, and 19.

21. Strategy
- First odd integer: n
  Second odd integer: $n + 2$
  Third odd integer: $n + 4$
- Three times the second number is one more than the sum of the first and third numbers.

Solution

$$
\begin{aligned}
3(n + 2) &= 1 + n + (n + 4) \\
3n + 6 &= 5 + 2n \\
n + 6 &= 5 \\
n &= -1 \\
n + 2 &= -1 + 2 = 1 \\
n + 4 &= -1 + 4 = 3
\end{aligned}
$$

The integers are -1, 1, and 3.

23. Strategy • First even integer: n
Second even integer: $n+2$
• Three times the first integer equals twice the second integer.

Solution
$$3n = 2(n+2)$$
$$3n = 2n+4$$
$$n = 4$$
$$n+2 = 4+2 = 6$$
The integers are 4 and 6.

25. Strategy • First odd integer: n
Second odd integer: $n+2$
• Seven times the first equals five times the second.

Solution
$$7n = 5(n+2)$$
$$7n = 5n+10$$
$$2n = 10$$
$$n = 5$$
$$n+2 = 5+2 = 7$$
The integers are 5 and 7.

Objective B Exercises

27. Strategy To find the processor speed, write and solve an equation using x to represent the speed of the newer personal computer.

Solution [3.2] is [three-fourths of the speed of the older model]

$$3.2 = \frac{3}{4}x$$
$$\frac{4}{3}(3.2) = \frac{4}{3}\left(\frac{3}{4}x\right)$$
$$4.2\overline{6} = x$$
The processor speed of the older personal computer is $4.2\overline{6}$ GHz.

29. Strategy To find the length of the sides of the triangle, write and solve an equation using x to represent the length of an equal side.

Solution [Perimeter of 23 ft] is [x ft + x ft + $(2x-1)$ft]

$$23 = x + x + (2x-1)$$
$$23 = 4x - 1$$
$$24 = 4x$$
$$\frac{1}{4} \cdot 24 = \frac{1}{4} \cdot 4x$$
$$6 = x$$
$$2x - 1 = 12 - 1 = 11$$
The length of the sides are 6 ft, 6 ft, and 11 ft.

31. Strategy To find the number of hours worked, write and solve an equation using h to represent the number of hours worked.

Solution [\$4.00 plus \$.15 for each hour worked] is [\$29.20]

$$4 + 0.15h = 29.20$$
$$4 - 4 + 0.15h = 29.20 - 4$$
$$0.15h = 25.20$$
$$\frac{0.15h}{0.15} = \frac{25.20}{0.15}$$
$$h = 168$$
The union member worked 168 h during March.

33. Strategy To find the number of hours, write and solve an equation using h to represent the number of hours to paint the inside of the house.

Solution [\$125 for material and \$33 per hour for labor] is [\$1,346]

$$125 + 33 \cdot h = 1{,}346$$
$$125 - 125 + 33h = 1{,}346 - 125$$
$$33h = 1{,}221$$
$$\frac{33h}{33} = \frac{1{,}221}{33}$$
$$h = 37$$
37 h of labor was required to paint the house.

35. Strategy To find the number of vertical pixels, write and solve an equation using x to represent the number of vertical pixels.

Solution [1280] is [768 less than twice the number of vertical pixels]

$$1280 = 2x - 768$$
$$1280 + 768 = 2x - 768 + 768$$
$$2048 = 2x$$
$$\frac{1}{2} \cdot 2048 = \frac{1}{2} \cdot 2x$$
$$1024 = x$$
There are 1024 vertical pixels.

37. Strategy To find the length and width of the rectangle, write and solve an equation using x to represent the width. Then the length is $2x - 3$.

Solution

Perimeter of 42 m	is	2 times the length plus 2 times the width

$$42 = 2(2x - 3) + 2x$$
$$42 = 4x - 6 + 2x$$
$$42 = 6x - 6$$
$$42 + 6 = 6x - 6 + 6$$
$$48 = 6x$$
$$\frac{1}{6} \cdot 48 = \frac{1}{6} \cdot 6x$$
$$8 = x$$
$$2x - 3 = 2 \cdot 8 - 3 = 13$$

The length is 13 m.
The width is 8 m.

39. Strategy To find the length of each piece, write and solve an equation using x to represent the shorter piece and $12 - x$ to represent the longer piece.

Solution

Twice the length of the shorter piece	is	three feet less than the longer piece

$$2x = (12 - x) - 3$$
$$2x = 9 - x$$
$$3x = 9$$
$$x = 3$$
$$12 - x = 12 - 3 = 9$$

The shorter piece is 3 ft.
The longer piece is 9 ft.

41. Strategy To find the amount of each scholarship, write and solve an equation using x to represent the smaller scholarship and $7000 - x$ to represent the larger scholarship.

Solution

Twice the smaller scholarship	is	1000 less than the larger scholarship

$$2x = (7000 - x) - 1000$$
$$2x = 7000 - x - 1000$$
$$2x + x = 6000 - x + x$$
$$3x = 6000$$
$$\frac{3x}{3} = \frac{6000}{3}$$
$$x = 2000$$
$$7000 - x = 7000 - 2000 = 5000$$

The larger scholarship is $5000.

Applying the Concepts

43. A possible problem for $6x = 123$ is "A student worked for 6 hours and earned $123. What was the student's hourly wage?" For the equation $8x + 100 = 300$, a possible problem is "A group of eight people spent $300 at an amusement park." This included $100 for lunch and admission tickets for the eight people. Find the cost of each ticket.

45. The problem states that a 4-quart mixture of fruit juice is made from apple juice and cranberry juice. There are 6 more quarts of apple juice than of cranberry juice. If we let $x =$ the number of quarts of cranberry juice, then $x + 6 =$ the number of quarts of apple juice. The total number of quarts is 4. Therefore, we can write the equation

$$x + (x + 6) = 4$$
$$2x + 6 = 4$$
$$2x = -2$$
$$x = -1$$

Because $x =$ the number of quarts of cranberry juice, there are -1 qt of cranberry juice in the mixture. We cannot add -1 qt to a mixture. The solution is not reasonable. We can see from the original problem that the answer will not be reasonable. If the total number of quarts in the mixture is 4, we cannot have more than 6 qt of apple juice in the mixture.

Section 3.5

Objective A Exercises

1. Strategy The sum of the measures of the three angles shown is 360°. To find $\angle a$, write an equation and solve for $\angle a$.

Solution

$$m\angle a + 76^\circ + 168^\circ = 360^\circ$$
$$m\angle a + 244^\circ = 360^\circ$$
$$m\angle a = 116^\circ$$

The measure of $\angle a$ is 116°.

3. Strategy The sum of the measures of the three angles shown is 180°. To find x, write an equation and solve for x.

Solution

$$3x + 4x + 2x = 180^\circ$$
$$9x = 180^\circ$$
$$x = 20^\circ$$

The measure of x is 20°.

5. Strategy The sum of the measures of the three angles shown is 180°. To find x, write an equation and solve for x.

Solution
$$5x + (x + 20^\circ) + 2x = 180^\circ$$
$$8x + 20^\circ = 180^\circ$$
$$8x = 160^\circ$$
$$x = 20^\circ$$
The measure of x is 20°.

7. Strategy The sum of the measures of the four angles shown is 360°. To find x, write an equation and solve for x.

Solution
$$3x + 4x + 6x + 5x = 360^\circ$$
$$18x = 360^\circ$$
$$x = 20^\circ$$
The measure of x is 20°.

9. Strategy The angles labeled are adjacent angles of intersecting lines and are, therefore, supplementary angles. To find x, write an equation and solve for x.

Solution
$$x + 74^\circ = 180^\circ$$
$$x = 106^\circ$$
The measure of x is 106°.

11. Strategy The angles labeled are vertical angles and are, therefore, equal. To find x, write an equation and solve for x.

Solution
$$5x = 3x + 22^\circ$$
$$2x = 22^\circ$$
$$x = 11^\circ$$
The measure of x is 11°.

13. Strategy
- To find the measure of $\angle a$, use the fact that corresponding angles of parallel lines are equal.
- To find the measure of $\angle b$, use the fact that adjacent angles of intersecting lines are supplementary.

Solution
$$m\angle a = 38^\circ$$
$$m\angle b + m\angle a = 180^\circ$$
$$m\angle b + 38^\circ = 180^\circ$$
$$m\angle b = 142^\circ$$
The measure of $\angle a$ is 38°.
The measure of $\angle b$ is 142°.

15. Strategy
- To find the measure of $\angle a$, use the fact that alternate interior angles of parallel lines are equal.
- To find the measure of $\angle b$, use the fact that adjacent angles of intersecting lines are supplementary.

Solution
$$m\angle a = 47^\circ$$
$$m\angle a + m\angle b = 180^\circ$$
$$47^\circ + m\angle b = 180^\circ$$
$$m\angle b = 133^\circ$$
The measure of $\angle a$ is 47°.
The measure of $\angle b$ is 133°.

17. Strategy

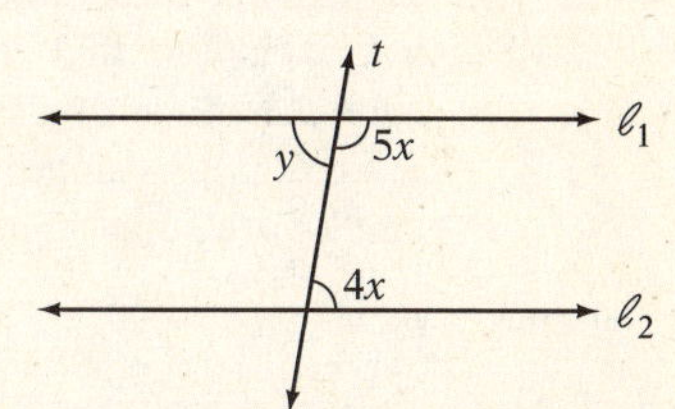

$4x = y$ because alternate interior angles have the same measure. $y + 5x = 180^\circ$ because adjacent angles of intersecting lines are supplementary. Substitute $4x$ for y and solve for x.

Solution
$$4x + 5x = 180^\circ$$
$$9x = 180^\circ$$
$$x = 20^\circ$$
The measure of x is 20°.

19. Strategy

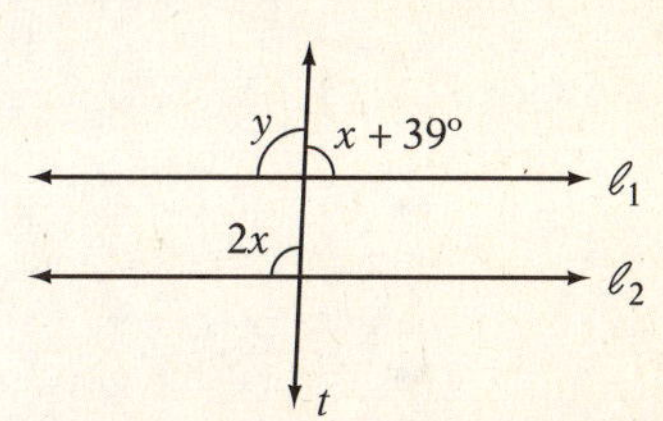

$y = 2x$ because corresponding angles have the same measure. $y + x + 39^\circ = 180^\circ$ because adjacent angles of intersecting lines are supplementary angles. Substitute $2x$ for y and solve for x.

Solution
$$2x + x + 39^\circ = 180^\circ$$
$$3x + 39^\circ = 180^\circ$$
$$3x = 141^\circ$$
$$x = 47^\circ$$
The measure of x is 47°.

21. Strategy

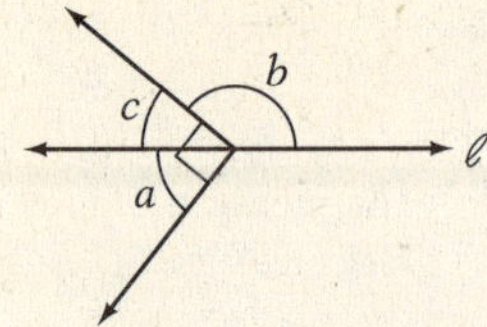

To find the measure of $\angle b$:
- Use the fact that $\angle a$ and $\angle c$ are complementary angles.
- Find $\angle b$ by using the fact that $\angle c$ and $\angle b$ are supplementary angles.

Solution

$$m\angle a + m\angle c = 90^\circ$$
$$51^\circ + m\angle c = 90^\circ$$
$$m\angle c = 39^\circ$$
$$m\angle b + m\angle c = 180^\circ$$
$$m\angle b + 39^\circ = 180^\circ$$
$$m\angle b = 141^\circ$$

The measure of $\angle b$ is 141°.

Objective B Exercises

23. Strategy

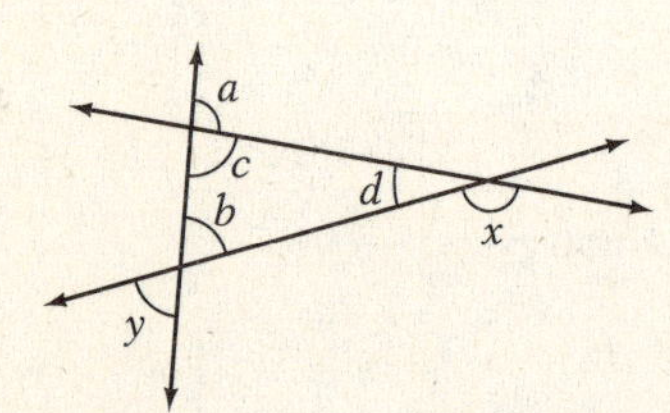

- To find the measure of angle y, use the fact that $\angle b$ and $\angle y$ are vertical angles.
- To find the measure of angle x: Find the measure of angle c by using the fact that the sum of an interior and exterior angle is 180°. Find the measure of angle d by using the fact that the sum of the interior angles of a triangle is 180°. Find the measure of angle x by using the fact that the sum of an interior and exterior angle is 180°.

Solution

$$m\angle y = m\angle b = 70^\circ$$
$$m\angle a + m\angle c = 180^\circ$$
$$95^\circ + m\angle c = 180^\circ$$
$$m\angle c = 85^\circ$$
$$m\angle b + m\angle c + m\angle d = 180^\circ$$
$$70^\circ + 85^\circ + m\angle d = 180^\circ$$
$$155^\circ + m\angle d = 180^\circ$$
$$m\angle d = 25^\circ$$
$$m\angle d + m\angle x = 180^\circ$$
$$25^\circ + m\angle x = 180^\circ$$
$$m\angle x = 155^\circ$$

The measure of $\angle x$ is 155°.
The measure of $\angle y$ is 70°.

25. Strategy

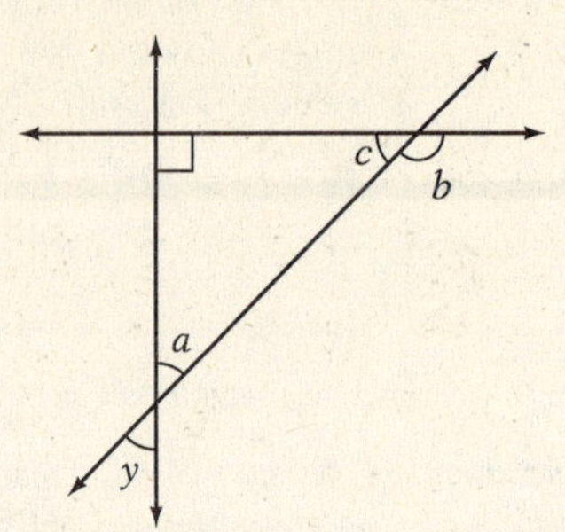

- To find the measure of angle a, use the fact that $\angle a$ and $\angle y$ are vertical angles.
- To find the measure of angle b: Find the measure of angle c by using the fact that the sum of the interior angles of a triangle is 180°. Find the measure of angle b by using the fact that the sum of an interior and exterior angle is 180°.

Solution

$$m\angle a = m\angle y = 45^\circ$$
$$m\angle a + m\angle c + 90^\circ = 180^\circ$$
$$45^\circ + m\angle c + 90^\circ = 180^\circ$$
$$m\angle c + 135^\circ = 180^\circ$$
$$m\angle c = 45^\circ$$
$$m\angle c + m\angle b = 180^\circ$$
$$45^\circ + m\angle b = 180^\circ$$
$$m\angle b = 135^\circ$$

The measure of $\angle a$ is 45°.
The measure of $\angle b$ is 135°.

27. Strategy To find the measure of $\angle BOC$, use the fact that the sum of the measure of the angles x, $\angle AOB$, and $\angle BOC$ is 180°. Since $AO \perp OB$, $m\angle AOB$ is 90°.

Solution

$$x + m\angle AOB + m\angle BOC = 180^\circ$$
$$x + 90^\circ + m\angle BOC = 180^\circ$$
$$m\angle BOC = 90^\circ - x$$

The measure of $\angle BOC$ is $90^\circ - x$.

29. Strategy To find the measure of the third angle, use the fact that the sum of the measures of the interior angles of a triangle is 180°. Write an equation using x to represent the measure of the third angle. Solve the equation of x.

Solution

$$x + 90^\circ + 30^\circ = 180^\circ$$
$$x + 120^\circ = 180^\circ$$
$$x = 60^\circ$$

The measure of the third angle is 60°.

31. Strategy To find the measure of the third angle, use the fact that the sum of the measures of the interior angles of a triangle is 180°. Write an equation using x to represent the measure of the third angle. Solve the equation for x.

Solution
$$x + 42^\circ + 103^\circ = 180^\circ$$
$$x + 145^\circ = 180^\circ$$
$$x = 35^\circ$$
The measure of the third angle is 35°.

33. Strategy To find the measure of the third angle, use the fact that the sum of the measures of the interior angles of a triangle is 180°. Write an equation using x to represent the measure of the third angle. Solve the equation of x.

Solution
$$x + 13^\circ + 65^\circ = 180^\circ$$
$$x + 78^\circ = 180^\circ$$
$$x = 102^\circ$$
The measure of the third angle is 102°.

Applying the Concepts

35. Strategy

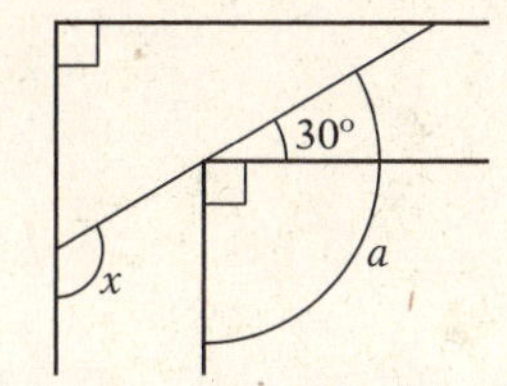

Using the fact that the lines representing the walls of the hallway are parallel, $\angle x$ and $\angle a$ are corresponding angles. So their measures are equal.

Solution
$$m\angle a = 30^\circ + 90^\circ = 120^\circ$$
$$m\angle x = m\angle a = 120^\circ$$
The measure of $\angle x$ is 120°.

37. Because the sum of the measures of an interior and an exterior angle is 180°.
$$m\angle a + m\angle y = 180^\circ$$
$$m\angle b + m\angle z = 180^\circ$$
$$m\angle c + m\angle x = 180^\circ$$
Adding these 3 equations results in
$$m\angle a + m\angle y + m\angle b + m\angle z + m\angle c + m\angle x = 180^\circ + 180^\circ + 180^\circ$$
$$m\angle a + m\angle y + m\angle b + m\angle z + m\angle c + m\angle x = 540^\circ$$
Because the sum of the measures of the interior angles of a triangle is 180°: $\angle a + \angle b + \angle c = 180^\circ$
$$(m\angle a + m\angle b + m\angle c) + (m\angle y + m\angle z + m\angle x) = 540^\circ$$
$$180^\circ + (m\angle y + m\angle z + m\angle x) = 540^\circ$$
$$m\angle y + m\angle z + m\angle x = 360^\circ$$
$$m\angle x + m\angle y + m\angle z = 360^\circ$$
The sum of the measures of angles x, y, and z is 360°.

Section 3.6

Objective A Exercises

1. Strategy • Amount of \$1 herb: x

	Amount	Cost	Value
\$1 herb	x	1	x
\$2 herb	30	2	2(30)
Mixture	$x + 30$	1.60	$1.60(x + 30)$

• The sum of the values before mixing equals the value after mixing.

Solution
$$x + 2(30) = 1.60(x + 30)$$
$$x + 60 = 1.60x + 48$$
$$x - 1.60x + 60 = 1.60x - 1.60x + 48$$
$$-0.60x + 60 = 48$$
$$-0.60x + 60 - 60 = 48 - 60$$
$$-0.60x = -12$$
$$\frac{-0.60x}{-0.60} = \frac{12}{-0.60}$$
$$x = 20$$
The amount of \$1 herbs is 20 oz.

3. Strategy • Cost per pound of the mixture: x

	Amount	Cost	Value
Beef	3	1.99	3(1.99)
Turkey	1	1.39	1(1.39)
Mixture	4	x	$4x$

• The sum of the values before mixing equals the value after mixing.

Solution
$$3(1.99) + 1(1.39) = 4x$$
$$5.97 + 1.39 = 4x$$
$$7.36 = 4x$$
$$\frac{7.36}{4} = \frac{4x}{4}$$
$$1.84 = x$$
The mixture will cost \$1.84 per pound.

5. Strategy • The amount of caramel: x

	Amount	Cost	Value
Popcorn	5	0.80	0.80(5)
Caramel	x	2.40	$2.40x$
Mixture	$5 + x$	1.40	$1.40(5 + x)$

• The sum of the values before mixing equals the value after mixing.

Solution
$$0.80(5) + 2.40x = 1.40(5 + x)$$
$$4 + 2.4x = 7 + 1.4x$$
$$4 + 2.4x - 1.4x = 7 + 1.4x - 1.4x$$
$$4 + x = 7$$
$$4 - 4 + x = 7 - 4$$
$$x = 3$$
The amount of caramel is 3 lb.

7. Strategy • Amount of olive oil: x
Amount of vinegar: $10 - x$

	Amount	Cost	Value
Olive oil	x	1.50	$1.50x$
Vinegar	$10 - x$	0.25	$0.25(10 - x)$
Blend	10	0.50	$0.50(10)$

• The sum of the values before mixing equals the value after mixing.

Solution

$$\begin{aligned} 1.50x + 0.25(10 - x) &= 0.50(10) \\ 1.50x + 2.5 - 0.25x &= 5 \\ 1.25x + 2.5 &= 5 \\ 1.25x + 2.5 - 2.5 &= 5 - 2.5 \\ 1.25x &= 2.5 \\ \frac{1.25x}{1.25} &= \frac{1.5}{1.25} \\ x &= 2 \\ 10 - x = 10 - 2 &= 8 \end{aligned}$$

The amount of olive oil is 2 c.
The amount of vinegar is 8 c.

9. Strategy • Cost per ounce of the mixture: x

	Amount	Cost	Value
\$5.50 type	200	5.50	$5.50(200)$
\$2.00 type	500	2.00	$2.00(500)$
Mixture	700	x	$700x$

• The sum of the values before mixing equals the value after mixing.

Solution

$$\begin{aligned} 5.50(200) + 2.00(500) &= 700x \\ 1100 + 1000 &= 700x \\ 2100 &= 700x \\ \frac{2100}{700} &= x \\ 3 &= x \end{aligned}$$

The cost of the mixture is \$3.00 per ounce.

11. Strategy • Amount of alloy: x
Amount of platinum: 20

	Amount	Cost	Value
Alloy	x	400	$400x$
Platinum	20	220	$220(20)$
Mixture	$20 + x$	300	$300(20 + x)$

• The sum of the values before mixing equals the value after mixing.

Solution

$$\begin{aligned} 220(20) + 400x &= 300(20 + x) \\ 4400 + 400x &= 6000 + 300x \\ 400x - 300x + 4400 &= 300x - 300x + 6000 \\ 100x + 4400 &= 6000 \\ 100x + 4400 - 4400 &= 6000 - 4400 \\ 100x &= 1600 \\ x &= \frac{1600}{100} \\ x &= 16 \end{aligned}$$

To make the mixture, 16 oz of the alloy are needed.

13. Strategy • Amount of almonds: x
Amount of walnuts: $100 - x$

	Amount	Cost	Value
Almonds	x	4.50	$4.50(x)$
Walnuts	$100 - x$	2.50	$2.50(100 - x)$
Mixture	100	3.24	$3.24(100)$

• The sum of the values before mixing equals the value after mixing.

Solution

$$\begin{aligned} 4.50x + 2.50(100 - x) &= 3.24(100) \\ 4.50x + 250 - 2.50x &= 324 \\ 2.00x + 250 &= 324 \\ 2.00x + 250 - 250 &= 324 - 250 \\ 2.00x &= 74 \\ x &= \frac{74}{2.00} \\ x &= 37 \end{aligned}$$

$100 - x = 100 - 37 = 63$

The amount of almonds is 37 lb.
The amount of walnuts is 63 lb.

15. Strategy • Number of adult tickets: x
Number of children's tickets: $370 - x$

	Number	Cost	Value
Adult tickets	x	6.00	$6.00x$
Children's ticket	$370 - x$	2.50	$2.50(370 - x)$

• The total receipts for the tickets were \$1723.

Solution

$$\begin{aligned} 6.00x + 2.50(370 - x) &= 1723 \\ 6.00x + 925 - 2.50x &= 1723 \\ 3.50x + 925 &= 1723 \\ 3.50x + 925 - 925 &= 1723 - 925 \\ 3.50x &= 798 \\ \frac{3.50x}{3.50} &= \frac{798}{3.50} \\ x &= 228 \end{aligned}$$

There were 228 adult tickets sold.

17. Strategy • Cost per pound of the sugar-coated cereal: x

	Amount	Cost	Value
Sugar	40	1.00	1.00(40)
Corn flakes	120	0.60	0.60(120)
Mixture	160	x	$160x$

• The sum of the values before mixing equals the value after mixing.

Solution

$$\begin{aligned} 1.00(40) + 0.60(120) &= 160x \\ 40.00 + 72.00 &= 160x \\ 112 &= 160x \\ \frac{112}{160} &= x \\ 0.70 &= x \end{aligned}$$

The cost per pound of the sugar-coated cereal is \$.70.

Objective B Exercises

19. Strategy • The percent concentration of gold in the mixture: x

	Amount	Percent	Quantity
30% gold	40	0.30	0.30(40)
20% gold	60	0.20	0.20(60)
Mixture	100	x	$100x$

• The sum of the quantities before mixing is equal to the quantity after mixing.

Solution

$$\begin{aligned} 0.30(40) + 0.20(60) &= 100x \\ 12 + 12 &= 100x \\ 24 &= 100x \\ 0.24 &= x \end{aligned}$$

The percent concentration of gold in the mixture is 24%.

21. Strategy • The amount of the 15% acid solution: x

	Amount	Percent	Quantity
15% acid solution	x	0.15	$0.15x$
20% acid solution	5	0.20	0.20(5)
Mixture	$5 + x$	0.16	$0.16(5 + x)$

• The sum of the quantities before mixing is equal to the quantity after mixing.

Solution

$$\begin{aligned} 0.15x + 0.20(5) &= 0.16(5 + x) \\ 0.15x + 1 &= 0.8 + 0.16x \\ -0.01x &= -0.2 \\ x &= 20 \end{aligned}$$

The amount of the 15% acid is 20 gal.

23. Strategy • Amount of 25% wool yarn: x

	Amount	Percent	Quantity
50% wool yarn	20	0.50	0.50(20)
25% wool yarn	x	0.25	$0.25x$
35% wool yarn	$20 + x$	0.35	$0.35(20 + x)$

• The sum of the quantities before mixing is equal to the sum after mixing.

Solution

$$\begin{aligned} 0.50(20) + 0.25x &= 0.35(20 + x) \\ 10 + 0.25x &= 7 + 0.35x \\ -0.10x &= -3 \\ x &= 30 \end{aligned}$$

The amount of the 25% wool yarn is 30 lb.

25. Strategy • Amount of plant food that is 9% nitrogen: x
Amount of plant food that is 25% nitrogen: $10 - x$

	Amount	Percent	Quantity
9% nitrogen	x	0.09	$0.09x$
25% nitrogen	$10 - x$	0.25	$0.25(10 - x)$
15% nitrogen	10	0.15	0.15(10)

• The sum of the quantities before mixing is equal to the quantity after mixing.

Solution

$$\begin{aligned} 0.09x + 0.25(10 - x) &= 0.15(10) \\ 0.09x + 2.5 - 0.25x &= 1.5 \\ -0.16x &= -1 \\ x &= 6.25 \end{aligned}$$

The amount of 9% nitrogen plant food is 6.25 gal.

27. Strategy • Percent concentration of sugar in the mixture: x

	Amount	Percent	Quantity
Sugar	5	1.00	1.00(5)
10% sugar	45	0.10	0.10(45)
Mixture	50	x	$50x$

• The sum of the quantities before mixing is equal to the quantity after mixing.

Solution

$$\begin{aligned} 1.00(5) + 0.10(45) &= 50x \\ 5 + 4.5 &= 50x \\ 9.5 &= 50x \\ 0.19 &= x \end{aligned}$$

The percent concentration of sugar in the mixture is 19%.

29. Strategy • Amount of 40% java beans: x
Amount of 30% java beans: 80

	Amount	Percent	Quantity
40% java beans	x	0.40	$0.40x$
30% java beans	80	0.30	0.30(80)
32% java beans (mix)	$x + 80$	0.32	$0.32(x + 80)$

• The sum of the quantities before mixing is equal to the quantity after mixing.

Solution

$$\begin{aligned} 0.40x + 0.30(80) &= 0.32(x + 80) \\ 0.40x + 24.00 &= 0.32x + 25.60 \\ 0.08x &= 1.60 \\ x &= 20 \end{aligned}$$

20 lb of 40% java bean coffee must be used.

31. Strategy • Amount of 7% hydrogen peroxide: x
Amount of 4% hydrogen peroxide: $300 - x$

	Amount	Percent	Quantity
7% hydrogen peroxide	x	0.07	$0.07x$
4% hydrogen peroxide	$300 - x$	0.04	$0.04(300 - x)$
Mixture	300	0.05	0.05(300)

• The sum of the quantities before mixing is equal to the quantity after mixing.

Solution

$$\begin{aligned} 0.07x + 0.04(300 - x) &= 0.05(300) \\ 0.07x + 12 - 0.04x &= 15 \\ 0.03x &= 3 \\ x &= 100 \\ 300 - x &= 200 \end{aligned}$$

The amount of the 7% solution is 100 ml.
The amount of the 4% solution is 200 ml.

33. Strategy • Amount of pure chocolate: x

	Amount	Percent	Quantity
Pure chocolate	x	1.00	$1.00x$
50% chocolate	150	0.50	0.50(150)
Mixture	$150 + x$	0.75	$0.75(150 + x)$

• The sum of the quantities before mixing is equal to the quantity after mixing.

Solution

$$\begin{aligned} 1.00x + 0.50(150) &= 0.75(150 + x) \\ 1.00x + 75 &= 112.5 + 0.75x \\ 0.25x &= 37.5 \\ x &= 150 \end{aligned}$$

150 oz of pure chocolate must be added.

35. Strategy • Percent concentration of the resulting alloy: x

	Amount	Percent	Quantity
Silver	30	1.00	1.00(30)
20% silver	50	0.20	0.20(50)
Resulting alloy	80	x	$80x$

• The sum of the quantities before mixing is equal to the quantity after mixing.

Solution

$$\begin{aligned} 1.00(30) + 0.20(50) &= 80x \\ 30 + 10 &= 80x \\ 40 &= 80x \\ 0.5 &= x \end{aligned}$$

The percent concentration of the resulting alloy is 50%.

Objective C Exercises

37. Strategy • Rate of the first plane: $r - 25$
Rate of the second plane: r

	Rate	Time	Distance
First plane	$r - 25$	2	$2(r - 25)$
Second plane	r	2	$2r$

• The total distance traveled by the two planes is 470 mi.

Solution
$$2(r - 25) + 2r = 470$$
$$2r - 50 + 2r = 470$$
$$4r = 520$$
$$r = 130$$
$$r - 25 = 130 - 25 = 105$$
The first plane is traveling at a rate of 105 mph. The second plane is traveling at a rate of 130 mph.

39. Strategy • Time that the planes will be flying: t

	Rate	Time	Distance
First plane	480	t	$480t$
Second plane	520	t	$520t$

• The sum of the distance that the planes fly is 3000 mi.

Solution
$$480t + 520t = 3000$$
$$1000t = 3000$$
$$t = 3$$
The planes fly for 3 hours.
Three hours after 8 A.M. is 11 A.M.
The planes will be 3000 km apart at 11 A.M.

41. Strategy • Time for the cabin cruiser: t
Time for the motorboat: $t + 2$

	Rate	Time	Distance
Motorboat	9	$t + 2$	$9(t + 2)$
Cruiser	18	t	$18t$

• The boats travel the same distance.

Solution
$$9(t + 2) = 18t$$
$$9t + 18 = 18t$$
$$18 = 9t$$
$$2 = t$$
In 2 h the cabin cruiser will be alongside the motorboat.

43. Strategy • Time to airport: t
Time in flight: $3 - t$

	Rate	Time	Distance
To airport	30	t	$30t$
In flight	60	$3 - t$	$60(3 - t)$

• The total distance was 150 mi.

Solution
$$30t + 60(3 - t) = 150$$
$$30t + 180 - 60t = 150$$
$$180 - 30t = 150$$
$$-30t = -30$$
$$t = 1$$
$$\text{Distance} = 60(3 - t) = 60(3 - 1)$$
$$= 60(2) = 120$$
The corporate offices are 120 mi from the airport.

45. Strategy • The average speed of the bus: r
The average speed of the car: $2r$

	Rate	Time	Distance
Bus	r	2	$2r$
Car	$2r$	2	$2(2r)$

• In 2 h the car is 68 mi ahead of the bus.

Solution
$$2(2r) - 2r = 68$$
$$4r - 2r = 68$$
$$2r = 68$$
$$r = 34$$
$$2r = 2(34) = 68$$
The rate of the car is 68 mph.

47. Strategy • Time of flight to the airport: t
Time of flight for the return trip: $5 - t$

	Rate	Time	Distance
To the airport	100	t	$100t$
The return trip	150	$(5 - t)$	$150(5 - t)$

• The flights are the same distance.

Solution
$$100t = 150(5 - t)$$
$$100t = 750 - 150t$$
$$250t = 750$$
$$t = 3$$
$$d = rt = 100(3) = 300$$
The distance to the airport was 300 mi.

49. Strategy • Time the Seattle plane is in the air: t
Time the Boston plane is in the air: $t+1$

	Rate	Time	Distance
Seattle plane	500	t	$500t$
Boston plane	500	$t+1$	$500(t+1)$

• The total distance that both planes flew is 3000 mi.

Solution
$$500t + 500(t+1) = 3000$$
$$500t + 500t + 500 = 3000$$
$$1000t = 2500$$
$$t = 2.5$$
The planes will pass each other in 2.5 h after the plane leaves Seattle.

51. Strategy • Time spent riding: t

	Rate	Time	Distance
1st rider	16	t	$16t$
2nd rider	18	t	$18t$

• The total distance is 45 mi.

Solution
$$16t + 18t = 51$$
$$34t = 51$$
$$t = 1.5$$
The cyclists will meet after 1.5 h.

53. Strategy • Time for the bus: t
Time for the car: $t+1$

	Rate	Time	Distance
Bus	60	t	$60t$
Car	45	$t+1$	$45(t+1)$

• The car and the bus travel the same distance.

Solution
$$60t = 45(t+1)$$
$$60t = 45t + 45$$
$$15t = 45$$
$$t = 3$$
$$60t = 60(3) = 180$$
The bus overtakes the car in 180 mi.

Applying the Concepts

55. Strategy • Amount of pure water: x

	Amount	Percent	Quantity
Pure water	x	0	$0x$
Pure acid	50	1.00	1.00(50)
Mixture	$x+50$	0.40	$0.40(x+50)$

• The sum of the quantities before mixing equals the quantity after mixing.

Solution
$$0x + 1.00(50) = 0.40(x+50)$$
$$50 = 0.40x + 20$$
$$30 = 0.40x$$
$$75 = x$$
Add 75 g of pure water.

57. Strategy • Amount of antifreeze drained: x
Amount of pure antifreeze added: x

	Amount	Percent	Quantity
20% antifreeze	$15-x$	0.20	$0.20(15-x)$
Pure antifreeze	x	1.00	$1.00x$
40% antifreeze	15	0.40	0.40(15)

• The sum of the quantities before mixing equals the quantity after mixing.

Solution
$$0.20(15-x) + 1.00x = 0.40(15)$$
$$3 - 0.20x + 1.00x = 6$$
$$3 + 0.80x = 6$$
$$0.80x = 3$$
$$x = 3.75$$
3.75 gal of 20% antifreeze must be drained.

59. Strategy • Find the total distance traveled and the total time.
• Divide total distance by total time to get the average speed.

	Rate	Time	Distance
Leaving	10	2	10(2) = 20
Returning	20	20/20 = 1	20
Total		3	40

$$\text{Average speed} = \frac{40}{3} = 13\frac{1}{3}$$

The bicyclist's average speed is $13\frac{1}{3}$ mph.

Chapter 3 Review Exercises

1. $$x + 3 = 24$$
$$x = 24 - 3$$
$$x = 21$$

2. $$x + 5(3x - 20) = 10(x - 4)$$
$$x + 15x - 100 = 10x - 40$$
$$16x - 100 = 10x - 40$$
$$16x - 10x = -40 + 100$$
$$6x = 60$$
$$\frac{1}{6}(6x) = \frac{1}{6}(60)$$
$$x = 10$$

3. $$\begin{aligned} 5x - 6 &= 29 \\ 5x &= 29 + 6 \\ 5x &= 35 \\ \frac{1}{5}(5x) &= \frac{1}{5}(35) \\ x &= 7 \end{aligned}$$

4. $$\begin{array}{r|l} 5x - 2 & = 4x + 5 \\ \hline 5(3) - 2 & 4(3) + 5 \\ 15 - 1 & 12 + 5 \\ \end{array}$$
$$13 \neq 17$$
No, 3 is not a solution.

5. $$\begin{aligned} \frac{3}{5}a &= 12 \\ a &= 12 \cdot \frac{5}{3} \\ a &= 20 \end{aligned}$$

6. $$\begin{aligned} 6x + 3(2x - 1) &= -27 \\ 6x + 6x - 3 &= -27 \\ 12x - 3 &= -27 \\ 12x &= -27 + 3 \\ 12x &= -24 \\ \frac{1}{12}(12x) &= \frac{1}{12}(-24) \\ x &= -2 \end{aligned}$$

7. $$\begin{aligned} P(12) &= 30 \\ P(12)\left(\frac{1}{12}\right) &= 30\left(\frac{1}{12}\right) \\ P &= 2.5 \end{aligned}$$
The percent is 250%.

8. $$\begin{aligned} 5x + 3 &= 10x - 17 \\ 3 + 17 &= 10x - 5x \\ 20 &= 5x \\ \frac{1}{5}(20) &= \frac{1}{5}(5x) \\ 4 &= x \end{aligned}$$

9. $$\begin{aligned} 7 - [4 + 2(x - 3)] &= 11(x + 2) \\ 7 - [4 + 2x - 6] &= 11x + 22 \\ 7 - [-2 + 2x] &= 11x + 22 \\ 7 + 2 - 2x &= 11x + 22 \\ 9 - 2x &= 11x + 22 \\ 9 - 22 &= 11x + 2x \\ -13 &= 13x \\ \frac{1}{13}(-13) &= \frac{1}{13}(13x) \\ -1 &= x \end{aligned}$$

10. $$\begin{aligned} -6x + 16 &= -2x \\ -6x + 2x &= -16 \\ -4x &= -16 \\ -\frac{1}{4}(-4x) &= -\frac{1}{4}(-16) \\ x &= 4 \end{aligned}$$

11. $$\begin{aligned} S &= C + rC \\ 1074 &= C + 60\%C \\ 1074 &= C + 0.60C \\ 1074 &= 1.60C \\ \frac{1074}{1.60} &= C \\ \$671.25 &= C \end{aligned}$$
The cost is $671.25.

12. **Strategy** To find x, use the fact that the sum of the measures of supplementary angles is 180°.

Solution $$\begin{aligned} (3x + 6) + (2x - 1) &= 180 \\ 5x + 5 &= 180 \\ 5x &= 175 \\ x &= 35^\circ \end{aligned}$$
The solution is $x = 35^\circ$.

13. **Strategy** To find x, use the fact that the measures of alternate exterior angles are equal.

Solution $$\begin{aligned} 4x + 7 &= 2x + 59 \\ 2x &= 52 \\ x &= 26^\circ \end{aligned}$$
The solution is $x = 26^\circ$.

14. $$\begin{aligned} F_1x &= F_2(d - x) \\ 120(2) &= F_2(10) \\ 240 &= 10F_2 \\ \frac{1}{10}(240) &= \frac{1}{10}(10F_2) \\ 24 &= F_2 \end{aligned}$$
The force is 24 lb.

15. **Strategy**
- Speed on winding road: r
  Speed on level road: $r + 20$

	Rate	Time	Distance
Winding road	r	3	$3r$
Level road	$r+20$	2	$2(r+20)$

- The total trip was 200 mi.

Solution $$\begin{aligned} 3r + 2(r + 20) &= 200 \\ 3r + 2r + 40 &= 200 \\ 5r + 40 &= 200 \\ 5r &= 160 \\ r &= 32 \end{aligned}$$
The average speed on the winding road was 32 mph.

16. $$\begin{aligned} S &= R - rR \\ 40 &= 60 - 60r \\ -20 &= -60r \\ \frac{1}{3} &= r \end{aligned}$$
The discount rate is $\frac{1}{3}$, or $33\frac{1}{3}\%$.

17. Strategy • To find the measure of angle c, use the fact that the sum of an interior and exterior angle is 180°.

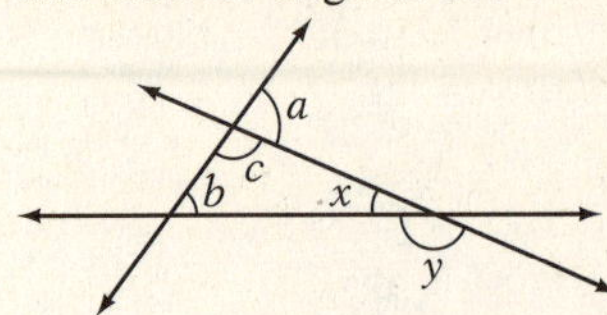

• To find the measure of angle x, use the fact that the sum of the measurements of the interior angles of a triangle is 180°.
• To find the measure of angle y, use the fact that the sum of an interior and exterior angle is 180°.

Solution

$$m\angle a + m\angle c = 180°$$
$$74° + m\angle c = 180°$$
$$m\angle c = 106°$$

$$m\angle b + m\angle c + m\angle x = 180°$$
$$52° + 106° + m\angle x = 180°$$
$$158° + m\angle x = 180°$$
$$m\angle x = 22°$$

$$m\angle x + m\angle y = 180°$$
$$22° + m\angle y = 180°$$
$$m\angle y = 158°$$

18. Strategy • Amount of cranberry juice: x
Amount of apple juice: $10 - x$

	Amount	Cost	Value
Cranberry juice	x	1.79	$1.79x$
Apple juice	$10 - x$	1.19	$1.19(10 - x)$
Mixture	10	1.61	1.61(10)

• The sum of the values before mixing equals the value after mixing.

Solution

$$1.79x + 1.19(10 - x) = 1.61(10)$$
$$1.79x + 11.90 - 1.19x = 16.10$$
$$0.60x = 4.2$$
$$x = 7$$
$$10 - x = 3$$

The amount of cranberry juice was 7 qt. The amount of apple juice was 3 qt.

19. Strategy • First integer: n
Second integer: $n + 1$
Third integer: $n + 2$
• Four times the second integer equals the sum of the first and third integers.

Solution

$$4(n + 1) = n + n + 2$$
$$4n + 4 = 2n + 2$$
$$2n = -2$$
$$n = -1$$

The integers are −1, 0, and 1.

20. Strategy • First angle: $x + 15$
Second angle: x
Third angle: $x - 15$
• The sum of the angles of a triangle is 180°.

Solution

$$x + 15 + x + x - 15 = 180$$
$$3x = 180$$
$$x = 60$$
$$x + 15 = 75$$
$$x - 15 = 45$$

The angles measure 75°, 60°, and 45°.

21.

$$n = \text{number}$$
$$5n - 4 = 16$$
$$5n = 16 + 4$$
$$5n = 20$$
$$\frac{1}{5}(5n) = \frac{1}{5}(20)$$
$$n = 4$$

22.

$$x = \text{height of Eiffel Tower}$$
$$1472 = 2x - 654$$
$$1472 + 654 = 2x$$
$$2126 = 2x$$
$$\frac{1}{2}(2126) = \frac{1}{2}(2x)$$
$$1063 = x$$

The height of the Eiffel Tower is 1063 ft.

23. Strategy To find $m\angle x$, use the fact that $\angle y$ and $\angle a$ are supplementary angles and that $\angle a$ and $\angle x$ are complementary angles.

Solution

$$m\angle y + m\angle a = 180°$$
$$115° + m\angle a = 180°$$
$$m\angle a = 65°$$
$$m\angle a + m\angle x = 90°$$
$$65° + m\angle x = 90°$$
$$m\angle x = 25°$$

The measure of angle x is 25°.

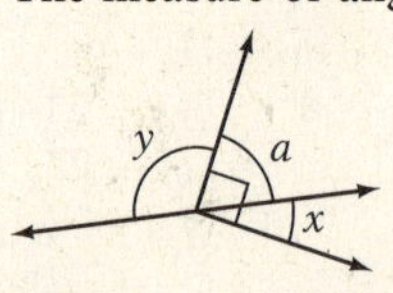

24. Strategy To find $m\angle y$, use the fact that the sum of the measures of $\angle y$, $\angle AOB$, and $\angle x$ is 180°.

Solution

$$m\angle y + m\angle AOB + m\angle x = 180°$$
$$m\angle y + 90° + 30° = 180°$$
$$m\angle y + 120° = 180°$$
$$m\angle y = 60°$$

The measure of $\angle y$ is 60°.

25. Strategy • Time for jet plane: t
Time for propeller-driven plane: $t+2$

	Rate	Time	Distance
Jet	600	t	$600t$
Propeller	200	$t+2$	$200(t+2)$

• The two planes travel the same distance.

Solution
$$600t = 200(t+2)$$
$$600t = 200t + 400$$
$$400t = 400$$
$$t = 1$$
Distance $= 600t = 600(1) = 600$
The jet overtakes the propeller-driven plane 600 mi from the starting point.

26. $x =$ smaller number
$21 - x =$ larger number
$$3x = 2(21-x) - 2$$
$$3x = 42 - 2x - 2$$
$$3x = 40 - 2x$$
$$3x + 2x = 40$$
$$5x = 40$$
$$\frac{1}{5}(5x) = \frac{1}{5}(40)$$
$$x = 8$$
$$21 - x = 13$$
The numbers are 8 and 13.

27. Strategy • Percent of butterfat in the mixture: x

	Amount	Percent	Quantity
Cream	5	0.30	0.30(5)
Milk	8	0.04	0.04(8)
Mixture	13	x	$13x$

• The sum of the quantities mixing is equal to the quantity after mixing.

Solution
$$0.30(5) + 0.04(8) = 13x$$
$$1.5 + 0.32 = 13x$$
$$1.82 = 13x$$
$$0.14 = x$$
The mixture is 14% butterfat.

Chapter 3 Test

1.
$$3x - 2 = 5x + 8$$
$$-2 - 8 = 5x - 3x$$
$$-10 = 2x$$
$$\frac{1}{2}(-10) = \frac{1}{2}(2x)$$
$$-5 = x$$

2.
$$x - 3 = -8$$
$$x = -8 + 3$$
$$x = -5$$

3.
$$3x - 5 = -14$$
$$3x = -14 + 5$$
$$3x = -9$$
$$\frac{1}{3}(3x) = \frac{1}{3}(-9)$$
$$x = -3$$

4.
$$4 - 2(3 - 2x) = 2(5 - x)$$
$$4 - 6 + 4x = 10 - 2x$$
$$-2 + 4x = 10 - 2x$$
$$4x + 2x = 10 + 2$$
$$6x = 12$$
$$\frac{6x}{6} = \frac{12}{6}$$
$$x = 2$$

5.

$x^2 - 3x$	$= 2x - 6$
$(-2)^2 - 3(-2)$	$2(-2) - 6$
$4 + 6$	$-4 - 6$

$$10 \neq -10$$
No, -2 is not a solution.

6.
$$7 - 4x = -13$$
$$-4x = -13 - 7$$
$$-4x = -20$$
$$-\frac{1}{4}(-4x) = -\frac{1}{4}(-20)$$
$$x = 5$$

7.
$$(0.005)(8) = A$$
$$0.04 = A$$
0.5% of 8 is 0.04.

8.
$$5x - 2(4x - 3) = 6x + 9$$
$$5x - 8x + 6 = 6x + 9$$
$$-3x + 6 = 6x + 9$$
$$6 - 9 = 6x + 3x$$
$$-3 = 9x$$
$$\frac{1}{9}(-3) = \frac{1}{9}(9x)$$
$$-\frac{3}{9} = x$$
$$-\frac{1}{3} = x$$

9.
$$5x + 3 - 7x = 2x - 5$$
$$-2x + 3 = 2x - 5$$
$$-2x - 2x = -5 - 3$$
$$-4x = -8$$
$$\frac{-4x}{-4} = \frac{-8}{-4}$$
$$x = 2$$

10.
$$\frac{3}{4}x = -9$$
$$\frac{4}{3}\left(\frac{3}{4}x\right) = \frac{4}{3}(-9)$$
$$x = -3(4)$$
$$x = -12$$

11. Strategy • Amount of rye flour: x
Amount of wheat flour: $15 - x$

	Amount	Cost	Value
Rye	x	0.70	$0.70x$
Wheat	$15 - x$	0.40	$0.40(15 - x)$
Blend	15	0.60	$0.60(15)$

• The sum of the values before mixing equals the value after mixing.

Solution
$$0.70x + 0.40(15 - x) = 0.60(15)$$
$$0.70x + 6 - 0.40x = 9$$
$$0.30x = 3$$
$$x = 10$$
$$15 - x = 5$$
The amount of rye flour is 10 lb.
The amount of wheat flour is 5 lb.

12. Strategy The sum of the measures of the three angles shown is 180°. To find x, write an equation and solve for x.

Solution
$$4x + 3x + (x + 28°) = 180°$$
$$8x + 28° = 180°$$
$$8x = 152°$$
$$x = 19°$$

The measure of x is 19°.

13.
$$S = R - rR$$
$$360 = 450 - 450r$$
$$-90 = -450r$$
$$\frac{90}{450} = r$$
$$0.20 = r$$
The discount rate is 20%.

14.
$$T = U \cdot N + F$$
$$5000 = 15N + 2000$$
$$5000 - 2000 = 15N$$
$$3000 = 15N$$
$$\frac{1}{15}(3000) = \frac{1}{15}(15N)$$
$$200 = N$$
200 calculators were produced.

15. Strategy • Measure of each equal angle: x
Measure of third angle: $x - 30$
• Use the equation
$A + B + C = 180°$.

Solution
$$x + x + x - 30 = 180$$
$$3x - 30 = 180$$
$$3x = 210$$
$$x = 70$$
The measure of one of the equal angles is 70°.

16. Strategy • First even integer: n
Second even integer: $n + 2$
Third even integer: $n + 4$
• The sum of the three integers is 36.

Solution
$$n + n + 2 + n + 4 = 36$$
$$3n + 6 = 36$$
$$3n = 30$$
$$n = 10$$
$$n + 2 = 10 + 2 = 12$$
$$n + 4 = 10 + 4 = 14$$
The numbers are 10, 12, and 14.

17. Strategy • Amount of pure water: x

	Amount	Percent	Quantity
Water	x	0.00	$0.00x$
20% solution	5	0.20	$0.20(5)$
Mixture	$x + 5$	0.16	$0.16(x + 5)$

• The sum of the quantities before mixing is equal to the quantity after mixing.

Solution
$$0.00x + 0.20(5) = 0.16(x + 5)$$
$$1 = 0.16x + 0.8$$
$$0.2 = 0.16x$$
$$1.25 = x$$
1.25 gal of water must be added.

18. Strategy $m\angle a = 138°$ because alternate interior angles of parallel lines are equal.
$m\angle a + m\angle b = 180°$ because adjacent angles of intersecting lines are supplementary.

Solution
$$m\angle a = 138°$$
$$m\angle a + m\angle b = 180°$$
$$138° + m\angle b = 180°$$
$$m\angle b = 42°$$
The measure of $\angle a$ is 138°.
The measure of $\angle b$ is 42°.

19. The unknown number: x
$$3x - 15 = 27$$
$$3x - 15 + 15 = 27 + 15$$
$$3x = 42$$
$$\frac{1}{3}(3x) = \frac{1}{3}(42)$$
$$x = 14$$
The number is 14.

20. Strategy • Rate of the skier: x
Rate of the snowmobile: $x+4$

	Rate	Time	Distance
Skier	x	3	$3x$
Snowmobile	$x+4$	1	$1(x+4)$

• The skier and the snowmobile travel the same distance.

Solution
$$3x = x+4$$
$$2x = 4$$
$$x = 2$$
$$x+4 = 6$$
The rate of the snowmobile is 6 mph.

21. Strategy Write and solve an equation letting x represent the number of 25-in. TVs and $140 - x$ represents the 15-in. TVs.

Solution
$$3(140 - x) = x - 20$$
$$420 - 3x = x - 20$$
$$440 = 4x$$
$$110 = x$$
The company makes 110 25-in. TVs each day.

22. Strategy The smaller number: x
The larger number: $18 - x$

Solution
$$4x - 7 = 2(18 - x) + 5$$
$$4x - 7 = 36 - 2x + 5$$
$$4x - 7 = 41 - 2x$$
$$4x + 2x - 7 = 41 - 2x + 2x$$
$$6x - 7 = 41$$
$$6x - 7 + 7 = 41 + 7$$
$$6x = 48$$
$$\frac{1}{6}(6x) = \frac{1}{6}(48)$$
$$x = 8$$
$$18 - x = 10$$
The smaller number is 8.
The larger number is 10.

23. Strategy • Time for flight out: t
Time for flight in: $7 - t$

	Rate	Time	Distance
Flight out	90	t	$90t$
Flight in	120	$7-t$	$120(7-t)$

• The distance traveled are the same.

Solution
$$90t = 120(7 - t)$$
$$90t = 840 - 120t$$
$$210t = 840$$
$$t = 4$$
Distance $= 90t = 90(4) = 360$
The distance to the airport is 360 mi.

24. Strategy

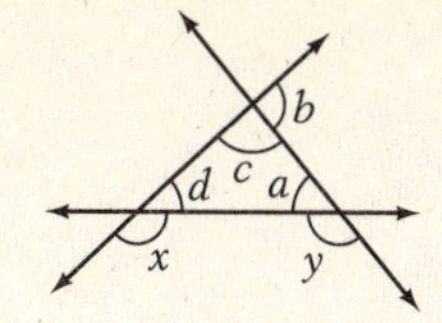

• To find $m\angle y$, use the fact that $\angle a$ and $\angle y$ are supplementary angles.
• To find $m\angle x$, find the measure of $\angle d$ and use the fact that $\angle d$ and $\angle x$ are supplementary angles. The measure of $\angle d$ is found using the fact that the sum of the measures of the angles of a triangle is 180°.

Solution
$$m\angle a + m\angle y = 180^\circ$$
$$50^\circ + m\angle y = 180^\circ$$
$$m\angle y = 130^\circ$$
$$m\angle c = 180^\circ - m\angle b$$
$$= 180^\circ - 92^\circ$$
$$= 88^\circ$$
$$m\angle a + m\angle c + m\angle d = 180^\circ$$
$$50^\circ + 88^\circ + m\angle d = 180^\circ$$
$$m\angle d = 42^\circ$$
$$m\angle x + m\angle d = 180^\circ$$
$$m\angle x + 42^\circ = 180^\circ$$
$$m\angle x = 138^\circ$$
The $m\angle x = 138^\circ$ and $m\angle y = 130^\circ$.

25.
$$m_1(T_1 - T) = m_2(T - T_2)$$
$$100(80 - T) = 50(T - 20)$$
$$8000 - 100T = 50T - 1000$$
$$-100T - 50T = -1000 - 8000$$
$$-150T = -9000$$
$$-\frac{1}{150}(-150)T = -\frac{1}{150}(-9000)$$
$$T = \frac{9000}{150}$$
$$T = 60$$
The final temperature is 60°C.

Cumulative Review Exercises

1. $-6 - (-20) - 8 = -6 + 20 - 8$
$= 14 - 8$
$= 6$

2. $(-2)(-6)(-4) = 12(-4)$
$= -48$

3. $-\frac{5}{6} - \left(-\frac{7}{16}\right) = -\frac{5}{6} + \frac{7}{16}$
$= -\frac{40}{48} + \frac{21}{48}$
$= -\frac{19}{48}$

4. $$-2\frac{1}{3} \div 1\frac{1}{6} = -\frac{7}{3} \div \frac{7}{6}$$
$$= -2$$

5. $$-4^2 \cdot \left(-\frac{3}{2}\right)^3 = -16 \cdot \left(-\frac{27}{8}\right)$$
$$= 54$$

6. $$\begin{aligned} 25 - 3\frac{(5-2)^2}{2^3+1} - (-2) &= 25 - \frac{3(3)^2}{8+1} + 2 \\ &= 25 - \frac{3(9)}{9} + 2 \\ &= 25 - 3 + 2 \\ &= 22 + 2 \\ &= 24 \end{aligned}$$

7. $$\begin{aligned} &3(a-c) - 2ab \\ 3(2-(-4)) - 2(2)(3) &= 3(2+4) - 4(3) \\ &= 3(6) - 4(3) \\ &= 18 - 12 \\ &= 6 \end{aligned}$$

8. $$\begin{aligned} 3x - 8x + (-12x) &= -5x + (-12x) \\ &= -17x \end{aligned}$$

9. $$\begin{aligned} 2a - (-3b) - 7a - 5b &= 2a + 3b - 7a - 5b \\ &= 2a - 7a + 3b - 5b \\ &= (2-7)a + (3-5)b \\ &= -5a - 2b \end{aligned}$$

10. $$(16x)\left(\frac{1}{8}\right) = 2x$$

11. $$-4(-9y) = 36y$$

12. $$\begin{aligned} -2(-x^2 - 3x + 2) &= -2(-x^2) - 2(-3x) - 2(2) \\ &= 2x^2 + 6x - 4 \end{aligned}$$

13. $$\begin{aligned} -2(x-3) + 2(4-x) &= -2x + 6 + 8 - 2x \\ &= -2x - 2x + 6 + 8 \\ &= -4x + 14 \end{aligned}$$

14. $$\begin{aligned} -3[2x - 4(x-3)] + 2 &= -3[2x - 4x + 12] + 2 \\ &= -3[-2x + 12] + 2 \\ &= 6x - 36 + 2 \\ &= 6x - 34 \end{aligned}$$

15. $$\begin{array}{r|l} x^2 + 6x + 9 & = x + 3 \\ \hline (-3)^2 + 6(-3) + 9 & -3 + 3 \\ 9 - 18 + 9 & 0 \\ \end{array}$$
$$0 = 0$$
Yes, -3 is a solution.

16. $$\begin{array}{r|l} 3 - 8x & = 12x - 2 \\ \hline 3 - 8\left(\frac{1}{2}\right) & 12\left(\frac{1}{2}\right) - 2 \\ 3 - 4 & 6 - 2 \\ \end{array}$$
$$-1 \neq 4$$
No, $\frac{1}{2}$ is not a solution.

17. $$\begin{aligned} (0.32)(60) &= A \\ 19.2 &= A \end{aligned}$$
32% of 60 is 19.2.

18. $$\begin{aligned} \frac{3}{5}x &= -15 \\ \frac{5}{3}\left(\frac{3}{5}x\right) &= \frac{5}{3}(-15) \\ x &= -25 \end{aligned}$$

19. $$\begin{aligned} 7x - 8 &= -29 \\ 7x - 8 + 8 &= -29 + 8 \\ 7x &= -21 \\ \frac{1}{7}(7x) &= \frac{1}{7}(-21) \\ x &= -3 \end{aligned}$$

20. $$\begin{aligned} 13 - 9x &= -14 \\ 13 - 13 - 9x &= -14 - 13 \\ -9x &= -27 \\ -\frac{1}{9}(-9x) &= -\frac{1}{9}(-27) \\ x &= 3 \end{aligned}$$

21. $$\begin{aligned} 8x - 3(4x - 5) &= -2x - 11 \\ 8x - 12x + 15 &= -2x - 11 \\ -4x + 15 &= -2x - 11 \\ -4x + 2x + 15 &= -2x + 2x - 11 \\ -2x + 15 &= -11 \\ -2x + 15 - 15 &= -11 - 15 \\ -2x &= -26 \\ -\frac{1}{2}(-2x) &= -\frac{1}{2}(-26) \\ x &= 13 \end{aligned}$$

22. $$\begin{aligned} 6 - 2(5x - 8) &= 3x - 4 \\ 6 - 10x + 16 &= 3x - 4 \\ -10x + 22 &= 3x - 4 \\ -10x - 3x + 22 &= 3x - 3x - 4 \\ -13x + 22 &= -4 \\ -13x + 22 - 22 &= -4 - 22 \\ -13x &= -26 \\ -\frac{1}{13}(-13x) &= -\frac{1}{13}(-26) \\ x &= 2 \end{aligned}$$

23. $$\begin{aligned} 5x - 8 &= 12x + 13 \\ 5x - 12x - 8 &= 12x - 12x + 13 \\ -7x - 8 &= 13 \\ -7x - 8 + 8 &= 13 + 8 \\ -7x &= 21 \\ -\frac{1}{7}(-7x) &= -\frac{1}{7}(21) \\ x &= -3 \end{aligned}$$

24. $$\begin{aligned} 11 - 4x &= 2x + 8 \\ 11 - 4x - 2x &= 2x - 2x + 8 \\ 11 - 6x &= 8 \\ 11 - 11 - 6x &= 8 - 11 \\ -6x &= -3 \\ -\frac{1}{6}(-6x) &= -\frac{1}{6}(-3) \\ x &= \frac{1}{2} \end{aligned}$$

25.

$$
\begin{aligned}
m_1(T_1 - T) &= m_2(T - T_2)\\
300(75 - T) &= 100(T - 15)\\
22{,}500 - 300T &= 100T - 1500\\
22{,}500 + 1500 - 300T &= 100T - 1500 + 1500\\
24{,}000 - 300T &= 100T\\
24{,}000 - 300T + 300T &= 100T + 300T\\
24{,}000 &= 400T\\
60 &= T
\end{aligned}
$$

The final temperature is 60°C.

26. The unknown number: x

$$
\begin{aligned}
12 - 5x &= -18\\
12 - 12 - 5x &= -18 - 12\\
-15x &= -30\\
-\frac{1}{5}(-5x) &= -\frac{1}{5}(-30)\\
x &= 6
\end{aligned}
$$

The number is 6.

27. Strategy The area of the garage: x

Solution

$$
\begin{aligned}
3x + 200 &= 2000\\
3x + 200 - 200 &= 2000 - 200\\
3x &= 1800\\
\frac{1}{3}(3x) &= \frac{1}{3}(1800)\\
x &= 600
\end{aligned}
$$

The area of the garage is 600 ft^2.

28. Strategy • Amount of oat flour: x

	Amount	Cost	Value
Oat	x	0.80	$0.80x$
Wheat	40	0.50	0.50(40)
Mixture	$x + 40$	0.60	$0.60(x + 40)$

• The sum of the values before mixing equals the value after mixing.

Solution

$$
\begin{aligned}
0.80x + 0.50(40) &= 0.60(x + 40)\\
0.80x + 20 &= 0.60x + 24\\
0.20x &= 4\\
x &= 20
\end{aligned}
$$

20 lb of oat flour are needed for the mixture.

29. Strategy • Amount of pure gold: x

	Amount	Cost	Value
Pure Gold	x	1.00	$1.00x$
Alloy	100	0.20	0.20(100)
Mixture	$x + 100$	0.36	$0.36(x + 100)$

• The sum of the quantities before mixing is equal to the quantity after mixing.

Solution

$$
\begin{aligned}
1.00x + 0.20(100) &= 0.36(x + 100)\\
1.00x + 20 &= 0.36x + 36\\
0.64x &= 16\\
x &= 25
\end{aligned}
$$

25 g of pure gold must be added.

30. Strategy • Width: x
Length: $x + 2$
• Use the equation for the perimeter of a rectangle.

Solution

$$
\begin{aligned}
2L + 2W &= P\\
2(x + 2) + 2(x) &= 44\\
2x + 4 + 2x &= 44\\
4x + 4 &= 44\\
4x &= 40\\
x &= 10\\
x + 2 &= 12
\end{aligned}
$$

The length is 12 ft.
The width is 10 ft.

31. Strategy To find the measure of angle x, use the fact that adjacent angles of intersecting lines are supplementary.

Solution

$$
\begin{aligned}
x + 49° &= 180°\\
x &= 131°
\end{aligned}
$$

The measure of x is 131°.

32. Strategy • Measure of one of the equal angles: x
• Use the equation $A + B + C = 180°$.

Solution

$$
\begin{aligned}
x + x + x &= 180°\\
3x &= 180\\
x &= 60
\end{aligned}
$$

The measure of one of the equal angles is 60°.

33. Strategy • Time running: t
Time jogging: $55 + t$

	Rate	Time	Distance
Running	8	t	$8t$
Jogging	3	$55 - t$	$3(55 - t)$

• The distance traveled is the same.

Solution

$$
\begin{aligned}
8t &= 3(55 - t)\\
8t &= 165 - 3t\\
11t &= 165\\
t &= 15
\end{aligned}
$$

Distance $= 8t = 8(15) = 120$
The length of the track is 120 m.

Chapter 4: Polynomials

Prep Test

1. $-2-(-3) = -2+3 = 1$

2. $-3(6) = -18$

3. $-\dfrac{24}{-36} = \dfrac{2}{3}$

4. $3n^4;\ n = -2$

$$\begin{aligned} 3(-2)^4 &= 3 \cdot 16 \\ &= 48 \end{aligned}$$

5. 0

6. No. $2x^2$ and $2x$ are not like terms since the exponent on x is different.

7. $$\begin{aligned} &3x^2 - 4x + 1 + 2x^2 - 5x - 7 \\ &= 3x^2 + 2x^2 - 4x - 5x + 1 - 7 \\ &= 5x^2 - 9x - 6 \end{aligned}$$

8. $-4y + 4y = 0$

9. $-3(2x-8) = -6x + 24$

10. $$\begin{aligned} 3xy - 4y - 2(5xy - 7y) &= 3xy - 4y - 10xy + 14y \\ &= -7xy + 10y \end{aligned}$$

Go Figure

Strategy To find the value of x and y:

- Set two of the three expressions equal and solve the equation for one of the variable.
- Use the third equation and the value of the variable found above to find the value of the second variable.

Solution

$$\begin{aligned} xy &= \frac{x}{y} \\ xy^2 &= x \\ y^2 &= 1,\ x \neq 0 \end{aligned}$$

So, $y = \pm 1$.

If $y = 1$, then

$$\begin{aligned} x + (1) &= x(1) \\ x + 1 &= x \\ 1 &= 0; \text{ a false statement. So } y \neq 1. \end{aligned}$$

If $y = -1$, then

$$\begin{aligned} x + (-1) &= x(-1) \\ x - 1 &= -x \\ 2x - 1 &= 0 \\ 2x &= 1 \\ x &= \frac{1}{2} \end{aligned}$$

So, $x = \dfrac{1}{2}$ and $y = -1$.

Section 4.1

Objective A Exercises

1. Yes

3. No

5. Yes

7. Yes

9. Binomial

11. Trinomial

13. None of these

15. Binomial

17. $$\begin{array}{r} x^2 + 7x \\ + \ -3x^2 - 4x \\ \hline -2x^2 + 3x \end{array}$$

19. $$\begin{array}{rrr} & y^2 + 4y & \\ + & -4y & -8 \\ \hline & y^2 & -8 \end{array}$$

21. $$\begin{array}{r} 2x^2 + 6x + 12 \\ + \ 3x^2 + x + 8 \\ \hline 5x^2 + 7x + 20 \end{array}$$

23. $$\begin{array}{r} x^3 \qquad\quad - 7x + 4 \\ + \qquad 2x^2 + x - 10 \\ \hline x^3 + 2x^2 - 6x - 6 \end{array}$$

25. $$\begin{array}{r} 2a^3 \qquad\quad - 7a + 1 \\ + \quad -3a^2 - 4a + 1 \\ \hline 2a^3 - 3a^2 - 11a + 2 \end{array}$$

27. $$\begin{aligned} &(4x^2 + 2x) + (x^2 + 6x) \\ &= 5x^2 + 8x \end{aligned}$$

29. $$\begin{aligned} &(4x^2 - 5xy) + (3x^2 + 6xy - 4y^2) \\ &= 7x^2 + xy - 4y^2 \end{aligned}$$

31. $$\begin{aligned} &(2a^2 - 7a + 10) + (a^2 + 4a + 7) \\ &= 3a^2 - 3a + 17 \end{aligned}$$

33. $$\begin{aligned} &(7x + 5x^3 - 7) + (10x^2 - 8x + 3) \\ &= 5x^3 + 10x^2 - x - 4 \end{aligned}$$

35. $$\begin{aligned} &(7 - 5r + 2r^2) + (3r^3 - 6r) \\ &= 3r^3 + 2r^2 - 11r + 7 \end{aligned}$$

Objective B Exercises

37. $$\begin{array}{r} x^2 - 6x \\ -x^2 - 10x \\ \hline \end{array} = \begin{array}{r} x^2 - 6x \\ + \ -x^2 + 10x \\ \hline 4x \end{array}$$

39. $$\begin{array}{r} 2y^2 - 4y \\ - \ -y^2 + 2 \\ \hline \end{array} = \begin{array}{r} 2y^2 - 4y \qquad \\ + \ y^2 \qquad\quad - 2 \\ \hline 3y^2 - 4y - 2 \end{array}$$

41. $\begin{array}{r} x^2 - 2x + 1 \\ \underline{-x^2 + 5x + 8} \end{array} = \begin{array}{r} x^2 - 2x + 1 \\ \underline{+ -x^2 - 5x - 8} \\ -7x - 7 \end{array}$

43. $\begin{array}{r} 4x^3 \qquad + 5x + 2 \\ \underline{- \quad -3x^2 + 2x + 1} \end{array} = \begin{array}{r} 4x^3 \qquad + 5x + 2 \\ \underline{+ \quad 3x^2 - 2x - 1} \\ 4x^3 + 3x^2 + 3x + 1 \end{array}$

45. $\begin{array}{r} 2y^3 + 6y^2 - 2y \\ \underline{-y^3 + y^2 \qquad + 4} \end{array} = \begin{array}{r} 2y^3 + 6y^2 - 2y \\ \underline{+ -y^3 - y^2 \qquad - 4} \\ y^3 + 5y^2 - 2y - 4 \end{array}$

47. $(y^2 - 10xy) - (2y^2 + 3xy)$
$= (y^2 - 10xy) + (-2y^2 - 3xy)$
$= -y^2 - 13xy$

49. $(3x^2 + x - 3) - (x^2 + 4x - 2)$
$= (3x^2 + x - 3) + (-x^2 - 4x + 2)$
$= 2x^2 - 3x - 1$

51. $(-2x^3 + x - 1) - (-x^2 + x - 3)$
$= (-2x^3 + x - 1) + (x^2 - x + 3)$
$= -2x^3 + x^2 + 2$

53. $(1 - 2a + 4a^3) - (a^3 - 2a + 3)$
$= (1 - 2a + 4a^3) + (-a^3 + 2a - 3)$
$= 3a^3 - 2$

55. $(-1 - y + 4y^3) - (3 - 3y - 2y^2)$
$= (-1 - y + 4y^3) + (-3 + 3y + 2y^2)$
$= 4y^3 + 2y^2 + 2y - 4$

Applying the Concepts

57. Subtract the given polynomial from the sum.
$(4x^2 + 3x - 2) - (3x^2 - 6x + 9)$
$= (4x^2 + 3x - 2) + (-3x^2 + 6x - 9)$
$= x^2 + 9x - 11$

59. A monomial is a one-term expression that consists of numbers and variables. An example is $5xy$. A binomial, such as $x^2 + y$, is the sum of two monomials. A trinomial is the sum of three monomials, such as $5x^2 + 3x + 4$. A polynomial has any number of terms, all of which are monomials. An example is $6xy + x - y - 4$. Monomials, binomials, and trinomials are specific types of polynomials. If a polynomial has more than three terms, it does not have a specific name. Be sure the student has not defined a monomial in terms of a polynomial and a polynomial in terms of a monomial; one of these terms should be defined independently of the other.

61. Yes, it is possible to add two polynomials, each of degree 3, and have the sum be a polynomial of degree 2. For example, $(3x^3 - 2x^2) + (-3x^3 - 6x) = -2x^2 - 6x$. Students should discover that for this to occur, the two terms of degree 3 in the two polynomials must be opposites.

Section 4.2

Objective A Exercises

1. Answers will vary. Students should include the fact that when monomials are multiplied, the coefficients are multiplied and the exponents on the like variables are added. Example: $(2x^3y^2)(3xy^4) = 6x^4y^6$

3. $(6x^2)(5x) = (6 \cdot 5)(x^2 \cdot x) = 30x^3$

5. $(7c^2)(-6c^4) = [7(-6)](c^2 \cdot c^4) = -42c^6$

7. $(-3a^3)(-3a^4) = (-3 \cdot -3)(a^3a^4) = 9a^7$

9. $(x^2)(xy^4) = (x^2 \cdot x)y^4 = x^3y^4$

11. $(-2x^4)(5x^5y) = (-2 \cdot 5)(x^4 \cdot x^5)y = -10x^9y$

13. $(-4x^2y^4)(-3x^5y^4) = (-4 \cdot -3)(x^2 \cdot x^5)(y^4 \cdot y^4)$
$= 12x^7y^8$

15. $(2xy)(-3x^2y^4) = (2 \cdot -3)(x \cdot x^2)(y \cdot y^4) = -6x^3y^5$

17. $(x^2yz)(x^2y^4) = (x^2 \cdot x^2)(y \cdot y^4)z = x^4y^5z$

19. $(-a^2b^3)(-ab^2c^4) = (-1 \cdot -1)(a^2 \cdot a)(b^3 \cdot b^2)c^4$
$= a^3b^5c^4$

21. $(-5a^2b^2)(6a^3b^6) = (-5 \cdot 6)(a^2 \cdot a^3)(b^2 \cdot b^6)$
$= -30a^5b^8$

23. $(-6a^3)(-a^2b) = (-6 \cdot -1)(a^3 \cdot a^2)b = 6a^5b$

25. $(-5y^4z)(-8y^6z^5) = (-5 \cdot -8)(y^4 \cdot y^6)(z \cdot z^5)$
$= 40y^{10}z^6$

27. $(x^2y)(yz)(xyz) = (x^2 \cdot x)(y \cdot y \cdot y)(z \cdot z) = x^3y^3z^2$

29. $(3ab^2)(-2abc)(4ac^2)$
$= (3 \cdot -2 \cdot 4)(a \cdot a \cdot a)(b^2 \cdot b)(c \cdot c^2)$
$= -24a^3b^3c^3$

31. $(4x^4z)(-yz^3)(-2x^3z^2)$
$= (4 \cdot -1 \cdot -2)(x^4 \cdot x^3)(y)(z \cdot z^3 \cdot x^2)$
$= 8x^7yz^6$

33. $(-2x^2y^3)(3xy)(-5x^3y^4)$
$= (-2 \cdot 3 \cdot -5)(x^2 \cdot x \cdot x^3)(y^3 \cdot y \cdot y^4)$
$= 30x^6y^8$

35. $(3a^2b)(-6bc)(2ac^2)$
$= (3 \cdot -6 \cdot 2)(a^2 \cdot a)(b \cdot b)(c \cdot c^2)$
$= -36a^3b^2c^3$

Objective B Exercises

37. $(x^3)^5 = x^{15}$

39. $(x^7)^2 = x^{14}$

41. $(-x^2)^4 = x^8$

43. $(-y^3)^4 = y^{12}$

45. $(-2x^2)^3 = (-2)^3x^6 = -8x^6$

47. $(x^2y^3)^2 = x^4y^6$

49. $(3x^2y)^2 = 3^2x^4y^2 = 9x^4y^2$

51. $(-3x^3y^2)^5 = (-3)^5x^{15}y^{10} = -243x^{15}y^{10}$

53. $(-2x)(2x^3)^2 = (-2x)(2^2)x^6 = -8x^7$

55. $(3x^2y)(2x^2y^2)^3 = (3x^2y)(2^3)x^6y^6 = 24x^8y^7$

57. $(ab^2)^2(ab)^2 = (a^2b^4)(a^2b^2) = a^4b^6$

59. $(-2x)^3(-2x^3y)^3 = (-2)^3x^3(-2)^3x^9y^3 = 64x^{12}y^3$

61. $(-2x)(-3xy^2)^2 = (-2x)(-3)^2x^2y^4 = -18x^3y^4$

63. $(ab^2)(-2a^2b)^3 = (ab^2)(-2)^3a^6b^3 = -8a^7b^5$

65. $(-2a^3)(3a^2b)^3 = (-2a^3)(3^3)a^6b^3 = -54a^9b^3$

Applying the Concepts

67. $3x^2 + (3x)^2 = 3x^2 + 9x^2 = 12x^2$

69. $2x^6y^2 + (3x^2y)^2 = 2x^6y^2 + 9x^4y^2$

71. $(2a^3b^2)^3 - 8a^9b^6 = 8a^9b^6 - 8a^9b^6 = 0$

73. $(x^2y^4)^2 + (2xy^2)^4 = x^4y^8 + 16x^4y^8$
$= 17x^4y^8$

75. True.

77. False. $(x^2)^5 = x^{2\cdot 5} = x^{10}$

79. No. $2^{(3^2)}$ is larger. $(2^3)^2 = 2^{3\cdot 2} = 2^6 = 64$ and $2^{(3^2)} = 2^9 = 512$

81. Answers will vary. Some examples include the area of a square, which can be expressed as s^2; the volume of a cube, which can be expressed as s^3; and the surface area of a cube, which can be expressed as $6s^2$.

Section 4.3

Objective A Exercises

1. $3(4x - 5) = (3)(4x) - (3)(5)$
$= 12x - 15$
3, 3, $12x - 15$

3. $x(x - 2) = x^2 - 2x$

5. $-x(x + 7) = -x^2 - 7x$

7. $3a^2(a - 2) = 3a^3 - 6a^2$

9. $-5x^2(x^2 - x) = -5x^4 + 5x^3$

11. $-x^3(3x^2 - 7) = -3x^5 + 7x^3$

13. $2x(6x^2 - 3x) = 12x^3 - 6x^2$

15. $(2x - 4)(3x) = 6x^2 - 12x$

17. $(3x + 4)x = 3x^2 + 4x$

19. $-xy(x^2 - y^2) = -x^3y + xy^3$

21. $x(2x^3 - 3x + 2) = 2x^4 - 3x^2 + 2x$

23. $-a(-2a^2 - 3a - 2) = 2a^3 + 3a^2 + 2a$

25. $x^2(3x^4 - 3x^2 - 2) = 3x^6 - 3x^4 - 2x^2$

27. $2y^2(-3y^2 - 6y + 7) = -6y^4 - 12y^3 + 14y^2$

29. $(a^2 + 3a - 4)(-2a) = -2a^3 - 6a^2 + 8a$

31. $-3y^2(-2y^2 + y - 2) = 6y^4 - 3y^3 + 6y^2$

33. $xy(x^2 - 3xy + y^2) = x^3y - 3x^2y^2 + xy^3$

Objective B Exercises

35.
$$\begin{array}{r} x^2 + 3x + 2 \\ \times \qquad x + 1 \\ \hline x^2 + 3x + 2 \\ x^3 + 3x^2 + 2x \quad \\ \hline x^3 + 4x^2 + 5x + 2 \end{array}$$

37.
$$\begin{array}{r} a^2 - 3a + 4 \\ \times \qquad a - 3 \\ \hline -3a^2 + 9a - 12 \\ a^3 - 3a^2 + 4a \quad \\ \hline a^3 - 6a^2 + 13a - 12 \end{array}$$

39.
$$\begin{array}{r} -2b^2 - 3b + 4 \\ \times \qquad b - 5 \\ \hline 10b^2 + 15b - 20 \\ -2b^3 - 3b^2 + 4b \quad \\ \hline -2b^3 + 7b^2 + 19b - 20 \end{array}$$

41.
$$\begin{array}{r} -2x^2 + 7x - 2 \\ \times \qquad 3x - 5 \\ \hline 10x^2 - 35x + 10 \\ -6x^3 + 21x^2 - 6x \quad \\ \hline -6x^3 + 31x^2 - 41x + 10 \end{array}$$

43.
$$\begin{array}{r} x^2 \qquad + 5 \\ \times \qquad x - 3 \\ \hline -3x^2 \qquad - 15 \\ x^3 \qquad + 5x \quad \\ \hline x^3 - 3x^2 + 5x - 15 \end{array}$$

45.
$$\begin{array}{r} x^3 \qquad - 3x + 2 \\ \times \qquad x - 4 \\ \hline -4x^3 \qquad + 12x - 8 \\ x^4 \qquad - 3x^2 + 2x \quad \\ \hline x^4 - 4x^3 - 3x^2 + 14x - 8 \end{array}$$

47.
$$\begin{array}{r} 5y^2 + 8y - 2 \\ \times \qquad 3y - 8 \\ \hline -40y^2 - 64y + 16 \\ 15y^3 + 24y^2 - 6y \quad \\ \hline 15y^3 - 16y^2 - 70y + 16 \end{array}$$

49.
$$\begin{array}{r} 5a^3 \qquad\quad - 5a + 2 \\ \times \qquad\qquad\qquad\quad a - 4 \\ \hline -20a^3 \qquad +20a - 8 \\ 5a^4 \qquad -5a^2 + 2a \qquad \\ \hline 5a^4 - 20a^3 - 5a^2 + 22a - 8 \end{array}$$

51.
$$\begin{array}{r} y^3 + 2y^2 - 3y + 1 \\ \times \qquad\qquad\quad y + 2 \\ \hline 2y^3 + 4y^2 - 6y + 2 \\ y^4 + 2y^3 - 3y^2 + y \qquad \\ \hline y^4 + 4y^3 + y^2 - 5y + 2 \end{array}$$

Objective C Exercises

53. $(x+1)(x+3) = x^2 + 3x + x + 3$
$= x^2 + 4x + 3$

55. $(a-3)(a+4) = a^2 + 4a - 3a - 12$
$= a^2 + a - 12$

57. $(y+3)(y-8) = y^2 - 8y + 3y - 24$
$= y^2 - 5y - 24$

59. $(y-7)(y-3) = y^2 - 3y - 7y + 21$
$= y^2 - 10y + 21$

61. $(2x+1)(x+7) = 2x^2 + 14x + x + 7$
$= 2x^2 + 15x + 7$

63. $(3x-1)(x+4) = 3x^2 + 12x - x - 4$
$= 3x^2 + 11x - 4$

65. $(4x-3)(x-7) = 4x^2 - 28x - 3x + 21$
$= 4x^2 - 31x + 21$

67. $(3y-8)(y+2) = 3y^2 + 6y - 8y - 16$
$= 3y^2 - 2y - 16$

69. $(3x+7)(3x+11) = 9x^2 + 33x + 21x + 77$
$= 9x^2 + 54x + 77$

71. $(7a-16)(3a-5) = 21a^2 - 35a - 48a + 80$
$= 21a^2 - 83a + 80$

73. $(3a-2b)(2a-7b) = 6a^2 - 21ab - 4ab + 14b^2$
$= 6a^2 - 25ab + 14b^2$

75. $(a-9b)(2a+7b) = 2a^2 + 7ab - 18ab - 63b^2$
$= 2a^2 - 11ab - 63b^2$

77. $(10a-3b)(10a-7b)$
$= 100a^2 - 70ab - 30ab + 21b^2$
$= 100a^2 - 100ab + 21b^2$

79. $(5x+12y)(3x+4y) = 15x^2 + 20xy + 36xy + 48y^2$
$= 15x^2 + 56xy + 48y^2$

81. $(2x-15y)(7x+4y) = 14x^2 + 8xy - 105xy - 60y^2$
$= 14x^2 - 97xy - 60y^2$

83. $(8x-3y)(7x-5y) = 56x^2 - 40xy - 21xy + 15y^2$
$= 56x^2 - 61xy + 15y^2$

Objective D Exercises

85. $(y-5)(y+5) = y^2 - 25$

87. $(2x+3)(2x-3) = 4x^2 - 9$

89. $(3x-7)(3x+7) = 9x^2 - 49$

91. $(4-3y)(4+3y) = 16 - 9y^2$

93. $(x+1)^2 = x^2 + 2x + 1$

95. $(3a-5)^2 = 9a^2 - 30a + 25$

97. $(2a+b)^2 = 4a^2 + 4ab + b^2$

99. $(x-2y)^2 = x^2 - 4xy + 4y^2$

101. $(5x+2y)^2 = 25x^2 + 20xy + 4y^2$

Objective E Exercises

103. **Strategy**
- Length: $5x$
  Width: $2x - 7$
- Use the formula for the area of a rectangle.

Solution $A = LW$
$A = 5x(2x-7)$
$A = 10x^2 - 35x$
The area of the rectangle is $(10x^2 - 35x)$ ft^2.

105. **Strategy**
- Side of a square: $2x + 1$
- Use the equation for the area of a square.

Solution
$$\begin{aligned} S^2 &= A \\ (2x+1)^2 &= A \\ (2x+1)(2x+1) &= A \\ 4x^2 + 4x + 1 &= A \end{aligned}$$
The area is $(4x^2 + 4x + 1)$ km^2.

107. **Strategy**
- Base: $4x$
  Height: $2x + 5$
- Use the equation for the area of a triangle.

Solution $A = \frac{1}{2}bh$
$A = \frac{1}{2}(4x)(2x+5)$
$A = 2x(2x+5)$
$A = 4x^2 + 10x$
The area of the triangle is $(4x^2 + 10x)$ m^2.

109. **Strategy**
- One side: 30
  Second side: $100 + 2w$
- Use the equation for the area of a rectangle.

Solution $A = LW$
$A = 30(100 + 2w)$
$A = 3000 + 60w$
The area is $(60w + 3000)$ yd^2.

Applying the Concepts

111. $(x^2+x-3)^2 = [x^2+(x-3)]^2$
$= (x^2)^2 + 2(x^2)(x-3) + (x-3)^2$
$= x^4 + 2x^3 - 6x^2 + x^2 - 6x + 9$
$= x^4 + 2x^3 - 5x^2 - 6x + 9$

113. Multiply the quotient by the divisor.
$(3x-4)(4x+5) = 12x^2 + 15x - 16x - 20$
$= 12x^2 - x - 20$
The polynomial is $12x^2 - x - 20$.

115. $(x^2+x+3)(x-4) - (4x^2-x-5)$
$= x^3 - 4x^2 + x^2 - 4x + 3x - 12 - 4x^2 + x + 5$
$= x^3 - 7x^2 - 7$

117. No, it is not possible to multiply a polynomial of degree 2 by a polynomial of degree 2 and have the product be a polynomial of degree 3. A polynomial of degree 2 contains the term ax^2, $a \neq 0$. Two polynomials of degree 2 will have the terms ax^2 and bx^2, $a \neq 0$, $b \neq 0$. Multiplying these terms yields abx^4, where $ab \neq 0$. Therefore, the product will have an x^4 term and will be of degree 4.

Section 4.4

Objective A Exercises

1. $\frac{x^7}{x^5} = x^{7-5} = x^2$

3. $5^{-2} = \frac{1}{5^2} = \frac{1}{25}$

5. $\frac{1}{8^{-2}} = 8^2 = 64$

7. $\frac{3^{-2}}{3} = \frac{1}{3^3} = \frac{1}{27}$

9. $\frac{2^{-2}}{2^{-3}} = 2^{-2-(-3)} = 2^1 = 2$

11. $x^{-2} = \frac{1}{x^2}$

13. $\frac{1}{a^{-6}} = a^6$

15. $4x^{-7} = 4 \cdot \frac{1}{x^7} = \frac{4}{x^7}$

17. $\frac{2}{3}z^{-2} = \frac{2}{3} \cdot \frac{1}{z^2} = \frac{2}{3z^2}$

19. $\frac{5}{b^{-8}} = 5 \cdot \frac{1}{b^{-8}} = 5b^8$

21. $\frac{1}{3x^{-2}} = \frac{1}{3} \cdot \frac{1}{x^{-2}} = \frac{1}{3} \cdot x^2 = \frac{x^2}{3}$

23. $(ab^5)^0 = 1$

25. $-(3p^2q^5)^0 = -(1) = -1$

27. $\frac{y^7}{y^3} = y^{7-3} = y^4$

29. $\frac{a^8}{a^5} = a^{8-5} = a^3$

31. $\frac{p^5}{p} = p^{5-1} = p^4$

33. $\frac{4x^8}{2x^5} = 2x^{8-5} = 2x^3$

35. $\frac{22k^5}{11k^4} = 2k^{5-4} = 2k$

37. $\frac{m^9n^7}{m^4n^5} = m^{9-4}n^{7-5} = m^5n^2$

39. $\frac{6r^4}{4r^2} = \frac{3r^{4-2}}{2} = \frac{3r^2}{2}$

41. $\frac{-16a^7}{24a^6} = \frac{-2a^{7-6}}{3} = -\frac{2a}{3}$

43. $\frac{y^3}{y^8} = y^{3-8} = y^{-5} = \frac{1}{y^5}$

45. $\frac{a^5}{a^{11}} = a^{5-11} = a^{-6} = \frac{1}{a^6}$

47. $\frac{4x^2}{12x^5} = \frac{x^{2-5}}{3} = \frac{x^{-3}}{3} = \frac{1}{3x^3}$

49. $\frac{-12x}{-18x^6} = \frac{2x^{1-6}}{3} = \frac{2x^{-5}}{3} = \frac{2}{3x^5}$

51. $\frac{x^6y^5}{x^8y} = x^{6-8}y^{5-1} = x^{-2}y^4 = \frac{y^4}{x^2}$

53. $\frac{2m^6n^2}{5m^9n^{10}} = \frac{2m^{6-9}n^{2-10}}{5} = \frac{2m^{-3}n^{-8}}{5} = \frac{2}{5m^3n^8}$

55. $\frac{pq^3}{p^4q^4} = p^{1-4}q^{3-4} = p^{-3}q^{-1} = \frac{1}{p^3q}$

57. $\frac{3x^4y^5}{6x^4y^8} = \frac{1x^{4-4}y^{5-8}}{2} = \frac{x^0y^{-3}}{2} = \frac{1}{2y^3}$

59. $\frac{14x^4y^6z^2}{16x^3y^9z} = \frac{7x^{4-3}y^{6-9}z^{2-1}}{8} = \frac{7xy^{-3}z}{8} = \frac{7xz}{8y^3}$

61. $\frac{15mn^9p^3}{30m^4n^9p} = \frac{1m^{1-4}n^{9-9}p^{3-1}}{2} = \frac{m^{-3}n^0p^2}{2}$
$= \frac{p^2}{2m^3}$

63. $(-2xy^{-2})^3 = -2^3x^3y^{-6} = -\frac{8x^3}{y^6}$

65. $(3x^{-1}y^{-2})^2 = 3^2x^{-2}y^{-4} = \frac{9}{x^2y^4}$

67. $(2x^{-1})(x^{-3}) = 2x^{-4} = \frac{2}{x^4}$

69. $(-5a^2)(a^{-5})^2$
$= (-5a^2)(a^{-10}) = -5a^{-8} = -\frac{5}{a^8}$

71. $(-2ab^{-2})(4a^{-2}b)^{-2}$
$= (-2ab^{-2})(4^{-2}a^4b^{-2})$
$= \frac{-2a^5}{4^2b^4} = \frac{-2a^5}{16b^4} = -\frac{a^5}{8b^4}$

73. $(-5x^{-2}y)(-2x^{-2}y^2) = 10x^{-4}y^3$
$= \frac{10y^3}{x^4}$

75. $\dfrac{3x^{-2}y^2}{6xy^2} = \dfrac{1}{2x^3}$

77. $\dfrac{3x^{-2}y}{xy} = \dfrac{3}{x^3}$

79. $\dfrac{2x^{-1}y^{-4}}{4xy^2} = \dfrac{1}{2x^2y^6}$

81. $\dfrac{(x^{-2}y)^2}{x^2y^3} = \dfrac{x^{-4}y^2}{x^2y^3} = \dfrac{1}{x^6y}$

83. $\dfrac{(a^{-2}y^3)^{-3}}{a^2y} = \dfrac{a^6y^{-9}}{a^2y} = \dfrac{a^4}{y^{10}}$

85. $\dfrac{-16xy^4}{96x^4y^4} = -\dfrac{1}{6x^3}$

87. $\dfrac{22a^2b^4}{-132b^3c^2} = -\dfrac{a^2b}{6c^2}$

89. $\dfrac{-(14ab^4)^2}{28a^4b^2} = -\dfrac{196a^2b^8}{28a^4b^2}$
$= -\dfrac{7b^6}{a^2}$

91. $\dfrac{(3^{-1}r^4s^{-3})^{-2}}{(6r^2t^{-2}s^{-1})^2} = \dfrac{3^2r^{-8}s^6}{6^2r^4t^{-4}s^{-2}}$
$= \dfrac{9t^4s^8}{36r^{12}}$
$= \dfrac{s^8t^4}{4r^{12}}$

93. $\left(\dfrac{15m^3n^{-2}p^{-1}}{25m^{-2}n^{-4}}\right)^{-3} = \dfrac{15^{-3}m^{-9}n^6p^3}{25^{-3}m^6n^{12}}$
$= \dfrac{25^3p^3}{15^3m^{15}n^6}$
$= \dfrac{5 \cdot 5 \cdot 5p^3}{3 \cdot 3 \cdot 3m^{15}n^6} = \dfrac{125p^3}{27m^{15}n^6}$

Objective B Exercises

95. Very large or very small numbers are written in scientific notation so that they are easier to read.

97. $0.00000000324 = 3.24 \times 10^{-9}$

99. $0.000000000000000003 = 3 \times 10^{-18}$

101. $32{,}000{,}000{,}000{,}000{,}000 = 3.2 \times 10^{16}$

103. $0.000000000000000000122 = 1.22 \times 10^{-19}$

105. $547{,}000{,}000 = 5.47 \times 10^8$

107. $1.67 \times 10^{-4} = 0.000167$

109. $6.8 \times 10^7 = 68{,}000{,}000$

111. $3.05 \times 10^{-5} = 0.0000305$

113. $1.02 \times 10^{-9} = 0.00000000102$

115. $602{,}300{,}000{,}000{,}000{,}000{,}000{,}000$
$= 6.023 \times 10^{23}$

117. $0.0000037 = 3.7 \times 10^{-6}$

119. $0.000000001 = 1 \times 10^{-9}$

121. $0.00000000000000000016 = 1.6 \times 10^{-19}$

Applying the Concepts

123.

x	2^x
-2	$2^{-2} = \frac{1}{4}$
-1	$2^{-1} = \frac{1}{2}$
0	$2^0 = 1$
1	$2^1 = 2$
2	$2^2 = 4$

125.

x	2^{-x}
-2	$2^{-(-2)} = 2^2 = 4$
-1	$2^{-(-1)^2} = 2^1 = 2$
0	$2^0 = 1$
1	$2^{-1} = \frac{1}{2}$
2	$2^{-2} = \frac{1}{4}$

127. False. $(2a)^{-3} = \dfrac{1}{8a^3}$

129. False. $(2+3)^{-1} = (5)^{-1} = \dfrac{1}{5}$

131. The expression x^{-2} is positive for all nonzero real numbers x. The reason is that $x^{-2} = \dfrac{1}{x^2}$, and x^2 is positive for all nonzero values of x.

Section 4.5

Objective A Exercises

1. $\dfrac{10a - 25}{5} = \dfrac{10a}{5} - \dfrac{25}{5} = 2a - 5$

3. $\dfrac{6y^2 + 4y}{y} = 6y + 4$

5. $\dfrac{3x^2 - 6x}{3x} = \dfrac{3x^2}{3x} - \dfrac{6x}{3x} = x - 2$

7. $\dfrac{5x^2 - 10x}{-5x} = \dfrac{5x^2}{-5x} + \dfrac{-10x}{-5x} = -x + 2$

9. $\dfrac{x^3 + 3x^2 - 5x}{x} = \dfrac{x^3}{x} + \dfrac{3x^2}{x} - \dfrac{5x}{x} = x^2 + 3x - 5$

11. $\dfrac{x^6 - 3x^4 - x^2}{x^2} = \dfrac{x^6}{x^2} - \dfrac{3x^4}{x^2} - \dfrac{x^2}{x^2} = x^4 - 3x^2 - 1$

13. $\dfrac{5x^2y^2 + 10xy}{5xy} = \dfrac{5x^2y^2}{5xy} + \dfrac{10xy}{5xy} = xy + 2$

15. $\dfrac{9y^6 - 15y^3}{-3y^3} = \dfrac{9y^6}{-3y^3} - \dfrac{15y^3}{-3y^3} = -3y^3 + 5$

17. $\dfrac{3x^2 - 2x + 1}{x} = \dfrac{3x^2}{x} - \dfrac{2x}{x} + \dfrac{1}{x} = 3x - 2 + \dfrac{1}{x}$

19. $\dfrac{-3x^2 + 7x - 6}{x} = \dfrac{-3x^2}{x} + \dfrac{7x}{x} - \dfrac{6}{x} = -3x + 7 - \dfrac{6}{x}$

21. $$\frac{16a^2b-20ab+24ab^2}{4ab}=\frac{16a^2b}{4ab}-\frac{20ab}{4ab}+\frac{24ab^2}{4ab}$$
$$=4a-5+6b$$

23. $$\frac{9x^2y+6xy-3xy^2}{xy}=\frac{9x^2y}{xy}+\frac{6xy}{xy}-\frac{3xy^2}{xy}$$
$$=9x+6-3y$$

Objective B Exercises

25. If $\frac{x^2-x-6}{x-3}=x+2$, then

$x^2-x-6=(x+2)(x-3)$

27.
$$\begin{array}{r} b-7 \\ b-7\overline{)b^2-14b+49} \\ \underline{b^2-7b} \\ -7b+49 \\ \underline{-7b+49} \\ 0 \end{array}$$
$(b^2-14b+49)\div(b-7)=b-7$

29.
$$\begin{array}{r} y-5 \\ y+7\overline{)y^2+2y-35} \\ \underline{y^2+7y} \\ -5y-35 \\ \underline{-5y-35} \\ 0 \end{array}$$
$(y^2+2y-35)\div(y+7)=y-5$

31.
$$\begin{array}{r} 2y-7 \\ y-3\overline{)2y^2-13y+21} \\ \underline{2y^2-6y} \\ -7y+21 \\ \underline{-7y+21} \\ 0 \end{array}$$
$(2y^2-13y+21)\div(y-3)=2y-7$

33.
$$\begin{array}{r} 2y+6 \\ y-3\overline{)2y^2+0+7} \\ \underline{2y^2-6y} \\ 6y+7 \\ \underline{6y-18} \\ 25 \end{array}$$
$(2y^2+7)\div(y-3)=2y+6+\frac{25}{y-3}$

35.
$$\begin{array}{r} x-2 \\ x+2\overline{)x^2+0+4} \\ \underline{x^2+2x} \\ -2x+4 \\ \underline{-2x-4} \\ 8 \end{array}$$
$(x^2+4)\div(x+2)=x-2+\frac{8}{x+2}$

37.
$$\begin{array}{r} 3y-5 \\ 2y+4\overline{)6y^2+2y+0} \\ \underline{6y^2+12y} \\ -10y+0 \\ \underline{-10y-20} \\ 20 \end{array}$$
$(6y^2+2y)\div(2y+4)=3y-5+\frac{20}{2y+4}$

39.
$$\begin{array}{r} 6x-12 \\ x+2\overline{)6x^2+0-5} \\ \underline{6x^2+12x} \\ -12x-5 \\ \underline{-12x-24} \\ 19 \end{array}$$
$(6x^2-5)\div(x+2)=6x-12+\frac{19}{x+2}$

41.
$$\begin{array}{r} b-5 \\ b-3\overline{)b^2-8b-9} \\ \underline{b^2-3b} \\ -5b-9 \\ \underline{-5b+15} \\ -24 \end{array}$$
$(b^2-8b-9)\div(b-3)=b-5-\frac{24}{b-3}$

43.
$$\begin{array}{r} 3x+17 \\ x-4\overline{)3x^2+5x-4} \\ \underline{3x^2-12x} \\ 17x-4 \\ \underline{17x-68} \\ 64 \end{array}$$
$(3x^2+5x-4)\div(x-4)=3x+17+\frac{64}{x-4}$

45.
$$\begin{array}{r} 5y+3 \\ 2y+3\overline{)10y^2+21y+10} \\ \underline{10y^2+15y} \\ 6y+10 \\ \underline{6y+9} \\ 1 \end{array}$$
$(10y^2+21y+10)\div(2y+3)=5y+3+\frac{1}{2y+3}$

47.
$$\begin{array}{r} 4a+1 \\ 3a-7\overline{)12a^2-25a-7} \\ \underline{12a^2-28a} \\ 3a-7 \\ \underline{3a-7} \\ 0 \end{array}$$
$(12a^2-25a-7)\div(3a-7)=4a+1$

49.
$$\begin{array}{r} 2a+9 \\ 3a-1\overline{)6a^2+25a+24} \\ \underline{6a^2-2a} \\ 27a+24 \\ \underline{27a-9} \\ 33 \end{array}$$
$(6a^2+25a+24)\div(3a-1)=2a+9+\frac{33}{3a-1}$

51.

$$\begin{array}{r} x^2 - 5x + 2 \\ x - 1\overline{)x^3 - 6x^2 + 7x - 2} \\ \underline{x^3 - x^2} \\ -5x^2 + 7x \\ \underline{-5x^2 + 5x} \\ 2x - 2 \\ \underline{2x - 2} \\ 0 \end{array}$$

$(x^3 - 6x^2 + 7x - 2) \div (x - 1) = x^2 - 5x + 2$

53.

$$\begin{array}{r} x^2 + 5 \\ x^2 + 0 - 2\overline{)x^4 + 0 + 3x^2 + 0 - 10} \\ \underline{x^4 + 0 - 2x^2} \\ 5x^2 + 0 - 10 \\ \underline{5x^2 + 0 - 10} \\ 0 \end{array}$$

$(x^4 + 3x^2 - 10) \div (x^2 - 2) = x^2 + 5$

Applying the Concepts

55. Divide the result by $4b$.

$\dfrac{12ab^2}{4b} = 3ab$

Chapter 4 Review Exercises

1. $(2b - 3)(4b + 5) = 8b^2 + 10b - 12b - 15$
$= 8b^2 - 2b - 15$

2. $(12y^2 + 17y - 4) + (9y^2 - 13y + 3)$

$$\begin{array}{r} 12y^2 + 17y - 4 \\ \underline{+\ 9y^2 - 13y + 3} \\ 21y^2 + 4y - 1 \end{array}$$

3. $(xy^5z^3)(x^3y^3z) = x^4y^8z^4$

4. $\dfrac{8x^{12}}{12x^9} = \dfrac{\overset{1}{\cancel{2}} \cdot \overset{1}{\cancel{2}} \cdot 2x^{12}}{\underset{1}{\cancel{2}} \cdot \underset{1}{\cancel{2}} \cdot 3x^9} = \dfrac{2x^{12-9}}{3} = \dfrac{2x^3}{3}$

5. $-2x(4x^2 + 7x - 9) = -2x(4x^2) - 2x(7x) - 2x(-9)$
$= -8x^3 - 14x^2 + 18x$

6. $\dfrac{3ab^4}{-6a^2b^4} = -\dfrac{1}{2a}$

7. $(-2u^3v^4)^4 = (-2^4)u^{12}v^{16} = 16u^{12}v^{16}$

8. $(2^3)^2 = 2^6 = 64$
or
$(8)^2 = 64$

9. $(5x^2 - 2x - 1) - (3x^2 - 5x + 7)$

$$\begin{array}{r} 5x^2 - 2x - 1 \\ \underline{+ -3x^2 + 5x - 7} \\ 2x^2 + 3x - 8 \end{array}$$

10. $\dfrac{a^{-1}b^3}{a^3b^{-3}} = \dfrac{b^6}{a^4}$

11. $(-2x^3)^2(-3x^4)^3 = (-2)^2x^6 \cdot (-3)^3x^{12}$
$= 4(-27)x^{18}$
$= -108x^{18}$

12. $(5y - 7)^2 = (5y - 7)(5y - 7)$
$= 25y^2 - 35y - 35y + 49$
$= 25y^2 - 70y + 49$

13. $(5a^7b^6)^2(4ab) = (5^2a^{14}b^{12})(4ab)$
$= (25a^{14}b^{12})(4ab)$
$= 100a^{15}b^{13}$

14. $\dfrac{12b^7 + 36b^5 - 3b^3}{3b^3} = \dfrac{12b^7}{3b^3} + \dfrac{36b^5}{3b^3} - \dfrac{3b^3}{3b^3}$
$= 4b^4 + 12b^2 - 1$

15. $-4^{-2} = -\dfrac{1}{4^2} = -\dfrac{1}{16}$

16. $(13y^3 - 7y - 2) - (12y^2 - 2y - 1)$

$$\begin{array}{r} 13y^3 \qquad\qquad - 7y - 2 \\ \underline{+ \qquad - 12y^2 + 2y + 1} \\ 13y^3 - 12y^2 - 5y - 1 \end{array}$$

17.

$$\begin{array}{r} -x + 2 \\ x + 3\overline{)-x^2 - x + 7} \\ \underline{-x^2 - 3x} \\ 2x + 7 \\ \underline{-2x + 6} \\ 1 \end{array}$$

$(-x^2 - x + 7) \div (x + 3) = -x + 2 + \dfrac{1}{x + 3}$

18. $(2a - b)(x - 2y) = 2ax - 4ay - bx + 2by$

19. $(3y^2 + 4y - 7)(2y + 3)$
$= 3y^2(2y + 3) + 4y(2y + 3) - 7(2y + 3)$

$$\begin{array}{r} 6y^3 + 9y^2 \qquad\qquad \\ + 8y^2 + 12y \qquad \\ - 14y - 21 \\ \hline 6y^3 + 17y^2 - 2y - 21 \end{array}$$

20.

$$\begin{array}{r} b^2 + 5b + 2 \\ b - 7\overline{)b^3 - 2b^2 - 33b - 7} \\ \underline{b^3 - 7b^2} \\ +5b^2 - 33b \\ \underline{5b^2 - 35b} \\ 2b - 7 \\ \underline{2b - 14} \\ 7 \end{array}$$

$(b^3 - 2b^2 - 33b - 7) \div (b - 7) = b^2 + 5b + 2 + \dfrac{7}{b - 7}$

21. $2ab^3(4a^2 - 2ab + 3b^2)$
$= 2ab^3(4a^2) + 2ab^3(-2ab) + 2ab^3(3b^2)$
$= 8a^3b^3 - 4a^2b^4 + 6ab^5$

22. $(2a - 5b)(2a + 5b) = 4a^2 + 10ab - 10ab - 25b^2$
$= 4a^2 - 25b^2$

23. $(6b^3 - 2b^2 - 5)(2b^2 - 1)$
$= 6b^3(2b^2 - 1) - 2b^2(2b^2 - 1) - 5(2b^2 - 1)$

$$\begin{array}{rrrrr} 12b^5 & & -6b^3 & & \\ & -4b^4 & & +2b^2 & \\ & & & -10b^2 & +5 \\ \hline 12b^5 & -4b^4 & -6b^3 & -8b^2 & +5 \end{array}$$

24. $(2x^3 + 7x^2 + x) + (2x^2 - 4x - 12)$

$$\begin{array}{rrrrr} & 2x^3 & +7x^2 & +x & \\ + & & 2x^2 & -4x & -12 \\ \hline & 2x^3 & +9x^2 & -3x & -12 \end{array}$$

25. $\dfrac{16y^2 - 32y}{-4y} = \dfrac{16y^2}{-4y} + \dfrac{-32y}{-4y} = -4y + 8$

26. $(a + 7)(a - 7) = a^2 - 7a + 7a - 49$
$= a^2 - 49$

27. $37{,}560{,}000{,}000 = 3.756 \times 10^{10}$

28. $1.46 \times 10^7 = 14{,}600{,}000$

29. $(2a^{12}b^3)(-9b^2c^6)(3ac) = -54a^{13}b^5c^7$

30.

$$\begin{array}{r} 2y - 9 \\ 3y - 4\overline{)6y^2 - 35y + 36} \\ \underline{6y^2 - 8y} \\ -27y + 36 \\ \underline{-27y + 36} \\ 0 \end{array}$$

$(6y^2 - 35y + 36) \div (3y + 4) = 2y - 9$

31. $(-3x^{-2}y^{-3})^{-2} = (-3)^{-2}x^4y^6 = \dfrac{x^4y^6}{(-3)^2} = \dfrac{x^4y^6}{9}$

32. $(5a - 7)(2a + 9) = 10a^2 + 45a - 14a - 63$
$= 10a^2 + 31a - 63$

33. $0.000000127 = 1.27 \times 10^{-7}$

34. $3.2 \times 10^{-12} = 0.0000000000032$

35. Width $= w$
Length $= 2w - 1$
Use the equation for the area of a rectangle.
$Lw = A$
$(2w - 1)w = 2w^2 - w$
The area is $(2w^2 - w)$ ft^2.

36. Side of a square: $3x - 2$
Use the equation for the area of a square.

$$\begin{aligned} S^2 &= A \\ (3x - 2)^2 &= A \\ 9x^2 - 6x - 6x + 4 &= A \\ 9x^2 - 12x + 4 &= A \end{aligned}$$

The area is $(9x^2 - 12x + 4)$ in^2.

Chapter 4 Test

1. $2x(2x^2 - 3x) = 2x(2x^2) - 2x(3x)$
$= 4x^3 - 6x^2$

2. $\dfrac{12x^3 - 3x^2 + 9}{3x^2} = \dfrac{12x^3}{3x^2} - \dfrac{3x^2}{3x^2} + \dfrac{9}{3x^2}$
$= 4x - 1 + \dfrac{3}{x^2}$

3. $\dfrac{12x^2}{-3x^8} = -\dfrac{4}{x^6}$

4. $(-2xy^2)(3x^2y^4) = -6x^3y^6$

5.

$$\begin{array}{r} x - 1 \\ x + 1\overline{)x^2 + 0x + 1} \\ \underline{x^2 + x} \\ -x + 1 \\ \underline{-x - 1} \\ 2 \end{array}$$

$(x^2 + 1) \div (x + 1) = x - 1 + \dfrac{2}{x + 1}$

6. $(x - 3)(x^2 - 4x + 5)$
$= x(x^2 - 4x + 5) - 3(x^2 - 4x + 5)$

$$\begin{array}{rrrr} = x^3 & -4x^2 & +5x & \\ & -3x^2 & +12x & -15 \\ \hline x^3 & -7x^2 & +17x & -15 \end{array}$$

7. $(-2a^2b)^3 = -2^3a^6b^3 = -8a^6b^3$

8. $\dfrac{(3x^{-2}y^3)^3}{3x^4y^{-1}} = \dfrac{27x^{-6}y^9}{3x^4y^{-1}} = \dfrac{9y^{10}}{x^{10}}$

9. $(a - 2b)(a + 5b) = a^2 + 5ab - 2ab - 10b^2$
$= a^2 + 3ab - 10b^2$

10. $\dfrac{16x^5 - 8x^3 + 20x}{4x} = \dfrac{16x^5}{4x} - \dfrac{8x^3}{4x} + \dfrac{20x}{4x}$
$= 4x^4 - 2x^2 + 5$

11.

$$\begin{array}{r} x + 7 \\ x - 1\overline{)x^2 + 6x - 7} \\ \underline{x^2 - x} \\ 7x - 7 \\ \underline{7x - 7} \\ 0 \end{array}$$

$(x^2 + 6x - 7) \div (x - 1) = x + 7$

12. $-3y^2(-2y^2 + 3y - 6)$
$= -3y^2(-2y^2) - 3y^2(3y) - 3y^2(-6)$
$= 6y^4 - 9y^3 + 18y^2$

13. $(-2x^3 + x^2 - 7)(2x - 3)$
$= -2x^3(2x - 3) + x^2(2x - 3) - 7(2x - 3)$
$= -4x^4 + 6x^3 + 2x^3 - 3x^2 - 14x + 21$
$= -4x^4 + 8x^3 - 3x^2 - 14x + 21$

14. $(4y - 3)(4y + 3) = 16y^2 + 12y - 12y - 9$
$= 16y^2 - 9$

15. $(ab^2)(a^3b^5) = a^4b^7$

16. $\dfrac{2a^{-1}b}{2^{-2}a^{-2}b^{-3}} = 2^3ab^4 = 8ab^4$

17. $\dfrac{20a-35}{5}=\dfrac{20a}{5}-\dfrac{35}{5}$
$= 4a-7$

18. $(3a^2-2a-7)-(5a^3+2a-10)$

$$\begin{array}{r} 3a^2-2a-7 \\ +\,-5a^3 \qquad -2a+10 \\ \hline -5a^3+3a^2-4a+3 \end{array}$$

19. $(2x-5)^2=(2x-5)(2x-5)$
$= 4x^2-10x-10x+25$
$= 4x^2-20x+25$

20.
$$\begin{array}{r} 2x+3 \\ 2x-3\overline{)4x^2+0x-7} \\ \underline{4x^2-6x} \\ 6x-7 \\ \underline{6x-9} \\ 2 \end{array}$$

$(4x^2-7)\div(2x-3)=2x+3+\dfrac{2}{2x-3}$

21. $\dfrac{-(2x^2y)^3}{4x^3y^3}=\dfrac{-8x^6y^3}{4x^3y^3}=-2x^3$

22. $(2x-7)(5x-4y)=10x^2-8xy-35xy+28y^2$
$= 10x^2-43xy+28y^2$

23. $(3x^3-2x^2-4)+(8x^2-8x+7)$

$$\begin{array}{r} 3x^3-2x^2 \qquad -4 \\ +\qquad 8x^2-8x+7 \\ \hline 3x^3+6x^2-8x+3 \end{array}$$

24. $0.00000000302=3.02\times10^{-9}$

25. $A=\pi r^2$
$= \pi(x-5)^2$
$= \pi(x^2-10x+25)$
$= \pi x^2-10\pi x+25\pi$
The area of the circle is $(\pi x^2-10\pi x+25\pi)$ m^2.

Cumulative Review Exercises

1. $\dfrac{3}{16}-\left(-\dfrac{5}{8}\right)-\dfrac{7}{9}=\dfrac{3}{16}+\dfrac{5}{8}-\dfrac{7}{9}$
$= \dfrac{27}{144}+\dfrac{90}{144}-\dfrac{112}{144}$
$= \dfrac{5}{144}$

2. $-3^2\left(\dfrac{2}{3}\right)^3\left(-\dfrac{5}{8}\right)$
$= -9\left(\dfrac{8}{27}\right)\left(-\dfrac{5}{8}\right)$
$= \dfrac{5}{3}$

3. $\left(-\dfrac{1}{2}\right)^3\div\left(\dfrac{3}{8}-\dfrac{5}{6}\right)+2=\left(-\dfrac{1}{8}\right)\div\left(\dfrac{9}{24}-\dfrac{20}{24}\right)+2$
$= \left(-\dfrac{1}{8}\right)\div\left(\dfrac{-11}{24}\right)+2$
$= \left(-\dfrac{1}{8}\right)\cdot\left(-\dfrac{24}{11}\right)+2$
$= \dfrac{3}{11}+2$
$= \dfrac{3}{11}+\dfrac{22}{11}$
$= \dfrac{25}{11}$

4. $\dfrac{b-(a-b)^2}{b^2}=\dfrac{3-(-2-3)^2}{3^2}=\dfrac{3-(-5)^2}{9}$
$= \dfrac{3-25}{9}=-\dfrac{22}{9}$

5. $-2x-(-xy)+7x-4xy=-2x+7x+xy-4xy$
$= 5x-3xy$

6. $(12x)\left(\dfrac{-3}{4}\right)=-9x$

7. $-2[3x-2(4-3x)+2]=-2[3x-8+6x+2]$
$= -2[9x-6]$
$= -18x+12$

8. $12=-\dfrac{3}{4}x$
$\dfrac{-4}{3}(12)=\dfrac{-3}{4}x\left(\dfrac{-4}{3}\right)$
$-16=x$

9. $2x-9=3x+7$
$-x=16$
$x=-16$

10. $2-3(4-x)=2x+5$
$2-12+3x=2x+5$
$3x-10=2x+5$
$x=15$

11. $P\times B=A$
$P\times160=35.2$
$P=\dfrac{35.2}{160}$
$P=0.22=22\%$

12. $(4b^3-7b^2-7)+(3b^2-8b+3)$

$$\begin{array}{r} 4b^3-7b^2 \qquad -7 \\ +\qquad 3b^2-8b+3 \\ \hline 4b^3-4b^2-8b-4 \end{array}$$

13. $(3y^3-5y+8)-(-2y^2+5y+8)$

$$\begin{array}{r} 3y^3 \qquad -5y+8 \\ +\qquad 2y^2-5y-8 \\ \hline 3y^3+2y^2-10y \end{array}$$

14. $(a^3b^5)^3=a^9b^{15}$

15. $(4xy^3)(-2x^2y^3)=-8x^3y^6$

16. $-2y^2(-3y^2 - 4y + 8)$
$= -2y^2(-3y^2) - 2y^2(-4y) - 2y^2(8)$
$= 6y^4 + 8y^3 - 16y^2$

17. $(2a - 7)(5a^2 - 2a + 3)$
$= 2a(5a^2 - 2a + 3) - 7(5a^2 - 2a + 3)$

$$\begin{array}{rrrr} 10a^3 & - 4a^2 & + 6a & \\ + & - 35a^2 & + 14a & - 21 \\ \hline 10a^3 & - 39a^2 & + 20a & - 21 \end{array}$$

18. $(3b - 2)(5b - 7) = 15b^2 - 21b - 10b + 14$
$= 15b^2 - 31b + 14$

19. $\frac{(-2a^2b^3)^2}{8a^4b^8} = \frac{(-2)^2a^4b^6}{8a^4b^8}$
$= \frac{4a^4b^6}{8a^4b^8} = \frac{1}{2b^2}$

20.
$$\begin{array}{r} a - 7 \\ a + 3\overline{)a^2 - 4a - 21} \\ \underline{a^2 + 3a} \\ -7a - 21 \\ \underline{-7a - 21} \\ 0 \end{array}$$

$(a^2 - 4a - 21) \div (a + 3) = a - 7$

21. $6.09 \times 10^{-5} = 0.0000609$

22. The unknown number: x
$8x - 2x = 18$
$6x = 18$
$x = 3$
The unknown number is 3.

23. **Strategy** • Percent concentration of orange juice in mixture: x

	Amount	Percent	Quantity
Pure juice	50	1.00	1.00(50)
10% juice	200	0.10	0.10(200)
Mixture	250	x	$250x$

• The sum of the quantities before mixing is equal to the quantity after mixing.

Solution $1.00(50) + 0.10(200) = 250x$
$50 + 20 = 250x$
$70 = 250x$
$0.28 = x$
The percent concentration of orange juice in the mixture is 28%.

24. **Strategy** • Time for car: x

	Rate	Time	Distance
Car	50	x	$50x$
Cyclist	10	$x + 2$	$10(x + 2)$

• The car and the cyclist travel the same distance.

Solution $50x = 10(x + 2)$
$50x = 10x + 20$
$40x = 20$
$x = \frac{1}{2}$
The car overtakes the cyclist 25 mi from the starting point.

25. **Strategy** • Length: x
Width: 40% x
• Use the equation for the perimeter of a rectangle.

Solution $2L + 2W = P$
$2(x) + 2(0.40x) = 42$
$2x + 0.80x = 42$
$2.8x = 42$
$x = 15$
$40\%\ x = 6$
The length is 15 m and the width is 6 m.

Chapter 5: Factoring

Prep Test

1. $30 = 2 \cdot 3 \cdot 5$

2. $-3(4y - 5) = -12y + 15$

3. $-(a - b) = -a + b$

4. $2(a - b) - 5(a - b) = 2a - 2b - 5a + 5b$
$= -3a + 3b$

5. $4x = 0$
$\frac{4x}{4} = \frac{0}{4}$
$x = 0$

6. $2x + 1 = 0$
$2x + 1 - 1 = 0 - 1$
$2x = -1$
$\frac{2x}{2} = \frac{-1}{2}$
$x = -\frac{1}{2}$

7. $(x + 4)(x - 6) = x^2 - 6x + 4x - 24$
$= x^2 - 2x - 24$

8. $(2x - 5)(3x + 2) = 6x^2 + 4x - 15x - 10$
$= 6x^2 - 11x - 10$

9. $\frac{x^5}{x^2} = x^{5-2} = x^3$

10. $\frac{6x^4y^3}{2xy^2} = 3x^3y$

Go Figure

$4^{54} \cdot 5^{100} = (2^2)^{54} \cdot 5^{100}$
$= 2^{108} \cdot 5^{100}$
$= 2^8 \cdot 2^{100} \cdot 5^{100}$
$= 2^8 \cdot (2 \cdot 5)^{100}$
$= 2^8 \cdot 10^{100}$
$= 256 \cdot 10^{10^2}$

There will be 103 digits in the product.

Section 5.1

Objective A Exercises

1. A common monomial factor is a monomial that is a factor of each term of a polynomial.

3. $5a + 5 = 5(a + 1)$

5. $16 - 8a^2 = 8(2 - a^2)$

7. $8x + 12 = 4(2x + 3)$

9. $30a - 6 = 6(5a - 1)$

11. $7x^2 - 3x = x(7x - 3)$

13. $3a^2 + 5a^5 = a^2(3 + 5a^3)$

15. $14y^2 + 11y = y(14y + 11)$

17. $2x^4 - 4x = 2x(x^3 - 2)$

19. $10x^4 - 12x^2 = 2x^2(5x^2 - 6)$

21. $8a^8 - 4a^5 = 4a^5(2a^3 - 1)$

23. $x^2y^2 - xy = xy(xy - 1)$

25. $3x^2y^4 - 6xy = 3xy(xy^3 - 2)$

27. $x^2y - xy^3 = xy(x - y^2)$

29. $5y^3 - 20y^2 + 5y = 5y(y^2 - 4y + 1)$

31. $3y^4 - 9y^3 - 6y^2 = 3y^2(y^2 - 3y - 2)$

33. $3y^3 - 9y^2 + 24y = 3y(y^2 - 3y + 8)$

35. $6a^5 - 3a^3 - 2a^2 = a^2(6a^3 - 3a - 2)$

37. $2a^2b - 5a^2b^2 + 7ab^2 = ab(2a - 5ab + 7b)$

39. $4b^5 + 6b^3 - 12b = 2b(2b^4 + 3b^2 - 6)$

41. $8x^2y^2 - 4x^2y + x^2 = x^2(8y^2 - 4y + 1)$

Objective B Exercises

43. $y(a + z) + 7(a + z) = (a + z)(y + 7)$

45. $3r(a - b) + s(a - b) = (a - b)(3r + s)$

47. $t(m - 7) + 7(7 - m) = t(m - 7) - 7(m - 7)$
$= (m - 7)(t - 7)$

49. $2y(4a - b) - (b - 4a) = 2y(4a - b) + 1(4a - b)$
$= (4a - b)(2y + 1)$

51. $x^2 + 2x + 2xy + 4y = x(x + 2) + 2y(x + 2)$
$= (x + 2)(x + 2y)$

53. $p^2 - 2p - 3rp + 6r = p(p - 2) - 3r(p - 2)$
$= (p - 2)(p - 3r)$

55. $ab + 6b - 4a - 24 = b(a + 6) - 4(a + 6)$
$= (a + 6)(b - 4)$

57. $2z^2 - z + 2yz - y = z(2z - 1) + y(2z - 1)$
$= (2z - 1)(z + y)$

59. $8v^2 - 12vy + 14v - 21y = 4v(2v - 3y) + 7(2v - 3y)$
$= (2v - 3y)(4v + 7)$

61. $2x^2 - 5x - 6xy + 15y = x(2x - 5) - 3y(2x - 5)$
$= (2x - 5)(x - 3y)$

63. $3y^2 - 6y - ay + 2a = 3y(y - 2) - a(y - 2)$
$= (y - 2)(3y - a)$

65. $3xy - y^2 - y + 3x = y(3x - y) + 1(-y + 3x)$
$= y(3x - y) + 1(3x - y)$
$= (3x - y)(y + 1)$

67. $3st + t^2 - 2t - 6s = t(3s + t) - 2(t + 3s)$
$= t(3s + t) - 2(3s + t)$
$= (3s + t)(t - 2)$

Applying the Concepts

69. **a.** 28 is a perfect number, because 28 has factors 1, 2, 7, and 14, and $1 + 2 + 4 + 7 + 14 = 28$.

b. 496 is a perfect number, because 496 has factors 1, 2, 4, 8, 16, 31, 62, 124, and 248, and their sum is 496.

71. **a.** $\text{Area} = \pi r^2 - 4\left(\frac{1}{2}bh\right)$
$= \pi r^2 - 4\left(\frac{1}{2}r \cdot r\right)$
$= \pi r^2 - 2r^2$
$= r^2(\pi - 2)$

b. $\text{Area} = bh - 2(\pi r^2)$
$= 4r(2r) - 2\pi r^2$
$= 8r^2 - 2\pi r^2$
$= 2r^2(4 - \pi)$

c. $\text{Area} = bh - \pi r^2$
$= 2r(2r) - \pi r^2$
$= 4r^2 - \pi r^2$
$= r^2(4 - \pi)$

Section 5.2

Objective A Exercises

1. In factoring a trinomial, if the constant term is positive, then the signs in both binomial factors will be the same.

3. $(x + \_)(x + \_)$

Factors	Sum
+1, +2	3

$(x + 1)(x + 2)$
$x^2 + 3x + 2 = (x + 1)(x + 2)$

5. $(x + \_)(x - \_)$

Factors	Sum
−1, +2	1
+1, −2	−1

$(x + 1)(x - 2)$
$x^2 - x - 2 = (x + 1)(x - 2)$

7. $(a + \_)(a - \_)$

Factors	Sum
−1, +12	11
+1, −12	−11
−2, +6	4
+2, −6	−4
−3, +4	1
+3, −4	−1

$(a + 4)(a - 3)$
$a^2 + a - 12 = (a + 4)(a - 3)$

9. $(a - \_)(a - \_)$

Factors	Sum
−1, −2	−3

$(a - 1)(a - 2)$
$a^2 - 3a + 2 = (a - 1)(a - 2)$

11. $(a + \_)(a - \_)$

Factors	Sum
−1, +2	1
+1, −2	−1

$(a + 2)(a - 1)$
$a^2 + a - 2 = (a + 2)(a - 1)$

13. $(b - \_)(b - \_)$

Factors	Sum
−1, −9	−10
−3, −3	−6

$(b - 3)(b - 3)$
$b^2 - 6b + 9 = (b - 3)(b - 3)$

15. $(b + \_)(b - \_)$

Factors	Sum
−1, +8	7
+1, −8	−7
−2, +4	2
+2, −4	−2

$(b + 8)(b - 1)$
$b^2 + 7b - 8 = (b + 8)(b - 1)$

17. $(y + \_)(y - \_)$

Factors	Sum
−1, +55	54
+1, −55	−54
−5, +11	6
+5, −11	−6

$(y + 11)(y - 5)$
$y^2 + 6y - 55 = (y + 11)(y - 5)$

19. $(y - \_)(y - \_)$

Factors	Sum
−1, −6	−7
−2, −3	−5

$(y - 2)(y - 3)$
$y^2 - 5y + 6 = (y - 2)(y - 3)$

21. $(z - \_)(z - \_)$

Factors	Sum
−1, −45	−46
−3, −15	−18
−5, −9	−14

$(z - 5)(z - 9)$
$z^2 - 14z + 45 = (z - 5)(z - 9)$

23. $(z + \_)(z - \_)$

Factors	Sum
−1, +160	159
+1, −160	−159
−2, +80	78
+2, −80	−78
−4, +40	36
+4, −40	−36
−5, +32	27
+5, −32	−27
−8, +20	12
+8, −20	−12
−10, +16	6
+10, −16	−6

$(z + 8)(z - 20)$
$z^2 - 12z - 160 = (z + 8)(z - 20)$

25. $(p + \_)(p + \_)$

Factors	Sum
+1, +27	28
+3, +9	12

$(p + 3)(p + 9)$
$p^2 + 12p + 27 = (p + 3)(p + 9)$

27. $(x + \_)(x + \_)$

Factors	Sum
+1, +100	101
+2, +50	52
+4, +25	29
+5, +20	25
+10, +10	20

$(x + 10)(x + 10)$
$x^2 + 20x + 100 = (x + 10)(x + 10)$

29. $(b + \_)(b + \_)$

Factors	Sum
+1, +20	21
+2, +10	12
+4, +5	9

$(b + 4)(b + 5)$
$b^2 + 9b + 20 = (b + 4)(b + 5)$

31. $(x + \_)(x - \_)$

Factors	Sum
−1, +42	41
+1, −42	−41
−2, +21	19
+2, −21	−19
−3, +14	11
+3, −14	−11
−6, +7	1
+6, −7	−1

$(x + 3)(x - 14)$
$x^2 - 11x - 42 = (x + 3)(x - 14)$

33. $(b + \_)(b - \_)$

Factors	Sum
−1, +20	19
+1, −20	−19
−2, +10	8
+2, −10	−8
−4, +5	1
+4, −5	−1

$(b + 4)(b - 5)$
$b^2 - b - 20 = (b + 4)(b - 5)$

35. $(y + \_)(y - \_)$

Factors	Sum
−1, +51	50
+1, −51	−50
−3, +17	14
+3, −17	−14

$(y + 3)(y - 17)$
$y^2 - 14y - 51 = (y + 3)(y - 17)$

37. $(p + \_)(p - \_)$

Factors	Sum
−1, +21	20
+1, −21	−20
−3, +7	4
+3, −7	−4

$(p + 3)(p - 7)$
$p^2 - 4p - 21 = (p + 3)(p - 7)$

39. $(y - \_)(y - \_)$

Factors	Sum
−1, −32	−33
−2, −16	−18
−4, −8	−12

$y^2 - 8y + 32$ is nonfactorable over the integers.

41. $(x - \_)(x - \_)$

Factors	Sum
−1, −75	−76
−3, −25	−28
−5, −15	−20

$(x - 5)(x - 15)$
$x^2 - 20x + 75 = (x - 5)(x - 15)$

43. $(p + \_)(p + \_)$

Factors	Sum
+1, +63	64
+3, +21	24

$(p + 3)(p + 21)$
$p^2 + 24p + 63 = (p + 3)(p + 21)$

45. $(x + \_)(x + \_)$

Factors	Sum
+1, +38	39
+2, +19	21

$(x + 2)(x + 19)$
$x^2 + 21x + 38 = (x + 2)(x + 19)$

47. $(x + \_)(x - \_)$

Factors	Sum
−1, +36	35
+1, −36	−35
−2, +18	16
+2, −18	−16
−3, +12	9
+3, −12	−9
−4, +9	5
+4, −9	−5
−6, +6	0

$(x+9)(x-4)$
$x^2 + 5x - 36 = (x+9)(x-4)$

49. $(a - \_)(a - \_)$

Factors	Sum
−1, +44	43
+1, −44	−43
−2, +22	20
+2, −22	−20
−4, +11	7
+4, −11	−7

$(a+4)(a-11)$
$a^2 - 7a - 44 = (a+4)(a-11)$

51. $(a - \_)(a - \_)$

Factors	Sum
−1, −54	−55
−2, −27	−29
−3, −18	−21
−6, −9	−15

$(a-3)(a-18)$
$a^2 - 21a + 54 = (a-3)(a-18)$

53. $(z + \_)(z - \_)$

Factors	Sum
−1, +147	146
+1, −147	−146
−3, +49	46
+3, −49	−46
−7, +21	14
+7, −21	−14

$(z+21)(z-7)$
$z^2 + 14z - 147 = (z+21)(z-7)$

55. $(c + \_)(c - \_)$

Factors	Sum
−1, +180	+179
+1, −180	−179
−2, +90	+88
+2, −90	−88
−3, +60	+57
+3, −60	−57
−4, +45	+41
+4, −45	−41
−5, +36	+31
+5, −36	−31
−6, +30	+24
+6, −30	−24
−9, +20	+11
+9, −20	−11
−10, +18	+8
+10, −18	−8
−12, +15	+3
+12, −15	−3

$(c+12)(c-15)$
$c^2 - 3c - 180 = (c+12)(c-15)$

57. $(p + \_)(p + \_)$

Factors	Sum
+1, +135	+136
+3, +45	+48
+5, +27	+32
+9, +15	+24

$(p+9)(p+15)$
$p^2 + 24p + 135 = (p+9)(p+15)$

59. $(c + \_)(c + \_)$

Factors	Sum
+1, +18	+19
+2, +9	+11
+3, +6	+9

$(c+2)(c+9)$
$c^2 + 11c + 18 = (c+2)(c+9)$

61. $(x + \_)(x - \_)$

Factors	Sum
−1, +75	+74
+1, −75	−74
−3, +25	+22
+3, −25	−22
−5, +15	+10
+5, −15	−10

$(x+15)(x-5)$
$x^2 + 10x - 75 = (x+15)(x-5)$

63. $(x + \_)(x - \_)$

Factors	Sum
−1, +100	+99
+1, −100	−99
−2, +50	+48
+2, −50	−48
−4, +25	+21
+4, −25	−21
−5, +20	+15
+5, −20	−15
−10, +10	0

$(x + 25)(x - 4)$
$x^2 + 21x - 100 = (x + 25)(x - 4)$

65. $(b - \_)(b - \_)$

Factors	Sum
−1, −72	−73
−2, −36	−38
−3, −24	−27
−4, −18	−22
−6, −12	−18
−8, −9	−17

$(b - 4)(b - 18)$
$b^2 - 22b + 72 = (b - 4)(b - 18)$

67. $(a + \_)(a - \_)$

Factors	Sum
−1, +135	+134
+1, −135	−134
−3, +45	+42
+3, −45	−42
−5, +27	+22
+5, −27	−22
−9, +15	+6
+9, −15	−6

$(a + 45)(a - 3)$
$a^2 + 42a - 135 = (a + 45)(a - 3)$

69. $(b - \_)(b - \_)$

Factors	Sum
−1, −126	−127
−2, −63	−65
−3, −43	−46
−6, −21	−27
−7, −18	−25

$(b - 7)(b - 18)$
$b^2 - 25b + 126 = (b - 7)(b - 18)$

71. $(z + \_)(z + \_)$

Factors	Sum
+1, +144	+145
+2, +72	+74
+3, +48	+51
+4, +36	+40
+6, +24	+30
+8, +18	+26
+9, +16	+25
+12, +12	+24

$(z + 12)(z + 12)$
$z^2 + 24z + 144 = (z + 12)(z + 12)$

73. $(x - \_)(x - \_)$

Factors	Sum
−1, −100	−101
−2, −50	−52
−4, −25	−29
−5, −20	−25
−10, −10	−20

$(x - 4)(x - 25)$
$x^2 - 29x + 100 = (x - 4)(x - 25)$

75. $(x + \_)(x - \_)$

Factors	Sum
−1, +112	+111
+1, −112	−111
−2, +56	+54
+2, −56	−54
−4, +28	+24
+4, −28	−24
−7, +16	+9
+7, −16	−9
−8, +14	+6
+8, −14	−6

$(x + 16)(x - 7)$
$x^2 + 9x - 112 = (x + 16)(x - 7)$

Objective B Exercises

77. The GCF is 3.
$3x^2 + 15x + 18 = 3(x^2 + 5x + 6)$
Factor the trinomial.
$3(x + \_)(x + \_)$

Factors	Sum
+1, +6	+7
+2, +3	+5

$3(x + 2)(x + 3)$
$3x^2 + 15x + 18 = 3(x + 2)(x + 3)$

79. The GCF is −1.
$-x^2 - 4x + 12 = -(x^2 + 4x - 12)$
Factor the trinomial.
$-(x + \_)(x - \_)$

Factors	Sum
−1, +12	+11
+1, −12	−11
−2, +6	+4
+2, −6	−4

$-(x - 2)(x + 6)$
$12 - 4x - x^2 = -(x - 2)(x + 6)$

81. The GCF is a.
$ab^2 + 7ab - 8a = a(b^2 + 7b - 8)$
Factor the trinomial.
$a(b + \_)(b - \_)$

Factors	Sum
−1, +8	+7
+1, −8	−7
−2, +4	+2
+2, −4	−2

$a(b + 8)(b - 1)$
$ab^2 + 7ab - 8a = a(b + 8)(b - 1)$

83. The GCF is x.
$xy^2 + 8xy + 15x = x(y^2 + 8y + 15)$
Factor the trinomial.
$x(y + \_)(y + \_)$

Factors	Sum
+1, +15	+16
+3, +5	+8

$x(y + 3)(y + 5)$
$xy^2 + 8xy + 15x = x(y + 3)(y + 5)$

85. The GCF is $-2a$.
$-2a^3 - 6a^2 - 4a = -2a(a^2 + 3a + 2)$
Factor the trinomial.
$-2a(a + \_)(a + \_)$

Factors	Sum
+1, +2	+3

$-2a(a + 1)(a + 2)$
$-2a^3 - 6a^2 - 4a = -2a(a + 1)(a + 2)$

87. The GCF is $4y$.
$4y^3 + 12y^2 - 72y = 4y(y^2 + 3y - 18)$
Factor the trinomial.
$4y(y + \_)(y - \_)$

Factors	Sum
−1, +18	+17
+1, −18	−17
−2, +9	+7
+2, −9	−7
−3, +6	+3
+3, −6	−3

$4y(y + 6)(y - 3)$
$4y^3 + 12y^2 - 72y = 4y(y + 6)(y - 3)$

89. The GCF is $2x$.
$2x^3 - 2x^2 + 4x = 2x(x^2 - x + 2)$
Factor the trinomial.
$2x(x - \_)(x - \_)$

Factors	Sum
−1, −2	−3

$2x^3 - 2x^2 + 4x = 2x(x^2 - x + 2)$

91. The GCF is 6.
$6z^2 + 12z - 90 = 6(z^2 + 2z - 15)$
Factor the trinomial.
$6(z + \_)(z - \_)$

Factors	Sum
−1, +15	+14
+1, −15	−14
−3, +5	+2
+3, −5	−2

$6(z + 5)(z - 3)$
$6z^2 + 12z - 90 = 6(z + 5)(z - 3)$

93. The GCF is $3a$.
$3a^3 - 9a^2 - 54a = 3a(a^2 - 3a - 18)$
Factor the trinomial.
$3a(a + \_)(a - \_)$

Factors	Sum
−1, +18	+17
+1, −18	−17
−2, +9	+7
+2, −9	−7
−3, +6	+3
+3, −6	−3

$3a(a + 3)(a - 6)$
$3a^3 - 9a^2 - 54a = 3a(a + 3)(a - 6)$

95. There is no common factor.
Factor the trinomial.
$(x + \_y)(x - \_y)$

Factors	Sum
−1, +21	+20
+1, −21	−20
−3, +7	+4
+3, −7	−4

$(x + 7y)(x - 3y)$
$x^2 + 4xy - 21y^2 = (x + 7y)(x - 3y)$

97. There is no common factor.
Factor the trinomial.
$(a - \_b)(a - \_b)$

Factors	Sum
−1,−50	−51
−2,−25	−27
−5,−10	−15

$(a - 5b)(a - 10b)$
$a^2 - 15ab + 50b^2 = (a - 5b)(a - 10b)$

99. There is no common factor.
Factor the trinomial.
$(s + \_t)(s - \_t)$

Factors	Sum
−1, +48	+47
+1, −48	−47
−2, +24	+22
+2, −24	−22
−3, +16	+13
+3, −16	−13
−4, +12	+8
+4, −12	−8
−6, +8	+2
+6, −8	−2

$(s + 8t)(s - 6t)$
$s^2 + 2st - 48t^2 = (s + 8t)(s - 6t)$

101. There is no common factor.
Factor the trinomial.
$(x + \_y)(x - \_y)$

Factors	Sum
+1, +36	+37
+2, +18	+20
+3, +12	+15
+4, +9	+13
+6, +6	+12

$x^2 + 85xy + 36y^2$ is nonfactorable over the integers.

103. The GCF is z^2.
$z^4 + 2z^3 - 80z^2 = z^2(z^2 + 2z - 80)$
Factor the trinomial.
$z^2(z + \_)(z - \_)$

Factors	Sum
−1, +80	+79
+1, −80	−79
−2, +40	+38
+2, −40	−38
−4, +20	+16
+4, −20	−16
−5, +16	+11
+5, −16	−11
−8, +10	+2
+8, −10	−2

$z^2(z + 10)(z - 8)$
$z^4 + 2z^3 - 80z^2 = z^2(z + 10)(z - 8)$

105. The GCF is b^2.
$b^4 - 3b^3 - 10b^2 = b^2(b^2 - 3b - 10)$
Factor the trinomial.
$b^2(b + \_)(b - \_)$

Factors	Sum
−1, +10	+9
+1, −10	−9
−2, +5	+3
+2, −5	−3

$b^2(b + 2)(b - 5)$
$b^4 - 3b^3 - 10b^2 = b^2(b + 2)(b - 5)$

107. The GCF is $3y^2$.
$3y^4 + 54y^3 + 135y^2 = 3y^2(y^2 + 18y + 45)$
Factor the trinomial.
$3y^2(y + \_)(y + \_)$

Factors	Sum
+1, +45	+46
+3, +15	+18
+5, +9	+14

$3y^2(y + 3)(y + 15)$
$3y^4 + 54y^3 + 135y^2 = 3y^2(y + 3)(y + 15)$

109. The GCF is $-x^2$.
$-x^4 + 11x^3 + 12x^2 = -x^2(x^2 - 11x - 12)$
Factor the trinomial.
$-x^2(x + \_)(x - \_)$

Factors	Sum
−1, +12	+11
+1, −12	−11
−2, +6	+4
+2, −6	−4
−3, +4	+1
+3, −4	−1

$-x^2(x + 1)(x - 12)$
$-x^4 + 11x^3 + 12x^2 = -x^2(x + 1)(x - 12)$

111. The GCF is $3y$.
$3x^2y - 6xy - 45y = 3y(x^2 - 2x - 15)$
Factor the trinomial.
$3y(x + \_)(x - \_)$

Factors	Sum
−1, +15	+14
+1, −15	−14
−3, +5	+2
+3, −5	−2

$3y(x + 3)(x - 5)$
$3x^2y - 6xy - 45y = 3y(x + 3)(x - 5)$

113. The GCF is $-3x$.
$-3x^3 + 36x^2 - 81x = -3x(x^2 - 12x + 27)$
Factor the trinomial.
$-3x(x - \_)(x - \_)$

Factors	Sum
−1, −27	−28
−3, −9	−12

$-3x(x - 3)(x - 9)$
$-3x^3 + 36x^2 - 81x = -3x(x - 3)(x - 9)$

115. There is no common factor.
Factor the trinomial.
$(x - \_y)(x - \_y)$

Factors	Sum
−1, −15	−16
−3, −5	−8

$(x - 3y)(x - 5y)$
$x^2 - 8xy + 15y^2 = (x - 3y)(x - 5y)$

117. There is no common factor.
Factor the trinomial.
$(a - \_b)(a - \_b)$

Factors	Sum
−1, −42	−43
−2, −21	−23
−3, −14	−17
−6, −7	−13

$(a - 6b)(a - 7b)$
$a^2 - 13ab + 42b^2 = (a - 6b)(a - 7b)$

119. There is no common factor.
Factor the trinomial.
$(y + \_z)(y + \_z)$

Factors	Sum
+1, +7	+8

$(y + 1z)(y + 7z)$
$y^2 + 8yz + 7z^2 = (y + z)(y + 7z)$

121. The GCF is $3y$.
$3x^2y + 60xy - 63y = 3y(x^2 + 20x - 21)$
Factor the trinomial.
$3y(x + \_)(x - \_)$

Factors	Sum
−1, +21	+20
+1, −21	−20
−3, +7	+4
+3, −7	−4

$3y(x + 21)(x - 1)$
$3x^2y + 60xy - 63y = 3y(x + 21)(x - 1)$

123. The GCF is $3x$.
$3x^3 + 3x^2 - 36x = 3x(x^2 + x - 12)$
Factor the trinomial.
$3x(x + \_)(x - \_)$

Factors	Sum
−1, +12	+11
+1, −12	−11
−2, +6	+4
+2, −6	−4
−3, +4	+1
+3, −4	−1

$3x(x + 4)(x - 3)$
$3x^3 + 3x^2 - 36x = 3x(x + 4)(x - 3)$

125. The GCF is $4z$.
$4z^3 + 32z^2 - 132z = 4z(z^2 + 8z - 33)$
Factor the trinomial.
$4z(z + \_)(z - \_)$

Factors	Sum
−1, +33	+32
+1, −33	−32
−3, +11	+8
+3, −11	−8

$4z(z + 11)(z - 3)$
$4z^3 + 32z^2 - 132z = 4z(z + 11)(z - 3)$

127. The GCF is $4x$.
$4x^3 + 8x^2 - 12x = 4x(x^2 + 2x - 3)$
Factor the trinomial.
$4x(x + \_)(x - \_)$

Factors	Sum
−1, +3	+2
+1, −3	−2

$4x(x + 3)(x - 1)$
$4x^3 + 8x^2 - 12x = 4x(x + 3)(x - 1)$

129. The GCF is 5.
$5p^2 + 25p - 420 = 5(p^2 + 5p - 84)$
Factor the trinomial.
$5(p + \_)(p - \_)$

Factors	Sum
−1, +84	+83
+1, −84	−83
−2, +42	+40
+2, −42	−40
−3, +28	+25
+3, −28	−25
−4, +21	+17
+4, −21	−17
−6, +14	+8
+6, −14	−8
−7, +12	+5
+7, −12	−5

$5(p + 12)(p - 7)$
$5p^2 + 25p - 420 = 5(p + 12)(p - 7)$

131. The GCF is p^2.
$p^4 + 9p^3 - 36p^2 = p^2(p^2 + 9p - 36)$
Factor the trinomial.
$p^2(p + \_)(p - \_)$

Factors	Sum
−1, +36	+35
+1, −36	−35
−2, +18	+16
+2, −18	−16
−3, +12	+9
+3, −12	−9
−4, +9	+5
+4, −9	−5
−6, +6	0

$p^2(p + 12)(p - 3)$
$p^4 + 9p^3 - 36p^2 = p^2(p + 12)(p - 3)$

133. There is no common factor.
Factor the trinomial.
$(t - \_s)(t - \_s)$

Factors	Sum
−1, −35	−36
−5, −7	−12

$(t - 5s)(t - 7s)$
$t^2 - 12ts + 35s^2 = (t - 5s)(t - 7s)$

135. There is no common factor.
Factor the trinomial.
$(a + \_b)(a - \_b)$

Factors	Sum
−1, +33	+32
−1, −33	−32
−3, +11	+8
+3, −11	−8

$(a + 3b)(a - 11b)$
$a^2 - 8ab - 33b^2 = (a + 3b)(a - 11b)$

Applying the Concepts

137. The GCF is y.

$x^2y - 54y - 3xy = y(x^2 - 54 - 3x)$
$= y(x^2 - 3x - 54)$

Factor the trinomial.

$y(x + \_)(x - \_)$

Factors	Sum
−1, +54	+53
−2, +27	+25
−3, +18	+15
−6, +9	+3
+6, −9	−3

$y(x + 6)(x - 9)$

$x^2y - 54y - 3xy = y(x + 6)(x - 9)$

139. $x^2 + kx + 35$. The factors of 35 must sum to k.

Factors	Sum
−1, −35	−36
+1, +35	+36
−5, −7	−12
+5, +7	+12

k can be −36, +36, −12, or +12.

141. $x^2 + kx + 21$. The factors of 21 must sum to $-k$.

Factors	Sum
−1, −21	−22
+1, +21	+22
−3, −7	−10
+3, +7	+10

k can be 22, −22, 10, or −10.

143. $z^2 + 7z + k, k > 0$

Find two positive integers that sum to 7. Their product is k.

Integers	Product
+1, +6	6
+2, +5	10
+3, +4	12

k can be 6, 10, or 12.

145. $c^2 - 7c + k, k > 0$

Find two negative integers that sum to −7. Their product is k.

Integers	Product
−1, −6	6
−2, −5	10
−3, −4	12

k can be 6, 10, or 12.

147. $y^2 + 5y + k, k > 0$

Find two positive integers that sum to 5. Their product is k.

Integers	Product
+2, +3	6
+1, +4	4

k can be 4 or 6.

Section 5.3

Objective A Exercises

1. $(\_x + \_)(\_x + \_)$

Factors of 2: 1, 2 — Factors of 1: +1, +1

Trial Factors	Middle Term
$(1x + 1)(2x + 1)$	$x + 2x = 3x$

$(x + 1)(2x + 1)$

$2x^2 + 3x + 1 = (x + 1)(2x + 1)$

3. $(\_y + \_)(\_y + \_)$

Factors of 2: 1, 2 — Factors of 3: +1, +3

Trial Factors	Middle Term
$(1y + 1)(2y + 3)$	$3y + 2y = 5y$
$(1y + 3)(2y + 1)$	$y + 6y = 7y$

$(y + 3)(2y + 1)$

$2y^2 + 7y + 3 = (y + 3)(2y + 1)$

5. $(\_a - \_)(\_a - \_)$

Factors of 2: 1, 2 — Factors of 1: −1, −1

Trial Factors	Middle Term
$(1a - 1)(2a - 1)$	$-a - 2a = -3a$

$(a - 1)(2a - 1)$

$2a^2 - 3a + 1 = (a - 1)(2a - 1)$

7. $(\_b - \_)(\_b - \_)$

Factors of 2: 1, 2 — Factors of 5: −1, −5

Trial Factors	Middle Term
$(1b - 1)(2b - 5)$	$-5b - 2b = -7b$
$(1b - 5)(2b - 1)$	$-b - 10b = -11b$

$(b - 5)(2b - 1)$

$2b^2 - 11b + 5 = (b - 5)(2b - 1)$

9. $(\_x + \_)(\_x - \_)$ or $(\_x - \_)(\_x + \_)$

Factors of 2: 1, 2 — Factors of −1: −1, +1

Trial Factors	Middle Term
$(1x - 1)(2x + 1)$	$x - 2x = -x$
$(1x + 1)(2x - 1)$	$-x + 2x = x$

$(x + 1)(2x - 1)$

$2x^2 + x - 1 = (x + 1)(2x - 1)$

11. $(\_x + \_)(\_x\_)$ or $(\_x - \_)(\_x + \_)$

Factors of 2: 1, 2 — Factors of −3: −1, +3; +1, −3

Trial Factors	Middle Term
$(1x - 1)(2x + 3)$	$3x - 2x = x$
$(1x + 3)(2x - 1)$	$-x + 6x = 5x$
$(1x + 1)(2x - 3)$	$-3x + 2x = -x$
$(1x - 3)(2x + 1)$	$x - 6x = -5x$

$(x - 3)(2x + 1)$

$2x^2 - 5x - 3 = (x - 3)(2x + 1)$

13. $(\_t + \_)(\_t - \_)$ or $(\_t - \_)(\_t + \_)$

Factors of 2: 1, 2

Factors of −10: $-1, +10$; $+1, -10$; $-2, +5$; $+2, -5$

Trial Factors	Middle Term
$(1t-1)(2t+10)$	Common factor
$(1t+10)(2t-1)$	$-t+20t=19t$
$(1t+1)(2t-10)$	Common factor
$(1t-10)(2t+1)$	$t-20t=-19t$
$(1t-2)(2t+5)$	$5t-4t=t$
$(1t+5)(2t-2)$	Common factor
$(1t+2)(2t-5)$	$-5t+4t=-t$
$(1t-5)(2t+2)$	Common factor

$(t+2)(2t-5)$

$2t^2-t-10=(t+2)(2t-5)$

15. $(\_p - \_)(\_p - \_)$

Factors of 3: 1, 3

Factors of 5: $-1, -5$

Trial Factors	Middle Term
$(1p-1)(3p-5)$	$-5p-3p=-8p$
$(1p-5)(3p-1)$	$-p-15p=-16p$

$(p-5)(3p-1)$

$3p^2-16p+5=(p-5)(3p-1)$

17. $(\_y - \_)(\_y - \_)$

Factors of 12: 1, 12; 2, 6; 3, 4

Factors of 1: $-1, -1$

Trial Factors	Middle Term
$(1y-1)(12y-1)$	$-y-12y=-13y$
$(2y-1)(6y-1)$	$-2y-6y=-8y$
$(3y-1)(4y-1)$	$-3y-4y=-7y$

$(3y-1)(4y-1)$

$12y^2-7y+1=(3y-1)(4y-1)$

19. $(\_z - \_)(\_z - \_)$

Factors of 6: 1, 6; 2, 3

Factors of 3: $-1, -3$

Trial Factors	Middle Term
$(1z-1)(6z-3)$	Common factor
$(1z-3)(6z-1)$	$-z-18z=-19z$
$(2z-1)(3z-3)$	Common factor
$(2z-3)(3z-1)$	$-2z-9z=-11z$

$6z^2-7z+3$ is nonfactorable over the integers.

21. $(\_t - \_)(\_t - \_)$

Factors of 6: 1, 6; 2, 3

Factors of 4: $-1, -4$; $-2, -2$

Trial Factors	Middle Term
$(1t-1)(6t-4)$	Common factor
$(1t-4)(6t-1)$	$-t-24t=-25t$
$(1t-2)(6t-2)$	Common factor
$(2t-1)(3t-4)$	$-8t-3t=-11t$
$(2t-4)(3t-1)$	Common factor
$(2t-2)(3t-2)$	Common factor

$(2t-1)(3t-4)$

$6t^2-11t+4=(2t-1)(3t-4)$

23. $(\_x + \_)(\_x + \_)$

Factors of 8: 1, 8; 2, 4

Factors of 4: $+1, +4$; $+2, +2$

Trial Factors	Middle Term
$(1x+1)(8x+4)$	Common factor
$(1x+4)(8x+1)$	$x+32x=33x$
$(1x+2)(8x+2)$	Common factor
$(2x+1)(4x+4)$	Common factor
$(2x+4)(4x+1)$	Common factor
$(2x+2)(4x+2)$	Common factor

$(x+4)(8x+1)$

$8x^2+33x+4=(x+4)(8x+1)$

25. $(\_x + \_)(\_x - \_)$ or $(\_x - \_)(\_x + \_)$

Factors of 5: 1, 5

Factors of −7: $-1, +7$; $+1, -7$

Trial Factors	Middle Term
$(1x-1)(5x+7)$	$7x-5x=2x$
$(1x+7)(5x-1)$	$-x+35x=34x$
$(1x+1)(5x-7)$	$-7x+5x=-2x$
$(1x-7)(5x+1)$	$x-35x=-34x$

$5x^2-62x-7$ is nonfactorable over the integers.

27. $(\_y + \_)(\_y + \_)$

Factors of 12: 1, 12; 2, 6; 3, 4

Factors of 5: $+1, +5$

Trial Factors	Middle Term
$(1y+1)(12y+5)$	$5y+12y=17y$
$(1y+5)(12y+1)$	$y+60y=61y$
$(2y+1)(6y+5)$	$10y+6y=16y$
$(2y+5)(6y+1)$	$2y+30y=32y$
$(3y+1)(4y+5)$	$15y+4y=19y$
$(3y+5)(4y+1)$	$3y+20y=23y$

$(3y+1)(4y+5)$

$12y^2+19y+5=(3y+1)(4y+5)$

29. $(\_a + \_)(\_a - \_)$ or $(\_a - \_)(\_a + \_)$

Factors of 7: 1, 7

Factors of -14: $-1, +14$; $+1, -14$; $-2, +7$; $+2, -7$

Trial Factors	Middle Term
$(1a - 1)(7a + 14)$	Common factor
$(1a + 14)(7a - 1)$	$-a + 98a = 97a$
$(1a + 1)(7a - 14)$	Common factor
$(1a - 14)(7a + 1)$	$a - 98a = -97a$
$(1a - 2)(7a + 7)$	Common factor
$(1a + 7)(7a - 2)$	$-2a + 49a = 47a$
$(1a + 2)(7a - 7)$	Common factor
$(1a - 7)(7a + 2)$	$2a - 49a = -47a$
$(a + 7)(7a - 2)$	

$7a^2 + 47a - 14 = (a + 7)(7a - 2)$

31. $(\_b - \_)(\_b - \_)$

Factors of 3: 1, 3

Factors of 16: $-1, -16$; $-2, -8$; $-4, -4$

Trial Factors	Middle Term
$(1b - 1)(3b - 16)$	$-16b - 3b = -19b$
$(1b - 16)(3b - 1)$	$-b - 48b = -49b$
$(1b - 2)(3b - 8)$	$-8b - 6b = -14b$
$(1b - 8)(3b - 2)$	$-2b - 24b = -26b$
$(1b - 4)(3b - 4)$	$-4b - 12b = -16b$
$(b - 4)(3b - 4)$	

$3b^2 - 16b + 16 = (b - 4)(3b - 4)$

33. $(\_z + \_)(\_z - \_)$ or $(\_z - \_)(\_z + \_)$

Factors of 2: 1, 2

Factors of -14: $-1, +14$; $+1, -14$; $-2, +7$; $+2, -7$

Trial Factors	Middle Term
$(1z - 1)(2z + 14)$	Common factor
$(1z + 14)(2z - 1)$	$-z + 28z = 27z$
$(1z + 1)(2z - 14)$	Common factor
$(1z - 14)(2z + 1)$	$z - 28z = -27z$
$(1z - 2)(2z + 7)$	$7z - 4z = 3z$
$(1z + 7)(2z + 2)$	Common factor
$(1z + 2)(2z - 7)$	$-7z + 4z = -3z$
$(1z - 7)(2z + 2)$	Common factor
$(z - 14)(2z + 1)$	

$2z^2 - 27z - 14 = (z - 14)(2z + 1)$

35. $(\_p + \_)(\_p - \_)$ or $(\_p - \_)(\_p + \_)$

Factors of 3: 1, 3

Factors of -16: $-1, +16$; $+1, -16$; $-2, +8$; $+2, -8$; $-4, +4$

Trial Factors	Middle Term
$(1p - 1)(3p + 16)$	$16p - 3p = 13p$
$(1p + 16)(3p - 1)$	$-p + 48p = 47p$
$(1p + 1)(3p - 16)$	$-16p + 3p = -13p$
$(1p - 16)(3p + 1)$	$p - 48p = -47p$
$(1p - 2)(3p + 8)$	$8p - 6p = 2p$
$(1p + 8)(3p - 2)$	$-2p + 24p = 22p$
$(1p + 2)(3p - 8)$	$-8p + 6p = -2p$
$(1p - 8)(3p + 2)$	$2p - 24p = -22p$
$(1p - 4)(3p + 4)$	$4p - 12p = -8p$
$(1p + 4)(3p - 4)$	$-4p + 12p = 8p$
$(p + 8)(3p - 2)$	

$3p^2 + 22p - 16 = (p + 8)(3p - 2)$

37. The GCF is 2.

$4x^2 + 6x + 2 = 2(2x^2 + 3x + 1)$

Factor the trinomial.

$2(\_x + \_)(\_x + \_)$

Factors of 2: 1, 2

Factors of 1: $+1, +1$

Trial Factors	Middle Term
$(1x + 1)(2x + 1)$	$x + 2x = 3x$
$2(x + 1)(2x + 1)$	

$4x^2 + 6x + 2 = 2(x + 1)(2x + 1)$

39. The GCF is 5.

$15y^2 - 50y + 35 = 5(3y^2 - 10y + 7)$

Factor the trinomial

$5(\_y - \_)(\_y - \_)$

Factors of 3: 1, 3

Factors of 7: $-1, -7$

Trial Factors	Middle Term
$(1y - 1)(3y - 7)$	$-7y - 3y = -10y$
$(1y - 7)(3y - 1)$	$-y - 21y = -22y$
$5(y - 1)(3y - 7)$	

$15y^2 - 50y + 35 = 5(y - 1)(3y - 7)$

41. The GCF is x.

$2x^3 - 11x^2 + 5x = x(2x^2 - 11x + 5)$

Factor the trinomial.

$x(\_x - \_)(\_x - \_)$

Factors of 2: 1, 2

Factors of 5: $-1, -5$

Trial Factors	Middle Term
$(1x - 1)(2x - 5)$	$-5x - 2x = -7x$
$(1x - 5)(2x - 1)$	$-x - 10x = -11x$
$x(x - 5)(2x - 1)$	

$2x^3 - 11x^2 + 5x = x(x - 5)(2x - 1)$

43. The GCF is b.

$3a^2b - 16ab + 16b = b(3a^2 - 16a + 16)$

Factor the trinomial.

$b(\_a - \_)(\_a - \_)$

Factors of 3: 1, 3 — Factors of 16: −1, −16; −2, −8; −4, −4

Trial Factors	Middle Term
$(1a-1)(3a-16)$	$-16a - 3a = -19a$
$(1a-16)(3a-1)$	$-a - 48a = -49a$
$(1a-2)(3a-8)$	$-8a - 6a = -14a$
$(1a-8)(3a-2)$	$-2a - 24a = -26a$
$(1a-4)(3a-4)$	$-4a - 12a = -16a$

$b(a-4)(3a-4)$

$3a^2b - 16ab + 16b = b(a-4)(3a-4)$

45. There is no common factor.

Factor the trinomial.

$(\_z + \_)(\_z + \_)$

Factors of 3: 1, 3 — Factors of 10: +1, +10; +2, +5

Trial Factors	Middle Term
$(1z+1)(3z+10)$	$10z + 3z = 13z$
$(1z+10)(3z+1)$	$z + 30z = 31z$
$(1z+2)(3z+5)$	$5z + 6z = 11z$
$(1z+5)(3z+2)$	$2z + 8z = 10z$

$3z^2 + 95z + 10$ is nonfactorable over the integers.

47. The GCF is $-3x$.

$36x - 3x^2 - 3x^3 = -3x(x^2 + x - 12)$

Factor the trinomial.

$-3x(x + \_)(x - \_)$

Factors	Sum
−1, +12	11
+1, −12	−11
−2, +6	4
+2, −6	−4
−3, +4	1
+3, −4	−1

$-3x(x+4)(x-3)$

$36x - 3x^2 - 3x^3 = -3x(x+4)(x-3)$

49. The GCF is 4.

$80y^2 - 36y + 4 = 4(20y^2 - 9y + 1)$

Factor the trinomial.

$4(\_y - \_)(\_y - \_)$

Factors of 20: 1, 20; 2, 10; 4, 5 — Factors of 1: −1, −1

Trial Factors	Middle Term
$(1y-1)(20y-1)$	$-y - 20y = -21y$
$(2y-1)(10y-1)$	$-2y - 10y = -12y$
$(4y-1)(5y-1)$	$-4y - 5y = -9y$

$4(4y-1)(5y-1)$

$80y^2 - 36y + 4 = 4(4y-1)(5y-1)$

51. The GCF is z.

$8z^3 + 14z^2 + 3z = z(8z^2 + 14z + 3)$

Factor the trinomial.

$z(\_z + \_)(\_z + \_)$

Factors of 8: 1, 8; 2, 4 — Factors of 3: +1, +3

Trial Factors	Middle Term
$(1z+1)(8z+3)$	$3z + 8z = 11z$
$(1z+3)(8z+1)$	$z + 24z = 25z$
$(2z+1)(4z+3)$	$6z + 4z = 10z$
$(2z+3)(4z+1)$	$2z + 12z = 14z$

$z(2z+3)(4z+1)$

$8z^3 + 14z^2 + 3z = z(2z+3)(4z+1)$

53. The GCF is y.

$6x^2y - 11xy - 10y = y(6x^2 - 11x - 10)$

Factor the trinomial.

$y(\_x + \_)(\_x - \_)$ or $y(\_x - \_)(\_x + \_)$

Factors of 6: 1, 6; 2, 3 — Factors of −10: −1, +10; +1, −10; −2, +5; +2, −5

Trial Factors	Middle Term
$(1x-1)(6x+10)$	Common factor
$(1x+10)(6x-1)$	$-x + 60x = 59x$
$(1x+1)(6x-10)$	Common factor
$(1x-10)(6x+1)$	$x - 60x = -59x$
$(1x-2)(6x+5)$	$5x - 12x = -7x$
$(1x+5)(6x-2)$	Common factor
$(1x+2)(6x-5)$	$-5x + 12x = 7x$
$(1x-5)(6x+2)$	Common factor
$(2x-1)(3x+10)$	$20x - 3x = 17x$
$(2x+10)(3x-1)$	Common factor
$(2x+1)(3x-10)$	$-20x + 3x = -17x$
$(2x-10)(3x+1)$	Common factor
$(2x-2)(3x+5)$	Common factor
$(2x+5)(3x-2)$	$-4x + 15x = 11x$
$(2x+2)(3x-5)$	Common factor
$(2x-5)(3x+2)$	$4x - 15x = -11x$

$y(2x-5)(3x+2)$

$6x^2y - 11xy - 10y = y(2x-5)(3x+2)$

55. The GCF is 5.

$10t^2 - 5t - 50 = 5(2t^2 - t - 10)$

Factor the trinomial.

$5(\_t + \_)(\_t - \_)$ or $5(\_t - \_)(\_t + \_)$

Factors of 2: 1, 2 — Factors of −10: −1, +10; +1, −10; −2, +5; +2, −5

Trial Factors	Middle Term
$(1t - 1)(2t + 10)$	Common factor
$(1t + 10)(2t - 1)$	$-t + 20t = 19t$
$(1t + 1)(2t - 10)$	Common factor
$(1t - 10)(2t + 1)$	$t - 20t = -19t$
$(1t - 2)(2t + 5)$	$5t - 4t = t$
$(1t + 5)(2t - 2)$	Common factor
$(1t + 2)(2t - 5)$	$-5t + 4t = -t$
$(1t - 5)(2t + 2)$	Common factor

$5(t + 2)(2t - 5)$

$10t^2 - 5t - 50 = 5(t + 2)(2t - 5)$

57. The GCF is p.

$3p^3 - 16p^2 + 5p = p(3p^2 - 16p + 5)$

Factor the trinomial.

$p(\_p - \_)(\_p - \_)$

Factors of 3: 1, 3 — Factors of 5: −1, −5

Trial Factors	Middle Term
$(1p - 1)(3p - 5)$	$-5p - 3p = -8p$
$(1p - 5)(3p - 1)$	$-p - 15p = -16p$

$p(p - 5)(3p - 1)$

$3p^3 - 16p^2 + 5p = p(p - 5)(3p - 1)$

59. The GCF is 2.

$26z^2 + 98z - 24 = 2(13z^2 + 49z - 12)$

Factor the trinomial.

$2(\_z + \_)(\_z - \_)$ or $2(\_z - \_)(\_z + \_)$

Factors of 13: 1, 13 — Factors of −12: −1, +12; +1, −12; −2, +6; +2, −6; −3, +4; +3, −4

Trial Factors	Middle Term
$(1z - 1)(13z + 12)$	$12z - 13z = -z$
$(1z + 12)(13z - 1)$	$-z + 156z = 155z$
$(1z + 1)(13z - 12)$	$-12z + 13z = z$
$(1z - 12)(13z + 1)$	$z - 156z = -155z$
$(1z - 2)(13z + 6)$	$6z - 26z = -20z$
$(1z + 6)(13z - 2)$	$-2z + 78z = 76z$
$(1z + 2)(13z - 6)$	$-6z + 26z = 20z$
$(1z - 6)(13z + 2)$	$2z - 78z = -76z$
$(1z - 3)(13z + 4)$	$4z - 39z = -35z$
$(1z + 4)(13z - 3)$	$-3z + 52z = 49z$
$(1z + 3)(13z - 4)$	$-4z + 49z = 35z$
$(1z - 4)(13z + 3)$	$3z - 52z = -49z$

$2(z + 4)(13z - 3)$

$26z^2 + 98z - 24 = 2(z + 4)(13z - 3)$

61. The GCF is $2y$.

$10y^3 - 44y^2 + 16y = 2y(5y^2 - 22y + 8)$

Factor the trinomial.

$2y(\_y - \_)(\_y - \_)$

Factors of 5: 1, 5 — Factors of 8: −1, −8; −2, −4

Trial Factors	Middle Term
$(1y - 1)(5y - 8)$	$-8y - 5y = -13y$
$(1y - 8)(5y - 1)$	$-y - 40y = -41y$
$(1y - 2)(5y - 4)$	$-4y - 10y = -14y$
$(1y - 4)(5y - 2)$	$-2y - 20y = -22y$

$2y(y - 4)(5y - 2)$

$10y^3 - 44y^2 + 16y = 2y(y - 4)(5y - 2)$

63. The GCF is yz.
$4yz^3 + 5yz^2 - 6yz = yz(4z^2 + 5z - 6)$
Factor the trinomial.
$yz(\_z + \_)(\_z - \_)$ or $yz(\_z - \_)(\_z + \_)$

Factors of 4: 1, 4; 2, 2

Factors of −6: −1, +6; +1, −6; −2, +3; +2, −3

Trial Factors	Middle Term
$(1z - 1)(4z + 6)$	Common factor
$(1z + 6)(4z - 1)$	$-z + 24z = 23z$
$(1z + 1)(4z - 6)$	Common factor
$(1z - 6)(4z + 1)$	$z - 24z = -23z$
$(1z - 2)(4z + 3)$	$3z - 8z = -5z$
$(1z + 3)(4z - 2)$	Common factor
$(1z + 2)(4z - 3)$	$-3z + 8z = 5z$
$(1z - 3)(4z + 2)$	Common factor
$(2z - 1)(2z + 6)$	Common factor
$(2z + 1)(2z - 6)$	Common factor
$(2z - 2)(2z + 3)$	Common factor
$(2z + 2)(2z - 3)$	Common factor

$yz(z + 2)(4z - 3)$
$4yz^3 + 5yz^2 - 6yz = yz(z + 2)(4z - 3)$

65. The GCF is $3a$.
$42a^3 + 45a^2 - 27a = 3a(14a^2 + 15a - 9)$
Factor the trinomial.
$3a(\_a + \_)(\_a - \_)$ or $3a(\_a - \_)(\_a + \_)$

Factors of 14: 1, 14; 2, 7

Factors of −9: −1, +9; +1, −9; −3, +3

Trial Factors	Middle Term
$(1a - 1)(14a + 9)$	$9a - 14a = -5a$
$(1a + 9)(14a - 1)$	$-a + 126a = 125a$
$(1a + 1)(14a - 9)$	$-9a + 14a = 5a$
$(1a - 9)(14a + 1)$	$a - 126a = -125a$
$(1a - 3)(14a + 3)$	$3a - 42a = -39a$
$(1a + 3)(14a - 3)$	$-3a + 42a = 39a$
$(2a - 1)(7a + 9)$	$18a - 7a = 11a$
$(2a + 9)(7a - 1)$	$-2a + 63a = 61a$
$(2a + 1)(7a - 9)$	$-18a + 7a = -11a$
$(2a - 9)(7a + 1)$	$2a - 63a = -61a$
$(2a - 3)(7a + 3)$	$6a - 21a = -15a$
$(2a + 3)(7a - 3)$	$-6a + 21a = 15a$

$3a(2a + 3)(7a - 3)$
$42a^3 + 45a^2 - 27a = 3a(2a + 3)(7a - 3)$

67. The GCF is y.
$9x^2y - 30xy^2 + 25y^3 = y(9x^2 - 30xy + 25y^2)$
Factor the trinomial.
$y(\_x - \_y)(\_x - \_y)$

Factors of 9: 1, 9; 3, 3

Factors of 25: −1, −25; −5, −5

Trial Factors	Middle Term
$(1x - 1y)(9x - 25y)$	$-25xy - 9xy = -34xy$
$(1x - 25y)(9x - 1y)$	$-xy - 225xy = -226xy$
$(1x - 5y)(9x - 5y)$	$-5xy - 45xy = -50xy$
$(3x - 1y)(3x - 25y)$	$-75xy - 3xy = -78xy$
$(3x - 5y)(3x - 5y)$	$-15xy - 15xy = -30xy$

$y(3x - 5y)(3x - 5y)$
$9x^2y - 30xy^2 + 25y^3 = y(3x - 5y)(3x - 5y)$

69. The GCF is xy.
$9x^3y - 24x^2y^2 + 16xy^3 = xy(9x^2 - 24xy + 16y^2)$
Factor the trinomial.
$xy(\_x - \_y)(\_x - \_y)$

Factors of 9: 1, 9; 3, 3

Factors of 16: −1, −16; −2, −8; −4, −4

Trial Factors	Middle Term
$(1x - 1y)(9x - 16y)$	$-16xy - 9xy = -25xy$
$(1x - 16y)(9x - 1y)$	$-xy - 144xy = -145xy$
$(1x - 2y)(9x - 8y)$	$-8xy - 18xy = -26xy$
$(1x - 8y)(9x - 2y)$	$-2xy - 72xy = -74xy$
$(1x - 4y)(9x - 4y)$	$-4xy - 36xy = -40xy$
$(3x - 1y)(3x - 16y)$	$-48xy - 3xy = -51xy$
$(3x - 2y)(3x - 8y)$	$-24xy - 6xy = -30xy$
$(3x - 4y)(3x - 4y)$	$-12xy - 12xy = -24xy$

$xy(3x - 4y)(3x - 4y)$
$9x^3y - 24x^2y^2 + 16xy^3 = xy(3x - 4y)(3x - 4y)$

Objective B Exercises

71. $6x^2 - 17x + 12$ | Factors of 72 whose sum is -17: -9 and -8
$6 \times 12 = 72$ | $6x^2 - 9x - 8x + 12 = 3x(2x - 3) - 4(2x - 3) = (2x - 3)(3x - 4)$

73. $5b^2 + 33b - 14$ | Factors of -70 whose sum is 33: 35 and -2
$5 \times (-14) = -70$ | $5b^2 + 35b - 2b - 14 = 5b(b + 7) - 2(b + 7) = (b + 7)(5b - 2)$

75. $6a^2 + 7a - 24$ | Factors of -144 whose sum is 7: 16 and -9
$6 \times (-24) = -144$ | $6a^2 + 16a - 9a - 24 = 2a(3a + 8) - 3(3a + 8) = (3a + 8)(2a - 3)$

77. $4z^2 + 11z + 6$ | Factors of 24 whose sum is 11: 8 and 3
$4 \times 6 = 24$ | $4z^2 + 8z + 3z + 6 = 4z(z + 2) + 3(z + 2) = (z + 2)(4z + 3)$

79. $22p^2 + 51p - 10$ | Factors of -220 whose sum is 51: 55 and -4
$22 \times (-10) = -220$ | $22p^2 + 55p - 4p - 10 = 11p(2p + 5) - 2(2p + 5) = (2p + 5)(11p - 2)$

81. $8y^2 + 17y + 9$ | Factors of 72 whose sum is 17: 9 and 8
$8 \times 9 = 72$ | $8y^2 + 8y + 9y + 9 = 8y(y + 1) + 9(y + 1) = (y + 1)(8y + 9)$

83. $18t^2 - 9t - 5$ | Factors of -90 whose sum is -9: -15 and 6
$18 \times (-5) = -90$ | $18t^2 - 15t + 6t - 5 = 3t(6t - 5) + 1(6t - 5) = (6t - 5)(3t + 1)$

85. $6b^2 + 71b - 12$ | Factors of -72 whose sum is 71: 72 and -1
$6 \times (-12) = -72$ | $6b^2 + 72b - b - 12 = 6b(b + 12) - 1(b + 12) = (b + 12)(6b - 1)$

87. $9x^2 + 12x + 4$ | Factors of 36 whose sum is 12: 6 and 6
$9 \times 4 = 36$ | $9x^2 + 6x + 6x + 4 = 3x(3x + 2) + 2(3x + 2) = (3x + 2)(3x + 2)$

89. $6b^2 - 13b + 6$ | Factors of 36 whose sum is -13: -9 and -4
$6 \times 6 = 36$ | $6b^2 - 9b - 4b + 6 = 3b(2b - 3) - 2(2b - 3) = (2b - 3)(3b - 2)$

91. $33b^2 + 34b - 35$ | Factors of -1155 whose sum is 34: 55 and -21
$33 \times (-35) = -1155$ | $33b^2 + 55b - 21b - 35 = 11b(3b + 5) - 7(3b + 5) = (3b + 5)(11b - 7)$

93. $18y^2 - 39y + 20$ | Factors of 360 whose sum is -39: -24 and -15
$18 \times 20 = 360$ | $18y^2 - 24y - 15y + 20 = 6y(3y - 4) - 5(3y - 4) = (3y - 4)(6y - 5)$

95. $15a^2 + 26a - 21$ | Factors of -315 whose sum is 26: 35 and -9
$15 \times (-21) = -315$ | $15a^2 + 35a - 9a - 21 = 5a(3a + 7) - 3(3a + 7) = (3a + 7)(5a - 3)$

97. $8y^2 - 26y + 15$ | Factors of 120 whose sum is -26: -20 and -6
$8 \times 15 = 120$ | $8y^2 - 20y - 6y + 15 = 4y(2y - 5) - 3(2y - 5) = (2y - 5)(4y - 3)$

99. $8z^2 + 2z - 15$ | Factors of -120 whose sum is 2: 12 and -10
$8 \times (-15) = -120$ | $8z^2 + 12z - 10z - 15 = 4z(2z + 3) - 5(2z + 3) = (2z + 3)(4z - 5)$

101. $15x^2 - 82x + 24$ | Factors of 360 whose sum is -82:
$15 \times 24 = 360$ | Nonfactorable over the integers

103. $10z^2 - 29z + 10$ | Factors of 100 whose sum is -29: -25 and -4
$10 \times 10 = 100$ | $10z^2 - 25z - 4z + 10 = 5z(2z - 5) - 2(2z - 5) = (2z - 5)(5z - 2)$

105. $36z^2 + 72z + 35$ | Factors of 1260 whose sum is 72: 30 and 42
$36 \times 35 = 1260$ | $36z^2 + 30z + 42z + 35 = 6z(6z + 5) + 7(6z + 5) = (6z + 5)(6z + 7)$

107. $3x^2 + xy - 2y^2$
$3 \times (-2) = -6$
Factors of -6 whose sum is 1: 3 and -2
$3x^2 + 3xy - 2xy - 2y^2 = 3x(x+y) - 2y(x+y) = (x+y)(3x-2y)$

109. $3a^2 + 5ab - 2b^2$
$3 \times (-2) = -6$
Factors of -6 whose sum is 5: 6 and -1
$3a^2 + 6ab - ab - 2b^2 = 3a(a+2b) - b(a+2b) = (a+2b)(3a-b)$

111. $4y^2 - 11yz + 6z^2$
$4 \times 6 = 24$
Factors of 24 whose sum is -11: -8 and -3
$4y^2 - 8yz - 3yz + 6z^2 = 4y(y-2z) - 3z(y-2z) = (y-2z)(4y-3z)$

113. $28 + 3z - z^2$
$-(z^2 - 3z - 28)$
$1 \times (-28) = -28$
Factors of -28 whose sum is -3: -7 and 4
$-(z^2 - 7z + 4z - 28) = -[z(z-7) + 4(z-7)] = -(z-7)(z+4)$

115. $8 - 7x - x^2$
$-(x^2 + 7x - 8)$
$1 \times (-8) = -8$
Factors of -8 whose sum is 7: -1 and 8
$-(x^2 - x + 8x - 8) = -[x(x-1) + 8(x-1)] = -(x-1)(x+8)$

117. $9x^2 + 33x - 60$
Common factor 3: $3(3x^2 + 11x - 20)$
$3 \times (-20) = -60$
Factors of -60 whose sum is 11: 15 and -4
$3(3x^2 + 15x - 4x - 20) = 3[3x(x+5) - 4(x+5)] = 3(x+5)(3x-4)$

119. $24x^2 - 52x + 24$
Common factor 4: $4(6x^2 - 13x + 6)$
$6 \times 6 = 36$
Factors of 36 whose sum is -13: -9 and -4
$4(6x^2 - 9x - 4x + 6) = 4[3x(2x-3) - 2(2x-3)] = 4(2x-3)(3x-2)$

121. $35a^4 + 9a^3 - 2a^2$
Common factor a^2: $a^2(35a^2 + 9a - 2)$
$35 \times (-2) = -70$
Factors of -70 whose sum is 9: 14 and -5
$a^2(35a^2 + 14a - 5a - 2) = a^2[7a(5a+2) - 1(5a+2)]$
$= a^2(5a+2)(7a-1)$

123. $15b^2 - 115b + 70$
Common factor 5: $5(3b^2 - 23b + 14)$
$3 \times 14 = 42$
Factors of 42 whose sum is -23: -21 and -2
$5(3b^2 - 21b - 2b + 14) = 5[3b(b-7) - 2(b-7)] = 5(b-7)(3b-2)$

125. $3x^2 - 26xy + 35y^2$
$3 \times 35 = 105$
Factors of 105 whose sum is -26: -21 and -5
$3x^2 - 21xy - 5xy + 35y^2 = 3x(x-7y) - 5y(x-7y) = (x-7y)(3x-5y)$

127. $216y^2 - 3y - 3$
Common factor 3: $3(72y^2 - y - 1)$
$72 \times (-1) = -72$
Factors of -72 whose sum is -1: -9 and 8
$3(72y^2 - 9y + 8y - 1) = 3[9y(8y-1) + 1(8y-1)] = 3(8y-1)(9y+1)$

129. $21 - 20x - x^2$
$-(x^2 + 20x - 21)$
$1 \times (-21) = -21$
Factors of -21 whose sum is 20: -1 and 21
$-(x^2 - x + 21x - 21) = -[x(x-1) + 21(x-1)] = -(x-1)(x+21)$

Applying the Concepts

131. Students should explain that the sign of the product of the last terms of the two binomial factors must be the same as the sign of the last term of the trinomial. Thus if the last term of the trinomial is positive, the last terms of the two binomial factors are either both positive or both negative, depending on the middle term of the trinomial. If the last term of the trinomial is negative, the last terms of the two binomial factors will have different signs.

133. $(x-2)^2 + 3(x-2) + 2$ Let $a = x - 2$
$a^2 + 3a + 2$
$(a+1)(a+2)$
$(x-2+1)(x-2+2)$
$(x-1)x$ or $x(x-1)$

135. $2(y+2)^2 - (y+2) - 3$ Let $a = y+2$

$2a^2 - a - 3$

$(2a-3)(a+1)$

$[2(y+2)-3][y+2+1]$

$(2y+4-3)(y+3)$

$(2y+1)(y+3)$

137. $4(y-1)^2 - 7(y-1) - 2$ Let $a = y-1$

$4a^2 - 7a - 2$

$(4a+1)(a-2)$

$[4(y-1)+1][y-1-2]$

$(4y-4+1)(y-3)$

$(4y-3)(y-3)$

139. $2x^2 + kx - 3$

$2 \times (-3) = -6$

Factors of –6	Sums
+1, −6	−5
−1, +6	+5
+2, −3	−1
−2, +3	+1

k can be -5, 5, -1, or 1.

141. $3x^2 + kx - 2$

$3 \times (-2) = -6$

Factors of –6	Sums
+1, −6	−5
−1, +6	+5
+2, −3	−1
−2, +3	+1

k can be -5, 5, -1, or 1.

143. $2x^2 + kx - 5$

$2 \times (-5) = -10$

Factors of –10	Sums
+1, −10	−9
−1, +10	+9
+2, −5	−3
−2, +5	+3

k can be -9, 9, -3, or 3.

Section 5.4

Objective A Exercises

1. **a.** Answers will vary. For instance, $x^2 - 25$.

b. Answers will vary. For instance, $x^2 + 6x + 9$.

3. $x^2 - 4 = x^2 - 2^2 = (x+2)(x-2)$

5. $a^2 - 81 = a^2 - 9^2 = (a+9)(a-9)$

7. $y^2 + 2y + 1$ $\sqrt{y^2} = y$ $2(y \cdot 1) = 2y$

$\sqrt{1} = 1$ The trinomial is a perfect square.

$y^2 + 2y + 1 = (y+1)^2$

9. $a^2 - 2a + 1$ $\sqrt{a^2} = a$ $-2(a \cdot 1) = -2a$

$\sqrt{1} = 1$ The trinomial is a perfect square.

$a^2 - 2a + 1 = (a-1)^2$

11. $4x^2 - 1 = (2x)^2 - 1^2 = (2x+1)(2x-1)$

13. $x^6 - 9 = (x^3)^2 - 3^2 = (x^3+3)(x^3-3)$

15. $x^2 + 8x - 16$ is nonfactorable over the integers.

$\sqrt{-16} =$ not real

17. $x^2+2xy+y^2$ $\sqrt{x^2}=x$ $\sqrt{y^2}=y$
$2(x\cdot y)=2xy$
The trinomial is a perfect square.
$x^2+2xy+y^2=(x+y)^2$

19. $4a^2+4a+1$ $\sqrt{4a^2}=2a$ $\sqrt{1}=1$
$2(2a\cdot 1)=4a$
The trinomial is a perfect square.
$4a^2+4a+1=(2a+1)^2$

21. $9x^2-1=(3x)^2-1^2=(3x+1)(3x-1)$

23. $1-64x^2=1^2-(8x)^2=(1+8x)(1-8x)$

25. x^2+64 is nonfactorable over integers.

27. $9a^2+6a+1$ $\sqrt{9a^2}=3a$ $\sqrt{1}=1$
$2(3a\cdot 1)=6a$
The trinomial is a perfect square.
$9a^2+6a+1=(3a+1)^2$

29. $b^4-16a^2=(b^2)^2-(4a)^2=(b^2+4a)(b^2-4a)$

31. $4a^2-20a+25$ $\sqrt{4a^2}=2a$ $\sqrt{25}=5$
$-2(2a\cdot 5)=-20a$
The trinomial is a perfect square.
$4a^2-20a+25=(2a-5)^2$

33. $9a^2-42a+49$ $\sqrt{9a^2}=3a$ $\sqrt{49}=7$
$-2(3a\cdot 7)=-42a$
The trinomial is a perfect square.
$9a^2-42a+49=(3a-7)^2$

35. $25z^2-y^2=(5z)^2-(y)^2=(5z+y)(5z-y)$

37. $a^2b^2-25=(ab)^2-5^2=(ab+5)(ab-5)$

39. $25x^2-1=(5x)^2-1^2=(5x+1)(5x-1)$

41. $4a^2-12ab+9b^2$ $\sqrt{4a^2}=2a$ $\sqrt{9b^2}=3b$
$-2(2a\cdot 3b)=-12ab$
The trinomial is a perfect square.
$4a^2-12ab+9b^2=(2a-3b)^2$

43. $4y^2-36yz+81z^2$ $\sqrt{4y^2}=2y$ $\sqrt{81z^2}=9z$
$-2(2y\cdot 9z)=-36yz$
The trinomial is a perfect square.
$4y^2-36yz+81z^2=(2y-9z)^2$

45. $\frac{1}{x^2}-4=\left(\frac{1}{x}\right)^2-2^2=\left(\frac{1}{x}+2\right)\left(\frac{1}{x}-2\right)$

47. $9a^2b^2-6ab+1$ $\sqrt{9a^2b^2}=3ab$ $\sqrt{1}=1$
$-2(3ab\cdot 1)=-6ab$
The trinomial is a perfect square.
$9a^2b^2-6ab+1=(3ab-1)^2$

Objective B Exercises

49. $8y^2-2$
The GCF is 2: $2(4y^2-1)=2\left[(2y)^2-1^2\right]=2(2y+1)(2y-1)$

51. $3a^3+6a^2+3a$ The GCF is $3a$: $3a(a^2+2a+1)$
$\sqrt{a^2}=a$ $\sqrt{1}=1$
$2(a\cdot 1)=2a$
The trinomial is a perfect square.
$3a(a+1)^2$

53. $m^4-256=(m^2)^2-16^2=(m^2+16)(m^2-16)=(m^2+16)(m+4)(m-4)$

55. $9x^2+13x+4$ $\sqrt{9x^2}=3x$ $\sqrt{4}=2$ Not a perfect square $9\times 4=36$ Factors of 36 whose sum is 13: 9 and 4
$9x^2+9x+4x+4=9x(x+1)+4(x+1)=(x+1)(9x+4)$

57. $16y^4 + 48y^3 + 36y^2$

The GCF is $4y^2$: $4y^2(4y^2 + 12y + 9)$

$\sqrt{4y^2} = 2y$ $\quad 2(2y \cdot 3) = 12y$

$\sqrt{9} = 3$ $\quad$ The trinomial is a perfect square.

$4y^2(2y + 3)^2$

59. $y^8 - 81 = (y^4)^2 - 9^2 = (y^4 + 9)(y^4 - 9) = (y^4 + 9)(y^2 + 3)(y^2 - 3)$

61. $25 - 20p + 4p^2$

$\sqrt{25} = 5$ $\quad -2(5 \cdot 2p) = -20p$

$\sqrt{4p^2} = 2p$ $\quad$ The trinomial is a perfect square.

$25 - 20p + 4p^2 = (5 - 2p)^2$

63. $(4x - 3)^2 - y^2 = (4x - 3 + y)(4x - 3 - y)$

65. $(x^2 - 4x + 4) - y^2 = (x - 2)^2 - y^2$

$= (x - 2 + y)(x - 2 - y)$

67. The GCF is 5.

$5x^2 - 5 = 5(x^2 - 1)$

Factor the difference of two squares.

$5(x + 1)(x - 1)$

$5x^2 - 5 = 5(x + 1)(x - 1)$

69. The GCF is x.

$x^3 + 4x^2 + 4x = x(x^2 + 4x + 4)$

Factor the perfect-square trinomial.

$x(x + 2)^2$

$x^3 + 4x^2 + 4x = x(x + 2)^2$

71. The GCF is x^2.

$x^4 + 2x^3 - 35x^2 = x^2(x^2 + 2x - 35)$

Factor the trinomial.

$x^2(x + \_)(x - \_)$

Factors	Sum
−1, +35	+34
+1, −35	−34
−5, +7	+2
+5, −7	−2

$x^2(x + 7)(x - 5)$

$x^4 + 2x^3 - 35x^2 = x^2(x + 7)(x - 5)$

73. The GCF is 5.

$5b^2 + 75b + 180 = 5(b^2 + 15b + 36)$

Factor the trinomial.

$5(b + \_)(b + \_)$

Factors	Sum
+1, +36	+37
+2, +18	+20
+3, +12	+15
+4, +9	+13
+6, +6	+12

$5(b + 3)(b + 12)$

$5b^2 + 75b + 180 = 5(b + 3)(b + 12)$

75. There is no common factor.

Factor the trinomial.

$(\_a + \_)(\_a + \_)$

Factors of 3: 1,3 $\quad$ Factors of 10: +1, +10

+2, +5

$3a^2 + 36a + 10$ is nonfactorable over the integers.

77. The GCF is $2y$.

$2x^2y + 16xy - 66y = 2y(x^2 + 8x - 33)$

Factor the trinomial.

$2y(x + \_)(x - \_)$

Factors	Sum
−1, +33	+32
+1, −33	−32
−3, +11	+8
+3, −11	−8

$2y(x + 11)(x - 3)$

$2x^2y + 16xy - 66y = 2y(x + 11)(x - 3)$

79. The GCF is x.

$x^3 - 6x^2 - 5x = x(x^2 - 6x - 5)$

Factor the trinomial.

$x(x + \_)(x - \_)$

Factors	Sum
−1, +5	+4
+1, −5	−4

The trinomial is nonfactorable.

$x^3 - 6x^2 - 5x = x(x^2 - 6x - 5)$

81. The GCF is 3.

$3y^2 - 36 = 3(y^2 - 12)$

$y^2 - 12$ is nonfactorable over the integers.

$3y^2 - 36 = 3(y^2 - 12)$

83. There is no common factor.

Factor the trinomial.

$(\_a + \_)(\_a + \_)$

Factors of 20: 1, 20 $\quad$ Factors of 1: +1, +1

2, 10

4, 5

$(2a + 1)(10a + 1)$

$20a^2 + 12a + 1 = (2a + 1)(10a + 1)$

85. The GCF is y^2.

$x^2y^2 - 7xy^2 - 8y^2 = y^2(x^2 - 7x - 8)$

Factor the trinomial.

$y^2(x + \_)(x - \_)$

Factors	Sum
−1, +8	+7
+1, −8	−7
−2, +4	+2
+2, −4	−2

$y^2(x + 1)(x - 8)$

$x^2y^2 - 7xy^2 - 8y^2 = y^2(x + 1)(x - 8)$

87. The GCF is 5.
$10a^2 - 5ab - 15b^2 = 5(2a^2 - ab - 3b^2)$
Factor the trinomial.
$5(\_a + \_b)(\_a - \_b)$ or $5(\_a - \_b)(\_a + \_b)$
Factors of 2: 1, 2 — Factors of −3: −1, +3; +1, −3
$5(a + b)(2a - 3b)$
$10a^2 - 5ab - 15b^2 = 5(a + b)(2a - 3b)$

89. The GCF is −2.
$50 - 2x^2 = -2(x^2 - 25)$
Factor the difference of two squares.
$-2(x + 5)(x - 5)$
$50 - 2x^2 = -2(x + 5)(x - 5)$

91. The GCF is b^2.
$a^2b^2 - 10ab^2 + 25b^2 = b^2(a^2 - 10a + 25)$
Factor the perfect-square trinomial.
$b^2(a - 5)^2$
$a^2b^2 - 10ab^2 + 25b^2 = b^2(a - 5)^2$

93. The GCF is ab.
$12a^3b - a^2b^2 - ab^3 = ab(12a^2 - ab - b^2)$
Factor the trinomial.
$ab(\_a + \_b)(\_a - \_b)$ or $ab(\_a - \_b)(\_a + \_b)$
Factors of 12: 1, 12; 2, 6; 3, 4 — Factors of −1: −1, +1
$ab(3a - b)(4a + b)$
$12a^3b - a^2b^2 - ab^3 = ab(3a - b)(4a + b)$

95. The GCF is $3a$.
$12a^3 - 12a^2 + 3a = 3a(4a^2 - 4a + 1)$
Factor the perfect-square trinomial.
$3a(2a - 1)^2$
$12a^3 - 12a^2 + 3a = 3a(2a - 1)^2$

97. The GCF is 3.
$243 + 3a^2 = 3(81 + a^2)$
$81 + a^2$ is nonfactorable over the integers.
$243 + 3a^2 = 3(81 + a^2)$

99. The GCF is $2a$.
$12a^3 - 46a^2 + 40a = 2a(6a^2 - 23a + 20)$
Factor the trinomial.
$2a(\_a - \_)(\_a - \_)$
Factors of 6: 1, 6; 2, 3 — Factors of 20: −1, −20; −2, −10; −4, −5
$2a(2a - 5)(3a - 4)$
$12a^3 - 46a^2 + 40a = 2a(2a - 5)(3a - 4)$

101. The GCF is a.
$4a^3 + 20a^2 + 25a = a(4a^2 + 20a + 25)$
Factor the perfect-square trinomial.
$a(2a + 5)^2$
$4a^3 + 20a^2 + 25a = a(2a + 5)^2$

103. The GCF is $3b$.
$27a^2b - 18ab + 3b = 3b(9a^2 - 6a + 1)$
Factor the perfect-square trinomial.
$3b(3a - 1)^2$
$27a^2b - 18ab + 3b = 3b(3a - 1)^2$

105. The GCF is −6.
$48 - 12x - 6x^2 = -6(x^2 + 2x - 8)$
Factor the trinomial.
$-6(x - \_)(x + \_)$

Factors of −8	Sum
+1, −8	−7
−1, +8	7
+2, −4	−2
−2, +4	2

$-6(x - 2)(x + 4)$
$48 - 12x - 6x^2 = -6(x - 2)(x + 4)$

107. The GCF is x^2.
$x^4 - x^2y^2 = x^2(x^2 - y^2)$
Factor the difference of two squares.
$x^2(x + y)(x - y)$
$x^4 - x^2y^2 = x^2(x + y)(x - y)$

109. The GCF is $2a$.
$18a^3 + 24a^2 + 8a = 2a(9a^2 + 12a + 4)$
Factor the perfect-square trinomial.
$2a(3a + 2)^2$
$18a^3 + 24a^2 + 8a = 2a(3a + 2)^2$

111. The GCF is $-b$.
$2b + ab - 6a^2b = -b(6a^2 - a - 2)$
Factor the trinomial.
$-b(\_a - \_)(\_a + \_)$
Factors of 6: 1, 6; 2, 3 — Factors of −2: −1, +2; +1, −2
$-b(3a - 2)(2a + 1)$
$2b + ab - 6a^2b = -b(3a - 2)(2a + 1)$

113. The GCF is $2x^2$.
$4x^4 - 38x^3 + 48x^2 = 2x^2(2x^2 - 19x + 24)$
Factor the trinomial.
$2x^2(\_x - \_)(\_x - \_)$
Factors of 2: 1, 2 — Factors of 24: −1, −24; −2, −5; −3, −8; −4, −6
$2x^2(x - 8)(2x - 3)$
$4x^4 - 38x^3 + 48x^2 = 2x^2(x - 8)(2x - 3)$

115. The GCF is x^2.
$x^4 - 25x^2 = x^2(x^2 - 25)$
Factor the difference of two squares.
$x^2(x + 5)(x - 5)$
$x^4 - 25x^2 = x^2(x + 5)(x - 5)$

117. There is no common factor.
$a^4 - 16 = (a^2 + 4)(a^2 - 4)$
Factor the difference of two squares.
$(a^2 + 4)(a + 2)(a - 2)$
$a^4 - 16 = (a^2 + 4)(a + 2)(a - 2)$

119. The GCF is $-3y^2$.
$45y^2 - 42y^3 - 24y^4 = -3y^2(8y^2 + 14y - 15)$
Factor the trinomial.
$-3y^2(\_y + \_)(\_y - \_)$

Factors of 8:	Factors of 15:
1, 8	−1, +15
2, 4	+1, −15
	−3, +5
	+3, −5

$-3y^2(2y + 5)(4y - 3)$
$45y^2 - 42y^3 - 24y^4 = -3y^2(2y + 5)(4y - 3)$

121. The common binomial factor is $x - 3$.
$4a(x - 3) - 2b(x - 3) = (x - 3)(4a - 2b)$
Factor 2 from the binomial factor.
$(x - 3)(4a - 2b) = 2(x - 3)(2a - b)$
$4a(x - 3) - 2b(x - 3) = 2(x - 3)(2a - b)$

123. The common binomial factor is $a - b$.
$y^2(a - b) - (a - b) = (a - b)(y^2 - 1)$
Factor the difference of two squares.
$(a - b)(y^2 - 1) = (a - b)(y + 1)(y - 1)$
$y^2(a - b) - (a - b) = (a - b)(y + 1)(y - 1)$

125. The common binomial factor is $a^2 - b^2$.
$x(a^2 - b^2) - y(a^2 - b^2) = (a^2 - b^2)(x - y)$
Factor the difference of two squares.
$(a^2 - b^2)(x - y) = (a + b)(a - b)(x - y)$
$x(a^2 - b^2) - y(a^2 - b^2) = (a + b)(a - b)(x - y)$

Applying the Concepts

127. $4x^2 - kx + 9$ $\quad \sqrt{4x^2} = 2x \quad \pm 2(2x \cdot 3) = -kx$
$\sqrt{9} = 3 \quad \pm 12x = -kx$
$k = 12$ or -12

129. $64x^2 + kxy + y^2$ $\quad \sqrt{64x^2} = 8x \quad \pm 2(8x \cdot y) = kxy$
$\sqrt{y^2} = y \quad \pm 16xy = kxy$
$k = 16$ or -16

131. $25x^2 - kx + 1$ $\quad \sqrt{25x^2} = 5x \quad \pm 2(5x \cdot 1) = -kx$
$\sqrt{1} = 1 \quad \pm 10 = -k$
$k = 10$ or -10

133. Choose the integer 7.
Square 7: $7^2 = 49$
Subtract 1: $49 - 1 = 48$
48 is evenly divisible by 8.
Let n be a natural number. Then $2n + 1$ is an odd integer greater than 1. Square $2n + 1$ and then subtract 1:
$(2n+1)^2 - 1$
$= (2n+1)(2n+1) - 1$
$= (4n^2 + 4n + 1) - 1$
$= 4n^2 + 4n$
$= 4n(n+1)$
If n is even, $n = 2, 4, 6, 8, \ldots$
and $4n = 8, 16, 24, 32, \ldots$, which is divisible by 8.
If n is odd, $n = 1, 3, 5, 7, \ldots$,
$n + 1 = 2, 4, 6, 8, \ldots$
and $4(n+1) = 8, 16, 24, 32, \ldots$, which is divisible by 8.
Therefore, $4n(n+1)$ is always divisible by 8. The procedure always produces a number that is divisible by 8.

Section 5.5

Objective A Exercises

1. The Principle of Zero Products states that if the product of two numbers equals zero, then one or both of the numbers is (are) zero.

3. $(y+3)(y+2) = 0$
$y + 3 = 0 \qquad y + 2 = 0$
$y = -3 \qquad y = -2$
The solutions are -3 and -2.

5. $(z-7)(z-3) = 0$
$z - 7 = 0 \qquad z - 3 = 0$
$z = 7 \qquad z = 3$
The solutions are 7 and 3.

7. $x(x-5) = 0$
$x = 0 \qquad x - 5 = 0$
$x = 5$
The solutions are 0 and 5.

9. $a(a-9) = 0$
$a = 0 \qquad a - 9 = 0$
$a = 9$
The solutions are 0 and 9.

11. $y(2y+3) = 0$
$y = 0 \qquad 2y + 3 = 0$
$2y = -3$
$y = -\frac{3}{2}$
The solutions are 0 and $-\frac{3}{2}$.

13. $2a(3a-2) = 0$
$2a = 0 \qquad 3a - 2 = 0$
$a = 0 \qquad 3a = 2$
$a = \frac{2}{3}$
The solutions are 0 and $\frac{2}{3}$.

15. $(b+2)(b-5) = 0$
$b + 2 = 0 \qquad b - 5 = 0$
$b = -2 \qquad b = 5$
The solutions are -2 and 5.

17. $x^2 - 81 = 0$
$(x+9)(x-9) = 0$
$x + 9 = 0 \qquad x - 9 = 0$
$x = -9 \qquad x = 9$
The solutions are -9 and 9.

19. $4x^2 - 49 = 0$
$(2x+7)(2x-7) = 0$
$2x + 7 = 0 \qquad 2x - 7 = 0$
$2x = -7 \qquad 2x = 7$
$x = -\frac{7}{2} \qquad x = \frac{7}{2}$
The solutions are $-\frac{7}{2}$ and $\frac{7}{2}$.

21. $9x^2 - 1 = 0$
$(3x+1)(3x-1) = 0$
$3x + 1 = 0 \qquad 3x - 1 = 0$
$3x = -1 \qquad 3x = 1$
$x = -\frac{1}{3} \qquad x = \frac{1}{3}$
The solutions are $-\frac{1}{3}$ and $\frac{1}{3}$.

23. $x^2 + 6x + 8 = 0$
$(x+2)(x+4) = 0$
$x + 2 = 0 \qquad x + 4 = 0$
$x = -2 \qquad x = -4$
The solutions are -2 and -4.

25. $z^2 + 5z - 14 = 0$
$(z+7)(z-2) = 0$
$z + 7 = 0 \qquad z - 2 = 0$
$z = -7 \qquad z = 2$
The solutions are -7 and 2.

27. $2a^2 - 9a - 5 = 0$
$(2a+1)(a-5) = 0$
$2a + 1 = 0 \qquad a - 5 = 0$
$2a = -1 \qquad a = 5$
$a = -\frac{1}{2}$
The solutions are $-\frac{1}{2}$ and 5.

29.
$$6z^2 + 5z + 1 = 0$$
$$(3z + 1)(2z + 1) = 0$$
$$3z + 1 = 0 \qquad 2z + 1 = 0$$
$$3z = -1 \qquad 2z = -1$$
$$z = -\frac{1}{3} \qquad z = -\frac{1}{2}$$
The solutions are $-\frac{1}{3}$ and $-\frac{1}{2}$.

31.
$$x^2 - 3x = 0$$
$$x(x - 3) = 0$$
$$x = 0 \quad x - 3 = 0$$
$$x = 3$$
The solutions are 0 and 3.

33.
$$x^2 - 7x = 0$$
$$x(x - 7) = 0$$
$$x = 0 \quad x - 7 = 0$$
$$x = 7$$
The solutions are 0 and 7.

35.
$$a^2 + 5a = -4$$
$$a^2 + 5a + 4 = 0$$
$$(a + 1)(a + 4) = 0$$
$$a + 1 = 0 \qquad a + 4 = 0$$
$$a = -1 \qquad a = -4$$
The solutions are -1 and -4.

37.
$$y^2 - 5y = -6$$
$$y^2 - 5y + 6 = 0$$
$$(y - 2)(y - 3) = 0$$
$$y - 2 = 0 \quad y - 3 = 0$$
$$y = 2 \qquad y = 3$$
The solutions are 2 and 3.

39.
$$2t^2 + 7t = 4$$
$$2t^2 + 7t - 4 = 0$$
$$(2t - 1)(t + 4) = 0$$
$$2t - 1 = 0 \quad t + 4 = 0$$
$$2t = 1 \qquad t = -4$$
$$t = \frac{1}{2}$$
The solutions are $\frac{1}{2}$ and -4.

41.
$$3t^2 - 13t = -4$$
$$3t^2 - 13t + 4 = 0$$
$$(3t - 1)(t - 4) = 0$$
$$3t - 1 = 0 \quad t - 4 = 0$$
$$3t = 1 \qquad t = 4$$
$$t = \frac{1}{3}$$
The solutions are $\frac{1}{3}$ and 4.

43.
$$x(x - 12) = -27$$
$$x^2 - 12x = -27$$
$$x^2 - 12x + 27 = 0$$
$$(x - 3)(x - 9) = 0$$
$$x - 3 = 0 \quad x - 9 = 0$$
$$x = 3 \qquad x = 9$$
The solutions are 3 and 9.

45.
$$y(y - 7) = 18$$
$$y^2 - 7y = 18$$
$$y^2 - 7y - 18 = 0$$
$$(y + 2)(y - 9) = 0$$
$$y + 2 = 0 \qquad y - 9 = 0$$
$$y = -2 \qquad y = 9$$
The solutions are -2 and 9.

47.
$$p(p + 3) = -2$$
$$p^2 + 3p = -2$$
$$p^2 + 3p + 2 = 0$$
$$(p + 1)(p + 2) = 0$$
$$p + 1 = 0 \qquad p + 2 = 0$$
$$p = -1 \qquad p = -2$$
The solutions are -1 and -2.

49.
$$y(y + 4) = 45$$
$$y^2 + 4y = 45$$
$$y^2 + 4y - 45 = 0$$
$$(y + 9)(y - 5) = 0$$
$$y + 9 = 0 \qquad y - 5 = 0$$
$$y = -9 \qquad y = 5$$
The solutions are -9 and 5.

51.
$$x(x + 3) = 28$$
$$x^2 + 3x = 28$$
$$x^2 + 3x - 28 = 0$$
$$(x + 7)(x - 4) = 0$$
$$x + 7 = 0 \qquad x - 4 = 0$$
$$x = -7 \qquad x = 4$$
The solutions are -7 and 4.

53.
$$(x + 8)(x - 3) = -30$$
$$x^2 + 5x - 24 = -30$$
$$x^2 + 5x + 6 = 0$$
$$(x + 2)(x + 3) = 0$$
$$x + 2 = 0 \qquad x + 3 = 0$$
$$x = -2 \qquad x = -3$$
The solutions are -2 and -3.

55.
$$(z - 5)(z + 4) = 52$$
$$z^2 - z - 20 = 52$$
$$z^2 - z - 72 = 0$$
$$(z + 8)(z - 9) = 0$$
$$z + 8 = 0 \qquad z - 9 = 0$$
$$z = -8 \qquad z = 9$$
The solutions are -8 and 9.

57.
$$(z - 6)(z + 1) = -10$$
$$z^2 - 5z - 6 = -10$$
$$z^2 - 5z + 4 = 0$$
$$(z - 1)(z - 4) = 0$$
$$z - 1 = 0 \quad z - 4 = 0$$
$$z = 1 \qquad z = 4$$
The solutions are 1 and 4.

59. $(a-4)(a+7)=-18$
$a^2+3a-28=-18$
$a^2+3a-10=0$
$(a+5)(a-2)=0$
$a+5=0 \quad a-2=0$
$a=-5 \quad a=2$
The solutions are -5 and 2.

Objective B Exercises

61. Strategy The positive number: x
The square of the positive number is six more than five times the positive number.

Solution
$x^2=5x+6$
$x^2-5x-6=0$
$(x-6)(x+1)=0$
$x-6=0 \quad x+1=0$
$x=6 \quad x=-1$
Because -1 is not a positive number, it is not a solution.
The number is 6.

63. Strategy One of the two numbers: x
The other number: $6-x$
The sum of the squares of the two numbers is 20.

Solution
$x^2+(6-x)^2=20$
$x^2+36-12x+x^2=20$
$2x^2-12x+36=20$
$2x^2-12x+16=0$
$2(x^2-6x+8)=0$
$2(x-4)(x-2)=0$
$x-4=0 \quad x-2=0$
$x=4 \quad x=2$
$x=4$: $\quad x=2$:
$6-x=6-4=2 \quad 6-x=6-2=4$
The numbers are 2 and 4.

65. Strategy First positive integer: x
Second positive integer: $x+1$
The sum of the squares of the two integers is forty-one.

Solution
$x^2+(x+1)^2=41$
$x^2+x^2+2x+1=41$
$2x^2+2x+1=41$
$2x^2+2x-40=0$
$2(x^2+x-20)=0$
$2(x+5)(x-4)=0$
$x+5=0 \quad x-4=0$
$x=-5 \quad x=4$
Because -5 is not a positive integer, it is not a solution.
$x=4$
$x+1=5$
The numbers are 4 and 5.

67. Strategy One of the numbers: x
The other number: $10-x$
The product of the two number is 21.

Solution
$x(10-x)=21$
$10x-x^2=21$
$0=x^2-10x+21$
$0=(x-3)(x-7)$
$x-3=0 \quad x-7=0$
$x=3 \quad x=7$
$x=3$: $\quad x=7$:
$10-x=10-3 \quad 10-x=10-7$
$=7 \quad =3$
The numbers are 3 and 7.

69.
$S=\frac{n^2+n}{2}$
$78=\frac{n^2+n}{2}$
$156=n^2+n$
$n^2+n-156=0$
$(n+13)(n-12)=0$
$n=12$
There will be 12 consecutive numbers.

71.
$N=\frac{t^2-t}{2}$
$15=\frac{t^2-t}{2}$
$30=t^2-t$
$t^2-t-30=0$
$t-6=0 \quad t+5=0$
$t=6 \quad t=-5$
There are 6 teams in the league.

73.
$s=vt+16t^2$
$192=16t+16t^2$
$16t^2+16t-192=0$
$16(t^2+t-12)=0$
$16(t-3)(t+4)=0$
$t-3=0 \quad t+4=0$
$t=3 \quad t=-4$
The object will hit the ground 3 s later.

75. $h=vt-16t^2$
$0=60t-16t^2$
$0=4t(15-4t)$
$t=0 \quad 15-4t=0$
$-4t=-15$
$t=3.75$
The time cannot be 0.
The golf ball will return to the ground 3.75 s later.

77. Strategy
- Width of the rectangle: x
  Length of the rectangle: $2x+5$
  The area of the rectangle is 75 in^2.
- The equation for the area of a rectangle is $A = L \times W$. Substitute in the equation and solve for x.

Solution
$$A = L \times W$$
$$75 = (2x+5)(x)$$
$$75 = 2x^2 + 5x$$
$$0 = 2x^2 + 5x - 75$$
$$0 = (2x+15)(x-5)$$
$$2x - 15 = 0 \qquad x - 5 = 0$$
$$x = -\frac{15}{2} \qquad x = 5$$
Because the width cannot be a negative number, $-\frac{15}{2}$ is not a solution.
$x = 5$
$2x + 5 = 2(5) + 5 = 10 + 5 = 15$
The length is 15 in.
The width is 5 in.

79. Strategy
- Base of the triangle: x
  Height of the triangle: $2x+4$
  The area of the triangle is 35 m^2.
- The equation of the area of a triangle is $A = \frac{1}{2}bh$. Substitute in the equation and solve for the height.

Solution
$$A = \frac{1}{2}bh$$
$$35 = \frac{1}{2}(x)(2x+4)$$
$$70 = 2x^2 + 4x$$
$$2x^2 + 4x - 70 = 0$$
$$2(x^2 + 2x - 35) = 0$$
$$2(x+7)(x-5) = 0$$
$$x + 7 = 0 \qquad x - 5 = 0$$
$$x = -7 \qquad x = 5$$
The height cannot be a negative number.
$2x + 4 = 2(5) + 4 = 14$
The height of the triangle is 14 m.

81. Strategy
- The width of the uniform border: x
  New width: $6 - 2x$
  New length: $9 - 2x$
- The required area is 28 in^2. The equation for the area of a rectangle is $A = LW$.

Solution
$$(6-2x)(9-2x) = 28$$
$$(3-x)(9-2x) = 14$$
$$27 - 15x + 2x^2 = 14$$
$$2x^2 - 15x + 13 = 0$$
$$(x-1)(2x-13) = 0$$
$$x = 1 \qquad x = \frac{13}{2}$$
(Not possible)
$$6 - 2x = 6 - 2 = 4$$
$$9 - 2x = 9 - 2 = 7$$
The dimensions of the type area are 4 in. by 7 in.

83. Strategy
- Radius of the original circular lawn: x
  Radius of the larger circular lawn: $3 + x$
- To find the radius of the original lawn, subtract the area of the small circle from the area of the larger circle and set equal to 100.

Solution
$$A_2 - A_1 = 100$$
$$\pi(3+x)^2 - \pi x^2 = 100$$
$$\pi(9 + 6x + x^2) - \pi x^2 = 100$$
$$9\pi + 6\pi x + \pi x^2 - \pi x^2 = 100$$
$$9\pi + 6\pi x = 100$$
$$6\pi x = 100 - 9\pi$$
$$x = \frac{100 - 9\pi}{6\pi}$$
$$x \approx 3.81$$
The radius of the original circular lawn was approximately 3.81 ft.

Applying the Concepts

85.
$$n(n+5) = -4$$
$$n^2 + 5n + 4 = 0$$
$$(n+1)(n+4) = 0$$
$$n + 1 = 0 \qquad n + 4 = 0$$
$$n = -1 \qquad n = -4$$
$$3n^2 = 3(-1)^2 \qquad 3n^2 = 3(-4)^2$$
$$= 3(1) \qquad = 3(16)$$
$$= 3 \qquad = 48$$
$3n^2 = 3$ or 48.

87.
$$2y(y+4) = -5(y+3)$$
$$2y^2 + 8y = -5y - 15$$
$$2y^2 + 13y + 15 = 0$$
$$(2y+3)(y+5) = 0$$
$$2y + 3 = 0 \qquad y + 5 = 0$$
$$2y = -3 \qquad y = -5$$
$$y = -\frac{3}{2}$$
The solutions are $-\frac{3}{2}$ and -5.

89. $p^3 = 9p^2$
$p^3 - 9p^2 = 0$
$p^2(p-9) = 0$
$p^2 = 0 \quad p - 9 = 0$
$p = 0 \quad p = 9$
The solutions are 0 and 9.

91. The error in the solution
$(x+2)(x-3) = 6$
$x+2 = 6 \quad x-3 = 6$
$x = 4 \quad x = 9$
occurs when the assumption is made that because
$(x+2)(x-3) = 6$,
$x+2 = 6$ and/or $x-3 = 6$.
In other words, it is an error to say that if the product of two numbers is 6, then at least one of the numbers must be 6. The correction solution is
$(x+2)(x-3) = 6$
$x^2 - x - 6 = 6$
$x^2 - x - 12 = 0$
$(x+3)(x-4) = 0$
$x = -3 \quad x = 4$

Chapter 5 Review Exercises

1. $b^2 - 13b + 30$
$1 \times 30 = 30$ Factors of 30 whose sum is -13: -10 and -3
$b^2 - 10b - 3b + 30$
$b(b-10) - 3(b-10) = (b-10)(b-3)$

2. $4x(x-3) - 5(3-x)$
$4x(x-3) + 5(x-3) = (x-3)(4x+5)$

3. $2x^2 - 5x + 6$

Factors of 2	Negative Factors of 6
1 and 2	-1 and -6
	-2 and -3

Trial Factors	Middle Term
$(x-1)(2x-6)$	$-6x - 2x = -8x$
$(x-2)(2x-3)$	$-3x - 4x = -7x$
$(2x-1)(x-6)$	$-12x - x = -13x$
$(2x-2)(x-3)$	$-6x - 2x = -8x$

$2x^2 - 5x + 6$ is nonfactorable over the integers.

4. $5x^3 + 10x^2 + 35x$
The GCF is $5x$: $5x(x^2 + 2x + 7)$

5. $14y^9 - 49y^6 + 7y^3$
The GCF is $7y^3$: $7y^3(2y^6 - 7y^3 + 1)$

6. $y^2 + 5y - 36$
$1 \times (-36) = -36$ Factors of -36 whose sum is 5: 9 and -4
$y^2 + 9y - 4y - 36$
$y(y+9) - 4(y+9) = (y+9)(y-4)$

7. $6x^2 - 29x + 28$

Factors of 6	Negative Factors of 28
2 and 3	−1 and −28
1 and 6	−4 and −7

Trial Factors	Middle Term
$(x-1)(6x-28)$	$-28x - 6x = -34x$
$(x-4)(6x-7)$	$-7x - 24x = -31x$
$(2x-1)(6x-28)$	$-56x - 6x = -62x$
$(2x-4)(3x-7)$	$-14x - 12x = -26x$
$(2x-7)(3x-4)$	$-8x - 21x = -29x$

$6x^2 - 29x + 28 = (2x-7)(3x-4)$

8. $12a^2b + 3ab^2$
The GCF is $3ab$: $3ab(4a + b)$

9. $a^6 - 100 = (a^3)^2 - 10^2 = (a^3 + 10)(a^3 - 10)$

10. $n^4 - 2n^3 - 3n^2$
The GCF is n^2: $n^2(n^2 - 2n - 3)$ Factors of −3 whose sum is −2: −3 and 1
$1 \times (-3) = -3$ $n^2(n^2 - 3n + n - 3)$
$n^2[n(n-3) + 1(n-3)] = n^2(n-3)(n+1)$

11. $12y^2 + 16y - 3$

Factors of 12	Negative Factors of −3
1 and 12	1 and −3
2 and 6	−1 and 3
3 and 4	

Trial Factors	Middle Term
$(3y+1)(4y-3)$	$-9y + 4y = -5y$
$(3y-1)(4y+3)$	$9y - 4y = 5y$
$(2y+1)(6y-3)$	$-6y + 6y = 0$
$(6y-1)(2y+3)$	$18y - 2y = 16y$

$12y^2 + 16y - 3 = (6y-1)(2y+3)$

12. $12b^3 - 58b^2 + 56b$
The GCF is $2b$: $2b(6b^2 - 29b + 28)$
$6 \times 28 = 168$ Factors of 168 whose sum is −29: −21 and −8
$2b(6b^2 - 21b - 8b + 28)$
$2b[3b(2b-7) - 4(2b-7)]$
$2b(2b-7)(3b-4)$

$12b^3 - 58b^2 + 56b = 2b(2b-7)(3b-4)$

13. $9y^4 - 25z^2 = (3y^2)^2 - (5z)^2 = (3y^2 + 5z)(3y^2 - 5z)$

14. $c^2 + 8c + 12$
$1 \times 12 = 12$ Factors of 12 whose sum is 8: 6 and 2
$c^2 + 6c + 2c + 12$
$c(c+6) + 2(c+6) = (c+6)(c+2)$

15. $18a^2 - 3a - 10$
$18 \times (-10) = -180$ Factors of −180 whose sum is −3: −15 and 12
$18a^2 - 15a + 12a - 10$
$3a(6a-5) + 2(6a-5) = (6a-5)(3a+2)$

$18a^2 - 3a - 10 = (6a-5)(3a+2)$

16.
$$4x^2 + 27x = 7$$
$$4x^2 + 27x - 7 = 0$$
$$(4x - 1)(x + 7) = 0$$
$$4x - 1 = 0 \quad x + 7 = 0$$
$$4x = 1 \quad x = -7$$
$$x = \frac{1}{4}$$
The solutions are $\frac{1}{4}$ and -7.

17. $4x^3 - 20x^2 - 24x$
The GCF is $4x$: $4x(x^2 - 5x - 6)$
$1 \times (-6) = -6$ Factors of -6 whose sum is -5: -6 and 1
$$4x(x^2 - 5x - 6)$$
$$4x(x^2 - 6x + x - 6)$$
$$4x[x(x - 6) + 1(x - 6)] = 4x(x - 6)(x + 1)$$

18. $3a^2 - 15a - 42$
The GCF is 3: $3(a^2 - 5a - 14)$
$1 \times (-14) = -14$ Factors of -14 whose sum is -5: -7 and 2
$$3(a^2 - 7a + 2a - 14)$$
$$3[a(a - 7) + 2(a - 7)] = 3(a - 7)(a + 2)$$

19. $2a^2 - 19a - 60$
$2 \times (-60) = -120$ Factors of -120 whose sum is -19: -24 and 5
$$2a^2 - 24a + 5a - 60$$
$$2a(a - 12) + 5(a - 12) = (a - 12)(2a + 5)$$
$2a^2 - 19a - 60 = (a - 12)(2a + 5)$

20.
$$(x + 1)(x - 5) = 16$$
$$x^2 - 4x - 5 = 16$$
$$x^2 - 4x - 21 = 0$$
$$(x - 7)(x + 3) = 0$$
$$x - 7 = 0 \quad x + 3 = 0$$
$$x = 7 \quad x = -3$$
The solutions are 7 and -3.

21. $21ax - 35bx - 10by + 6ay$
$7x(3a - 5b) + 2y(-5b + 3a)$
$7x(3a - 5b) + 2y(3a - 5b) = (3a - 5b)(7x + 2y)$

22. $a^2b^2 - 1 = (ab)^2 - 1^2 = (ab + 1)(ab - 1)$

23. $10x^2 + 25x + 4xy + 10y$
$5x(2x + 5) + 2y(2x + 5) = (2x + 5)(5x + 2y)$

24. $5x^2 - 5x - 30$
The GCF is 5: $5(x^2 - x - 6)$
$1 \times (-6) = -6$ Factors of -6 whose sum is -1: -3 and 2
$$5(x^2 - 3x + 2x - 6)$$
$$5[x(x - 3) + 2(x - 3)] = 5(x - 3)(x + 2)$$
$5x^2 - 5x - 30 = 5(x - 3)(x + 2)$

25. $3x^2 + 36x + 108$
The GCF is 3: $3(x^2 + 12x + 36)$
$\sqrt{x^2} = x$ $\quad 2(x \cdot 6) = 12x$
$\sqrt{36} = 6$ $\quad$ The trinomial is a perfect square.
$$3(x + 6)^2$$

26. $3x^2 - 17x + 10$
$3 \times 10 = 30$ Factors of 30 whose sum is -17: -15 and -2
$$3x^2 - 15x - 2x + 10$$
$$3x(x - 5) - 2(x - 5) = (x - 5)(3x - 2)$$
$3x^2 - 17x + 10 = (x - 5)(3x - 2)$

27. Strategy
- Width: x
  Length: $2x - 20$
- Use the equation for the area of a rectangle: $A = LW$.

Solution

$$LW = A$$
$$x(2x - 20) = 6000$$
$$2x^2 - 20x - 6000 = 0$$
$$2(x^2 - 10x - 3000) = 0$$
$$2(x - 60)(x + 50) = 0$$
$$x - 60 = 0 \qquad x + 50 = 0$$
$$x = 60 \qquad x = -50$$

$2x - 20 = 120 - 20 = 100$
Because the width of a rectangle cannot be a negative number, -50 is not a solution.
The length is 100 yd.
The width is 60 yd.

28.
$$S = d^2$$
$$400 = d^2$$
$$\sqrt{400} = d$$
$$20 = d$$
The distance is 20 ft.

29. Strategy
- Width of the picture frame: x
  New width: $12 + 2x$
  New length: $15 + 2x$
- Use the equation for the area of a rectangle.

Solution

$$LW = A$$
$$(12 + 2x)(15 + 2x) = 270$$
$$180 + 24x + 30x + 4x^2 = 270$$
$$4x^2 + 54x - 90 = 0$$
$$2(2x^2 + 27x - 45) = 0$$
$$2(2x - 3)(x + 15) = 0$$
$$2x - 3 = 0 \qquad x + 15 = 0$$
$$2x = 3 \qquad x = -15$$
$$x = 1.5$$

Because the width of a picture frame cannot be a negative number, -15 is not a solution.
The width of the frame is 1.5 in.

30. Strategy
- Side of original square: x
  Side of new square: $x + 4$
- Use the equation for the area of a square.

Solution

$$s^2 = A$$
$$(x + 4)^2 = 576$$
$$x^2 + 8x + 16 = 576$$
$$x^2 + 8x - 560 = 0$$
$$(x + 28)(x - 20) = 0$$
$$x = 20$$

The side of the original garden plot was 20 ft.

Chapter 5 Test

1. $ab + 6a - 3b - 18$
$a(b + 6) - 3(b + 6)$
$(b + 6)(a - 3)$

2. $2y^4 - 14y^3 - 16y^2$
$2y^2(y^2 - 7x - 8)$
$2y^2(y^2 - 8y + y - 8)$
$2y^2[y(y - 8) + 1(y - 8)]$
$2y^2(y - 8)(y + 1)$

3. $8x^2 + 20x - 48$
The GCF is 4: $4(2x^2 + 5x - 12)$
$2 \times (-12) = -24$ Factors of -24 whose sum is 5: 8 and -3
$4(2x^2 + 8x - 3x - 12)$
$4(2x(x + 4) - 3(x + 4))$
$4(x + 4)(2x - 3)$

$8x^2 + 20x - 48 = 4(x + 4)(2x - 3)$

4. $6x^2 + 19x + 8$

Factors of 6:	Factors of 8:
1 and 6	1 and 8
2 and 3	2 and 4

Trial Factors	Middle Term
$(x + 1)(6x + 8)$	Common factor
$(x + 2)(6x + 4)$	Common factor
$(2x + 1)(3x + 8)$	$16x + 3x = 19x$

$6x^2 + 19x + 8 = (2x + 1)(3x + 8)$

5. $a^2 - 19a + 48$
$a^2 - 16a - 3a + 48$
$a(a - 16) - 3(a - 16)$
$(a - 16)(a - 3)$

6. $6x^3 - 8x^2 + 10x$
The GCF is $2x$: $2x(3x^2 - 4x + 5)$

7. $x^2 + 2x - 15$
$x^2 + 5x - 3x - 15$
$x(x + 5) - 3(x + 5)$
$(x + 5)(x - 3)$

8. $4x^2 - 1 = 0$
$(2x + 1)(2x - 1) = 0$
$$2x + 1 = 0 \qquad 2x - 1 = 0$$
$$2x = -1 \qquad 2x = 1$$
$$x = -\frac{1}{2} \qquad x = \frac{1}{2}$$

The solutions are $-\frac{1}{2}$ and $\frac{1}{2}$.

9. $5x^2 - 45x - 15$
$5(x^2 - 9x - 3)$

10. $p^2 + 12p + 36$ $\quad \sqrt{p^2} = p$ $\quad 2(p \cdot 6) = 12p$
$\sqrt{36} = 6$ The trinomial is a perfect square.

$p^2 + 12p + 36 = (p + 6)^2$

11. $$\begin{aligned} x(x-8) &= -15 \\ x^2 - 8x + 15 &= 0 \\ (x-3)(x-5) &= 0 \end{aligned}$$
$x - 3 = 0 \quad x - 5 = 0$
$x = 3 \quad x = 5$
The solutions are 3 and 5.

12. $$\begin{aligned} 3x^2 + 12xy + 12y^2 &= 3(x^2 + 4xy + 4y^2) \\ &= 3(x + 2y)^2 \end{aligned}$$

13. $b^2 - 16 = b^2 - 4^2 = (b+4)(b-4)$

14. $6x^2y^2 + 9xy^2 + 3y^2$
The GCF is $3y^2 : 3y^2(2x^2 + 3x + 1)$
$2 \times 1 = 2$ Factors of 2 whose sum is 3: 1 and 2
$3y^2(2x^2 + x + 2x + 1)$
$3y^2(x(2x+1) + 1(2x+1))$
$3y^2(2x+1)(x+1)$
$6x^2y^2 + 9xy^2 + 3y^2 = 3y^2(2x+1)(x+1)$

15. $p^2 + 5p + 6$
$p^2 + 3p + 2p + 6$
$p(p+3) + 2(p+3)$
$(p+3)(p+2)$

16. $a(x-2) + b(x-2)$
$(x-2)(a+b)$

17. $x(p+1) - 1(p+1)$
$(p+1)(x-1)$

18. $3a^2 - 75 = 3(a^2 - 25) = 3(a+5)(a-5)$

19. $2x^2 + 4x - 5$
Factors of 2: 1 and 2
Factors of -5: 1 and -5
-1 and 5

$(x-1)(2x+5) \quad 5x - 2x = 3x$
$(x+1)(2x-5) \quad -5x + 2x = -3x$
Nonfactorable over the integers.

20. $x^2 - 9x - 36$
$x^2 - 12x + 3x - 36$
$x(x-12) + 3(x-12)$
$(x-12)(x+3)$

21. $4a^2 - 12ab + 9b^2 \quad \sqrt{4a^2} = 2a$
$2(2a)(3b) = 12ab \quad \sqrt{9b^2} = 3b$ The trinomial is a perfect square.

$4a^2 - 12ab + 9b^2 = (2a - 3b)^2$

22. $4x^2 - 49y^2 = (2x)^2 - (7y)^2 = (2x + 7y)(2x - 7y)$

23. $(2a-3)(a+7) = 0$
$2a - 3 = 0 \quad a + 7 = 0$
$2a = 3 \quad a = -7$
$a = \frac{3}{2}$
The solutions are $\frac{3}{2}$ and -7.

24. Strategy
- One number: x
  Other number: $10 - x$
- The sum of the squares of the two numbers is 58.

Solution
$$\begin{aligned} x^2 + (10-x)^2 &= 58 \\ x^2 + 100 - 20x + x^2 &= 58 \\ 2x^2 - 20x + 42 &= 0 \\ 2(x^2 - 10x + 21) &= 0 \\ 2(x-7)(x-3) &= 0 \end{aligned}$$
$x - 7 = 0 \quad x - 3 = 0$
$x = 7 \quad x = 3$
The two numbers are 7 and 3.

25. Strategy
- Width: x
  Length: $2x + 3$
- Use the equation for the area of a rectangle.

Solution
$$\begin{aligned} LW &= A \\ x(2x+3) &= 90 \\ 2x^2 + 3x - 90 &= 0 \\ (2x+15)(x-6) &= 0 \\ x - 6 &= 0 \\ x &= 6 \\ 2x + 3 = 12 + 3 &= 15 \end{aligned}$$
The length is 15 cm.
The width is 6 cm.

Cumulative Review Exercises

1. $$\begin{aligned} -2 - (-3) - 5 - (-11) &= -2 + 3 - 5 + 11 \\ &= -2 - 5 + 3 + 11 \\ &= -7 + 14 \\ &= 7 \end{aligned}$$

2. $$\begin{aligned} (3-7)^2 \div (-2) - 3 \cdot (-4) &= (-4)^2 \div (-2) - 3 \cdot (-4) \\ &= 16 \div (-2) - (-12) \\ &= -8 + 12 \\ &= 4 \end{aligned}$$

3. $$\begin{aligned} -2a^2 \div (2b) - c &= -2(-4)^2 \div 2(2) - (-1) \\ &= -2(16) \div 4 + 1 \\ &= -32 \div 4 + 1 \\ &= -8 + 1 \\ &= -7 \end{aligned}$$

4. $\left(\frac{-3}{\cancel{4}_{1}}\right)\left(\overset{-5}{\cancel{-20}}x^2\right) = 15x^2$

5. $$\begin{aligned} -2[4x - 2(3-2x) - 8x] &= -2[4x - 6 + 4x - 8x] \\ &= -2[-6] \\ &= 12 \end{aligned}$$

6. $-\frac{5}{7}x = \frac{-10}{21}$

$x = \frac{\overset{-2}{\cancel{-10}}}{\underset{3}{\cancel{21}}} \times \frac{\overset{-1}{\cancel{-7}}}{\underset{1}{\cancel{5}}} = \frac{2}{3}$

7. $3x - 2 = 12 - 5x$
$3x + 5x = 12 + 2$
$8x = 14$
$x = \frac{14}{8} = \frac{7}{4}$

8. $-2 + 4[3x - 2(4 - x) - 3] = 4x + 2$
$-2 + 4[3x - 8 + 2x - 3] = 4x + 2$
$-2 + 4[5x - 11] = 4x + 2$
$-2 + 20x - 44 = 4x + 2$
$20x - 46 = 4x + 2$
$16x = 48$
$x = 3$

9. $P \times B = A$
$120\% \times B = 54$
$1.2B = 54$
$B = \frac{54}{1.2}$
$B = 45$
The number is 45.

10. $(-3a^3b^2)^2 = (-3)^2a^6b^4$
$= 9a^6b^4$

11. $(x+2)(x^2 - 5x + 4)$
$x(x^2 - 5x + 4) + 2(x^2 - 5x + 4)$

$$\begin{array}{r} x^3 - 5x^2 + 4x \phantom{{}+8} \\ + \quad 2x^2 - 10x + 8 \\ \hline x^3 - 3x^2 - 6x + 8 \end{array}$$

12.

$$\begin{array}{r} 4x + 8 \\ 2x - 3 \overline{)8x^2 + 4x - 3} \\ \underline{8x^2 - 12x} \phantom{{}-3} \\ 16x - 3 \\ \underline{16x - 24} \\ 21 \end{array}$$

$(8x^2 + 4x - 3) \div (2x - 3) = 4x + 8 + \frac{21}{2x - 3}$

13. $(x^{-4}y^3)^2 = x^{-8}y^6 = \frac{y^6}{x^8}$

14. $3a - 3b - ax + bx = 3(a - b) - x(a - b)$
$= (a - b)(3 - x)$

15. $15xy^2 - 20xy^4$
The GCF is $5xy^2$: $5xy^2(3 - 4y^2)$

16. $x^2 - 5xy - 14y^2 = x^2 - 7xy + 2xy - 14y^2$
$= x(x - 7y) + 2y(x - 7y)$
$= (x - 7y)(x + 2y)$

17. $p^2 - 9p - 10 = p^2 - 10p + p - 10$
$= p(p - 10) + 1(p - 10)$
$= (p - 10)(p + 1)$

18. $18a^3 + 57a^2 + 30a = 3a(6a^2 + 19a + 10)$
$= 3a(3a + 2)(2a + 5)$

19. $36a^2 - 49b^2 = (6a)^2 - (7b)^2$
$= (6a + 7b)(6a - 7b)$

20. $4x^2 + 28xy + 49y^2$
$\sqrt{4x^2} = 2x \quad 2(2x \cdot 7y) = 28xy$
$\sqrt{49y^2} = 7y$
The trinomial is a perfect square.
$4x^2 + 28xy + 49y^2 = (2x + 7y)^2$

21. $9x^2 + 15x - 14$
$9(-14) = -126$
Factors of -126 whose sum is 15: 21 and -6
$9x^2 + 21x - 6x - 14 = 3x(3x + 7) - 2(3x - 7)$
$= (3x + 7)(3x - 2)$

22. $18x^2 - 48xy + 32y^2$
The GCF is 2: $2(9x^2 - 24xy + 16y^2)$
$\sqrt{9x^2} = 3x \quad -2(3x \cdot 4y) = -24y$
$\sqrt{16y^2} = 4y$
The trinomial is a perfect square.
$18x^2 - 48xy + 32y^2 = 2(3x - 4y)^2$

23. $3y(x - 3) - 2(x - 3)$
$(x - 3)(3y - 2)$

24. $3x^2 + 19x - 14 = 0$
$(3x - 2)(x + 7) = 0$
$3x - 2 = 0 \quad x + 7 = 0$
$3x = 2 \quad x = -7$
$x = \frac{2}{3}$
The solutions are $\frac{2}{3}$ and -7.

25. Strategy
- Shorter pieces: x
  Longer pieces: $10 - x$
- Four times the length of the shorter pieces is 2 ft less than three times the length of the longer piece.

Solution
$4x = 3(10 - x) - 2$
$4x = 30 - 3x - 2$
$4x = 28 - 3x$
$7x = 28$
$x = 4$
$10 - x = 10 - 4 = 6$
The shorter piece is 4 ft long.
The longer piece is 6 ft long.

26. Strategy Given: $S = \$99$ $R = \$165$
- Unknown: r
- Use the equation $S = R - rR$.

Solution
$S = R - rR$
$99 = 165 - 165r$
$-66 = -165r$
$\frac{-66}{-165} = r$
$0.4 = r$
The discount rate is 40%.

27. Strategy • To find the measure of $\angle a$, use the fact that $\angle a$ and the 72° angle are the alternate interior angles of parallel lines.
• To find the measure of $\angle b$, use the fact that $\angle a$ and $\angle b$ are supplementary angles.

Solution
$$m\angle a = 72^\circ$$
$$m\angle a + m\angle b = 180^\circ$$
$$72^\circ + m\angle b = 180^\circ$$
$$m\angle b = 108^\circ$$
$$m\angle a = 72^\circ$$
$$m\angle b = 108^\circ$$

28. Strategy • Time driving to resort: x

	Rate	Time	Distance
To resort	42	x	$42x$
From resort	56	$7-x$	$56(7-x)$

• The distances are the same.

Solution
$$42x = 56(7-x)$$
$$42x = 392 - 56x$$
$$98x = 392$$
$$x = 4$$
$$42x = 168$$
The distance to the resort is 168 mi.

29. Strategy • First integer: n
Middle integer: $n+2$
Third integer: $n+4$
• Five times the middle integer is twelve more than twice the sum of the first and third.

Solution
$$5(n+2) = 2(n+n+4) + 12$$
$$5n + 10 = 2(2n+4) + 12$$
$$5n + 10 = 4n + 8 + 12$$
$$5n + 10 = 4n + 20$$
$$n = 10$$
$$n + 2 = 12$$
$$n + 4 = 14$$
The integers are 10, 12, and 14.

30. Strategy • Height: x
Base: $3x$
• Use the equation for the area of a triangle.

Solution
$$\frac{1}{2}bh = A$$
$$\frac{1}{2}(3x)(x) = 24$$
$$\frac{3}{2}x^2 = 24$$
$$x^2 = \overset{8}{\cancel{24}} \times \frac{2}{\cancel{3}}$$
$$x^2 = 16$$
$$x = 4$$
$$3x = 12$$
The length of the base of the triangle is 12 in.

Chapter 6: Rational Expressions

Prep Test

1. $12 = 2 \cdot 2 \cdot 3$
$18 = 2 \cdot 3 \cdot 3$
$LCM = 2 \cdot 2 \cdot 3 \cdot 3 = 36$

2. $\dfrac{9x^3y^4}{3x^2y^7} = \dfrac{3x^{3-2}}{y^{7-4}} = \dfrac{3x}{y^3}$

3. $\dfrac{3}{4} - \dfrac{8}{9} = \dfrac{3 \cdot 9}{4 \cdot 9} - \dfrac{8 \cdot 4}{9 \cdot 4}$
$= \dfrac{27}{36} - \dfrac{32}{36}$
$= -\dfrac{5}{36}$

4. $\left(-\dfrac{8}{11}\right) \div \dfrac{4}{5} = \left(-\dfrac{\overset{2}{\cancel{8}}}{11}\right) \cdot \dfrac{5}{\underset{1}{\cancel{4}}} = -\dfrac{10}{11}$

5. No. $\dfrac{0}{a} = 0$, but $\dfrac{a}{0}$ is undefined.

6. $\dfrac{2}{3}x - \dfrac{3}{4} = \dfrac{5}{6}$
$12\left(\dfrac{2}{3}x - \dfrac{3}{4}\right) = \left(\dfrac{5}{6}\right) \cdot 12$
$8x - 9 = 10$
$8x = 19$
$x = \dfrac{19}{8}$

7. $130°$

8. $x^2 - 4x - 12 = (x-6)(x+2)$

9. $2x^2 - x - 3 = (2x-3)(x+1)$

10. **Strategy**
- Anthony's time running: t
- Jean's time running: $t - 10$

	Rate	Time	Distance
Anthony	9	t	$9t$
Jean	12	$t - 10$	$12(t-10)$

- The distance traveled is the same.

Solution
$9t = 12(t - 10)$
$9t = 12t - 120$
$-3t = -120$
$t = 40$ min
Jean will catch up with Anthony at 9:40 A.M.

Go Figure

Strategy
- Find the difference between the distances each mouse travels.

	Rate	Time	Distance
mouse 1	2	24	2(24)
mouse 2	3	18	3(18)

Solution $54 - 48 = 6$
The mice are 6 ft apart 18 seconds after the second mouse starts.

Section 6.1

Objective A Exercises

1. To write a rational expression in simplest form, factor the numerator and denominator. Then divide by the common factors.

3. $\dfrac{9x^3}{12x^4} = \dfrac{\overset{1}{\cancel{3}} \cdot 3x^3}{2 \cdot 2 \cdot \underset{1}{\cancel{3}}x^4} = \dfrac{3}{4x}$

5. $\dfrac{(x+3)^2}{(x+3)^3} = \dfrac{1}{(x+3)^{3-2}} = \dfrac{1}{x+3}$

7. $\dfrac{3n-4}{4-3n} = \dfrac{-(4-3n)}{4-3n} = -1$

9. $\dfrac{6y(y+2)}{9y^2(y+2)} = \dfrac{2 \cdot \overset{1}{\cancel{3}}y\overset{1}{\cancel{(y+2)}}}{\underset{1}{\cancel{3}} \cdot 3y^2\underset{1}{\cancel{(y+2)}}} = \dfrac{2}{3y}$

11. $\dfrac{6x(x-5)}{8x^2(5-x)} = \dfrac{\overset{1}{\cancel{2}} \cdot 3x\overset{-1}{\cancel{(x-5)}}}{\underset{1}{\cancel{2}} \cdot 2 \cdot 2x^2\underset{1}{\cancel{(5-x)}}} = -\dfrac{3}{4x}$

13. $\dfrac{a^2+4a}{ab+4b} = \dfrac{a\overset{1}{\cancel{(a+4)}}}{b\underset{1}{\cancel{(a+4)}}} = \dfrac{a}{b}$

15. $\dfrac{4-6x}{3x^2-2x} = \dfrac{2\overset{-1}{\cancel{(2-3x)}}}{x\underset{1}{\cancel{(3x-2)}}} = -\dfrac{2}{x}$

17. $\dfrac{y^2-3y+2}{y^2-4y+3} = \dfrac{\overset{1}{\cancel{(y-1)}}(y-2)}{\underset{1}{\cancel{(y-1)}}(y-3)} = \dfrac{y-2}{y-3}$

19. $\dfrac{x^2+3x-10}{x^2+2x-8} = \dfrac{(x+5)\overset{1}{\cancel{(x-2)}}}{(x+4)\underset{1}{\cancel{(x-2)}}} = \dfrac{x+5}{x+4}$

21. $\dfrac{x^2+x-12}{x^2-6x+9} = \dfrac{(x+4)\overset{1}{\cancel{(x-3)}}}{(x-3)\underset{1}{\cancel{(x-3)}}} = \dfrac{x+4}{x-3}$

23. $\dfrac{x^2-3x-10}{25-x^2} = \dfrac{(x+2)\overset{-1}{\cancel{(x-5)}}}{(5+x)\underset{1}{\cancel{(5-x)}}} = -\dfrac{x+2}{x+5}$

25. $\dfrac{2x^3+2x^2-4x}{x^3+2x^2-3x}=\dfrac{2x(x^2+x-2)}{x(x^2+2x-3)}$

$=\dfrac{2x(x+2)\overset{1}{\cancel{(x-1)}}}{x(x+3)\underset{1}{\cancel{(x-1)}}}$

$=\dfrac{2(x+2)}{x+3}$

27. $\dfrac{6x^2-7x+2}{6x^2+5x-6}=\dfrac{(2x-1)\overset{1}{\cancel{(3x-2)}}}{(2x+3)\underset{1}{\cancel{(3x-2)}}}=\dfrac{2x-1}{2x+3}$

29. $\dfrac{x^2-3x-28}{24-2x-x^2}=\dfrac{(x+7)\overset{-1}{\cancel{(x-4)}}}{(6+x)\underset{1}{\cancel{(4-x)}}}=-\dfrac{x+7}{x+6}$

Objective B Exercises

31. $\dfrac{14a^2b^3}{15x^5y^2}\cdot\dfrac{25x^3y}{16ab}=\dfrac{\overset{1}{\cancel{2}}\cdot7\cdot a^2b^3\cdot\overset{1}{\cancel{5}}\cdot5x^3y}{3\cdot\underset{1}{\cancel{5}}x^5y^2\cdot\underset{1}{\cancel{2}}\cdot2\cdot2\cdot2ab}=\dfrac{35ab^2}{24x^2y}$

33. $\dfrac{18a^4b^2}{25x^2y^3}\cdot\dfrac{50x^5y^6}{27a^6b^2}=\dfrac{2\cdot\overset{1}{\cancel{3}}\cdot\overset{1}{\cancel{3}}a^4b^2\cdot2\cdot\overset{1}{\cancel{5}}\cdot\overset{1}{\cancel{5}}x^5y^6}{\underset{1}{\cancel{5}}\cdot\underset{1}{\cancel{5}}x^2y^3\cdot\underset{1}{\cancel{3}}\cdot\underset{1}{\cancel{3}}\cdot3a^6b^2}=\dfrac{4x^3y^3}{3a^2}$

35. $\dfrac{8x-12}{14x+7}\cdot\dfrac{42x+21}{32x-48}=\dfrac{4(2x-3)}{7(2x+1)}\cdot\dfrac{21(2x+1)}{16(2x-3)}=\dfrac{\overset{1}{\cancel{2}}\cdot\overset{1}{\cancel{2}}\overset{1}{\cancel{(2x-3)}}\cdot3\cdot\overset{1}{\cancel{7}}\overset{1}{\cancel{(2x+1)}}}{\underset{1}{\cancel{7}}\underset{1}{\cancel{(2x+1)}}\cdot\underset{1}{\cancel{2}}\cdot\underset{1}{\cancel{2}}\cdot2\cdot2\underset{1}{\cancel{(2x-3)}}}=\dfrac{3}{4}$

37. $\dfrac{4a^2x-3a^2}{2by+5b}\cdot\dfrac{2b^3y+5b^3}{4ax-3a}=\dfrac{a^2(4x-3)}{b(2y+5)}\cdot\dfrac{b^3(2y+5)}{a(4x-3)}=\dfrac{a^2\overset{1}{\cancel{(4x-3)}}b^3\overset{1}{\cancel{(2y+5)}}}{b\underset{1}{\cancel{(2y+5)}}a\underset{1}{\cancel{(4x-3)}}}=ab^2$

39. $\dfrac{x^2+x-2}{xy^2}\cdot\dfrac{x^3y}{x^2+5x+6}=\dfrac{(x+2)(x-1)}{xy^2}\cdot\dfrac{x^3y}{(x+2)(x+3)}=\dfrac{\overset{1}{\cancel{(x+2)}}(x-1)x^3y}{xy^2\underset{1}{\cancel{(x+2)}}(x+3)}=\dfrac{x^2(x-1)}{y(x+3)}$

41. $\dfrac{x^5y^3}{x^2+13x+30}\cdot\dfrac{x^2+2x-3}{x^7y^2}=\dfrac{x^5y^3}{(x+3)(x+10)}\cdot\dfrac{(x+3)(x-1)}{x^7y^2}=\dfrac{x^5y^3\overset{1}{\cancel{(x+3)}}(x-1)}{\underset{1}{\cancel{(x+3)}}(x+10)x^7y^2}=\dfrac{y(x-1)}{x^2(x+10)}$

43. $\dfrac{3a^3+4a^2}{5ab-3b}\cdot\dfrac{3b^3-5ab^3}{3a^2+4a}=\dfrac{a^2(3a+4)}{b(5a-3)}\cdot\dfrac{b^3(3-5a)}{a(3a+4)}=\dfrac{a^2\overset{1}{\cancel{(3a+4)}}b^3\overset{-1}{\cancel{(3-5a)}}}{b\underset{1}{\cancel{(5a-3)}}a\underset{1}{\cancel{(3a+4)}}}=-ab^2$

45. $\dfrac{x^2-8x+7}{x^2+3x-4}\cdot\dfrac{x^2+3x-10}{x^2-9x+14}=\dfrac{(x-1)(x-7)}{(x+4)(x-1)}\cdot\dfrac{(x+5)(x-2)}{(x-2)(x-7)}=\dfrac{\overset{1}{\cancel{(x-1)}}\overset{1}{\cancel{(x-7)}}(x+5)\overset{1}{\cancel{(x-2)}}}{(x+4)\underset{1}{\cancel{(x-1)}}\underset{1}{\cancel{(x-2)}}\underset{1}{\cancel{(x-7)}}}=\dfrac{x+5}{x+4}$

47. $\dfrac{y^2+y-20}{y^2+2y-15}\cdot\dfrac{y^2+4y-21}{y^2+3y-28}=\dfrac{(y+5)(y-4)}{(y+5)(y-3)}\cdot\dfrac{(y+7)(y-3)}{(y+7)(y-4)}=\dfrac{\overset{1}{\cancel{(y+5)}}\overset{1}{\cancel{(y-4)}}\overset{1}{\cancel{(y+7)}}\overset{1}{\cancel{(y-3)}}}{\underset{1}{\cancel{(y+5)}}\underset{1}{\cancel{(y-3)}}\underset{1}{\cancel{(y+7)}}\underset{1}{\cancel{(y-4)}}}=1$

49. $\dfrac{25-n^2}{n^2-2n-35}\cdot\dfrac{n^2-8n-20}{n^2-3n-10}=\dfrac{(5+n)(5-n)}{(n+5)(n-7)}\cdot\dfrac{(n+2)(n-10)}{(n+2)(n-5)}=\dfrac{\overset{1}{\cancel{(5+n)}}\overset{-1}{\cancel{(5-n)}}\overset{1}{\cancel{(n+2)}}(n-10)}{\underset{1}{\cancel{(n+5)}}(n-7)\underset{1}{\cancel{(n+2)}}\underset{1}{\cancel{(n-5)}}}=-\dfrac{n-10}{n-7}$

51. $\dfrac{8x^3+4x^2}{x^2-3x+2}\cdot\dfrac{x^2-4}{16x^2+8x}=\dfrac{4x^2(2x+1)}{(x-1)(x-2)}\cdot\dfrac{(x+2)(x-2)}{8x(2x+1)}=\dfrac{\overset{1}{\cancel{2}}\cdot\overset{1}{\cancel{2}}x^2\overset{1}{\cancel{(2x+1)}}(x+2)\overset{1}{\cancel{(x-2)}}}{(x-1)\underset{1}{\cancel{(x-2)}}\cdot\underset{1}{\cancel{2}}\cdot\underset{1}{\cancel{2}}\cdot2x\underset{1}{\cancel{(2x+1)}}}=\dfrac{x(x+2)}{2(x-1)}$

53. $\dfrac{x^2-11x+28}{x^2-13x+42}\cdot\dfrac{x^2+7x+10}{20-x-x^2}=\dfrac{(x-4)(x-7)}{(x-6)(x-7)}\cdot\dfrac{(x+2)(x+5)}{(5+x)(4-x)}=\dfrac{\overset{-1}{\cancel{(x-4)}}\overset{1}{\cancel{(x-7)}}(x+2)\overset{1}{\cancel{(x+5)}}}{(x-6)\underset{1}{\cancel{(x-7)}}\underset{1}{\cancel{(5+x)}}\underset{1}{\cancel{(4-x)}}}=-\dfrac{x+2}{x-6}$

55. $\dfrac{x^2-4x-32}{x^2-8x-48}\cdot\dfrac{3x^2+17x+10}{3x^2-22x-16}=\dfrac{(x+4)(x-8)}{(x+4)(x-12)}\cdot\dfrac{(3x+2)(x+5)}{(3x+2)(x-8)}=\dfrac{(x+4)(x-8)(3x+2)(x+5)}{(x+4)(x-12)(3x+2)(x-8)}=\dfrac{x+5}{x-12}$

Objective C Exercises

57. To divide rational expressions, multiply by the reciprocal of the divisor.

59. $\dfrac{9x^3y^4}{16a^4b^2}\div\dfrac{45x^4y^2}{14a^7b}=\dfrac{9x^3y^4}{16a^4b^2}\cdot\dfrac{14a^7b}{45x^4y^2}=\dfrac{3\cdot 3x^3y^4\cdot 2\cdot 7a^7b}{2\cdot 2\cdot 2\cdot 2a^4b^2\cdot 3\cdot 3\cdot 5x^4y^2}=\dfrac{7a^3y^2}{40bx}$

61. $\dfrac{28x+14}{45x-30}\div\dfrac{14x+7}{30x-20}=\dfrac{28x+14}{45x-30}\cdot\dfrac{30x-20}{14x+7}=\dfrac{14(2x+1)}{15(3x-2)}\cdot\dfrac{10(3x-2)}{7(2x+1)}=\dfrac{2\cdot 7(2x+1)\cdot 2\cdot 5(3x-2)}{3\cdot 5(3x-2)\cdot 7(2x+1)}=\dfrac{4}{3}$

63. $\dfrac{5a^2y+3a^2}{2x^3+5x^2}\div\dfrac{10ay+6a}{6x^3+15x^2}=\dfrac{5a^2y+3a^2}{2x^3+5x^2}\cdot\dfrac{6x^3+15x^2}{10ay+6a}=\dfrac{a^2(5y+3)}{x^2(2x+5)}\cdot\dfrac{3x^2(2x+5)}{2a(5y+3)}=\dfrac{3a}{2}$

65. $\dfrac{x^3y^2}{x^2-3x-10}\div\dfrac{xy^4}{x^2-x-20}=\dfrac{x^3y^2}{x^2-3x-10}\cdot\dfrac{x^2-x-20}{xy^4}=\dfrac{x^3y^2}{(x+2)(x-5)}\cdot\dfrac{(x+4)(x-5)}{xy^4}=\dfrac{x^2(x+4)}{y^2(x+2)}$

67. $\dfrac{x^2y^5}{x^2-11x+30}\div\dfrac{xy^6}{x^2-7x+10}=\dfrac{x^2y^5}{x^2-11x+30}\cdot\dfrac{x^2-7x+10}{xy^6}=\dfrac{x^2y^5}{(x-5)(x-6)}\cdot\dfrac{(x-2)(x-5)}{xy^6}=\dfrac{x(x-2)}{y(x-6)}$

69. $\dfrac{3x^2y-9xy}{a^2b}\div\dfrac{3x^2-x^3}{ab^2}=\dfrac{3x^2y-9xy}{a^2b}\cdot\dfrac{ab^2}{3x^2-x^3}=\dfrac{3xy(x-3)}{a^2b}\cdot\dfrac{ab^2}{x^2(3-x)}=-\dfrac{3by}{ax}$

71. $\dfrac{x^2+3x-40}{x^2+2x-35}\div\dfrac{x^2+2x-48}{x^2+3x-18}=\dfrac{x^2+3x-40}{x^2+2x-35}\cdot\dfrac{x^2+3x-18}{x^2+2x-48}=\dfrac{(x+8)(x-5)}{(x+7)(x-5)}\cdot\dfrac{(x+6)(x-3)}{(x+8)(x-6)}=\dfrac{(x+6)(x-3)}{(x+7)(x-6)}$

73. $\dfrac{y^2-y-56}{y^2+8y+7}\div\dfrac{y^2-13y+40}{y^2-4y-5}=\dfrac{y^2-y-56}{y^2+8y+7}\cdot\dfrac{y^2-4y-5}{y^2-13y+40}=\dfrac{(y+7)(y-8)}{(y+1)(y+7)}\cdot\dfrac{(y+1)(y-5)}{(y-5)(y-8)}=1$

75. $\dfrac{x^2-x-2}{x^2-7x+10}\div\dfrac{x^2-3x-4}{40-3x-x^2}=\dfrac{x^2-x-2}{x^2-7x+10}\cdot\dfrac{40-3x-x^2}{x^2-3x-4}=\dfrac{(x+1)(x-2)}{(x-2)(x-5)}\cdot\dfrac{(8+x)(5-x)}{(x+1)(x-4)}=-\dfrac{x+8}{x-4}$

77. $\dfrac{6n^2+13n+6}{4n^2-9}\div\dfrac{6n^2+n-2}{4n^2-1}=\dfrac{6n^2+13n+6}{4n^2-9}\cdot\dfrac{4n^2-1}{6n^2+n-2}=\dfrac{(2n+3)(3n+2)}{(2n+3)(2n-3)}\cdot\dfrac{(2n+1)(2n-1)}{(3n+2)(2n-1)}=\dfrac{2n+1}{2n-3}$

Applying the Concepts

79. Choosing a value of y very close to 3 makes $y-3$ very close to 0. Dividing 1 by a number close to 0 produces a very large number. For instance, when $y = 3.00000001$, $\dfrac{1}{y-3}$ is greater than 10,000,000.

81. $\dfrac{\frac{1}{2}(2x)(x+4)}{(x+8)(x+4)}=\dfrac{x}{x+8}$

83. $\dfrac{n}{n+3} \div ? = \dfrac{n}{n-2}$

$\dfrac{n}{n+3} = \dfrac{n}{n-2} \cdot ?$

$? = \dfrac{n}{n+3} \div \dfrac{n}{n-2}$

$= \dfrac{\overset{1}{\cancel{n}}}{n+3} \cdot \dfrac{n-2}{\underset{1}{\cancel{n}}}$

$= \dfrac{n-2}{n+3}$

Section 6.2

Objective A Exercises

1. $8x^3y = 2 \cdot 2 \cdot 2 \cdot x \cdot x \cdot x \cdot y$
$12xy^2 = 2 \cdot 2 \cdot 3 \cdot x \cdot y \cdot y$
$\text{LCM} = 2 \cdot 2 \cdot 2 \cdot 3 \cdot x \cdot x \cdot x \cdot y \cdot y = 24x^3y^2$

3. $10x^4y^2 = 2 \cdot 5 \cdot x \cdot x \cdot x \cdot x \cdot y \cdot y$
$15x^3y = 3 \cdot 5 \cdot x \cdot x \cdot x \cdot y$
$\text{LCM} = 2 \cdot 3 \cdot 5 \cdot x \cdot x \cdot x \cdot x \cdot y \cdot y = 30x^4y^2$

5. $8x^2 = 2 \cdot 2 \cdot 2 \cdot x \cdot x$
$4x^2 + 8x = 4x(x+2) = 2 \cdot 2 \cdot x(x+2)$
$\text{LCM} = 2 \cdot 2 \cdot 2 \cdot x \cdot x(x+2) = 8x^2(x+2)$

7. $2x^2y = 2 \cdot x \cdot x \cdot y$
$3x^2 + 12x = 3x(x+4)$
$\text{LCM} = 2 \cdot 3 \cdot x \cdot x \cdot y(x+4) = 6x^2y(x+4)$

9. $9x(x+2) = 3 \cdot 3 \cdot x(x+2)$
$12(x+2)^2 = 2 \cdot 2 \cdot 3(x+2)(x+2)$
$\text{LCM} = 2 \cdot 2 \cdot 3 \cdot 3 \cdot x(x+2)(x+2) = 36x(x+2)^2$

11. $3x + 3 = 3(x+1)$
$2x^2 + 4x + 2 = 2(x^2 + 2x + 1) = 2(x+1)(x+1)$
$\text{LCM} = 2 \cdot 3(x+1)(x+1) = 6(x+1)^2$

13. $(x-1)(x+2)$
$(x-1)(x+3)$
$\text{LCM} = (x-1)(x+2)(x+3)$

15. $(2x+3)^2 = (2x+3)(2x+3)$
$(2x+3)(x-5)$
$\text{LCM} = (2x+3)(2x+3)(x-5) = (2x+3)^2(x-5)$

17. $(x-1)$
$(x-2)$
$(x-1)(x-2)$
$\text{LCM} = (x-1)(x-2)$

19. $x^2 - x - 6 = (x+2)(x-3)$
$x^2 + x - 12 = (x-3)(x+4)$
$\text{LCM} = (x+2)(x-3)(x+4)$

21. $x^2 + 5x + 4 = (x+1)(x+4)$
$x^2 - 3x - 28 = (x+4)(x-7)$
$\text{LCM} = (x+1)(x+4)(x-7)$

23. $x^2 - 2x - 24 = (x+4)(x-6)$
$x^2 - 36 = (x+6)(x-6)$
$\text{LCM} = (x+4)(x-6)(x+6)$

25. $x^2 - 7x - 30 = (x+3)(x-10)$
$x^2 - 5x - 24 = (x+3)(x-8)$
$\text{LCM} = (x+3)(x-10)(x-8)$

27. $3x^2 - 11x + 6 = (x-3)(3x-2)$
$3x^2 + 4x - 4 = (x+2)(3x-2)$
$\text{LCM} = (x-3)(3x-2)(x+2)$

29. $6 + x - x^2 = (2+x)(3-x)$
$x+2$
$x-3$
$\text{LCM} = (x+2)(x-3)$

31. $5 + 4x - x^2 = (1+x)(5-x)$
$x-5$
$x+1$
$\text{LCM} = (x+1)(x-5)$

33. $x^2 - 5x + 6 = (x-2)(x-3)$
$1-x$
$x-6$
$\text{LCM} = (x-1)(x-2)(x-3)(x-6)$

Objective B Exercises

35. The LCM is ab^2.
$\dfrac{5}{ab^2} = \dfrac{5}{ab^2}$
$\dfrac{6}{ab} = \dfrac{6}{ab} \cdot \dfrac{b}{b} = \dfrac{6b}{ab^2}$

37. The LCM is $18x^2y$.
$\dfrac{5y}{6x^2} = \dfrac{5y}{6x^2} \cdot \dfrac{3y}{3y} = \dfrac{15y^2}{18x^2y}$
$\dfrac{7}{9xy} = \dfrac{7}{9xy} \cdot \dfrac{2x}{2x} = \dfrac{14x}{18x^2y}$

39. The LCM is $y^2(y+5)$.
$\dfrac{a}{y^2} = \dfrac{a}{y^2} \cdot \dfrac{y+5}{y+5} = \dfrac{ay+5a}{y^2(y+5)}$
$\dfrac{6}{y(y+5)} = \dfrac{6}{y(y+5)} \cdot \dfrac{y}{y} = \dfrac{6y}{y^2(y+5)}$

41. The LCM is $y(y+7)^2$.
$\dfrac{a^2}{y(y+7)} = \dfrac{a^2}{y(y+7)} \cdot \dfrac{y+7}{y+7} = \dfrac{a^2y+7a^2}{y(y+7)^2}$
$\dfrac{a}{(y+7)^2} = \dfrac{a}{(y+7)^2} \cdot \dfrac{y}{y} = \dfrac{ay}{y(y+7)^2}$

43. $\dfrac{b^2}{4-y} = \dfrac{b^2}{-(y-4)} = -\dfrac{b^2}{y-4}$
The LCM is $y(y-4)$.
$\dfrac{b}{y(y-4)} = \dfrac{b}{y(y-4)}$
$\dfrac{b^2}{4-y} = -\dfrac{b^2}{y-4} \cdot \dfrac{y}{y} = -\dfrac{b^2y}{y(y-4)}$

45. $\dfrac{3}{7-y}=\dfrac{3}{-(y-7)}=-\dfrac{3}{y-7}$

The LCM is $(y-7)^2$.

$\dfrac{3}{7-y}=-\dfrac{3}{y-7}\cdot\dfrac{y-7}{y-7}=-\dfrac{3y-21}{(y-7)^2}$

$\dfrac{2}{(y-7)^2}=\dfrac{2}{(y-7)^2}$

47. The LCM is $y^2(y-3)$.

$\dfrac{2}{y-3}=\dfrac{2}{y-3}\cdot\dfrac{y^2}{y^2}=\dfrac{2y^2}{y^2(y-3)}$

$\dfrac{3}{y^3-3y^2}=\dfrac{3}{y^2(y-3)}$

49. The LCM is $(2x-1)(x+4)$.

$\dfrac{x^2}{2x-1}=\dfrac{x^2}{2x-1}\cdot\dfrac{x+4}{x+4}=\dfrac{x^3+4x^2}{(2x-1)(x+4)}$

$\dfrac{x+1}{x+4}=\dfrac{x+1}{x+4}\cdot\dfrac{2x-1}{2x-1}=\dfrac{2x^2+x-1}{(2x-1)(x+4)}$

51. The LCM is $(x+5)(x-5)$.

$\dfrac{3x}{x-5}=\dfrac{3x}{x-5}\cdot\dfrac{x+5}{x+5}=\dfrac{3x^2+15x}{(x+5)(x-5)}$

$\dfrac{4}{x^2-25}=\dfrac{4}{(x+5)(x-5)}$

53. The LCM is $(x-3)(x+5)(x+1)$.

$$\begin{aligned}\frac{x-1}{x^2+2x-15}&=\frac{x-1}{(x-3)(x+5)}\cdot\frac{x+1}{x+1}\\&=\frac{x^2-1}{(x-3)(x+5)(x+1)}\\\frac{x}{x^2+6x+5}&=\frac{x}{(x+5)(x+1)}\cdot\frac{x-3}{x-3}\\&=\frac{x^2-3x}{(x-3)(x+5)(x+1)}\end{aligned}$$

Applying the Concepts

55. The LCM is 10^5.

$\dfrac{8}{10^3}=\dfrac{8}{10^3}\cdot\dfrac{10^2}{10^2}=\dfrac{800}{10^5}$

$\dfrac{9}{10^5}=\dfrac{9}{10^5}$

57. The LCM is x^2-1.

$x=\dfrac{x}{1}\cdot\dfrac{x^2-1}{x^2-1}=\dfrac{x^3-x}{x^2-1}$

$\dfrac{x}{x^2-1}=\dfrac{x}{x^2-1}$

59. The LCM is $3(6c+d)(c+d)(c-d)$.

$$\frac{c}{6c^2+7cd+d^2}=\frac{c}{(6c+d)(c+d)}\cdot\frac{3(c-d)}{3(c-d)}=\frac{3c^2-3cd}{3(6c+d)(c+d)(c-d)}$$

$$\frac{d}{3c^2-3d^2}=\frac{d}{3(c+d)(c-d)}\cdot\frac{6c+d}{6c+d}=\frac{6cd+d^2}{3(6c+d)(c+d)(c-d)}$$

Section 6.3

Objective A Exercises

1. $\frac{3}{y^2}+\frac{8}{y^2}=\frac{3+8}{y^2}=\frac{11}{y^2}$

3. $\frac{3}{x+4}-\frac{10}{x+4}=\frac{3-10}{x+4}=-\frac{7}{x+4}$

5. $\frac{3x}{2x+3}+\frac{5x}{2x+3}=\frac{3x+5x}{2x+3}=\frac{8x}{2x+3}$

7. $\frac{2x+1}{x-3}+\frac{3x+6}{x-3}=\frac{2x+1+(3x+6)}{x-3}=\frac{5x+7}{x-3}$

9. $\frac{5x-1}{x+9}-\frac{3x+4}{x+9}=\frac{5x-1-(3x+4)}{x+9}=\frac{5x-1-3x-4}{x+9}=\frac{2x-5}{x+9}$

11. $\frac{x-7}{2x+7}-\frac{4x-3}{2x+7}=\frac{x-7-(4x-3)}{2x+7}=\frac{x-7-4x+3}{2x+7}=\frac{-3x-4}{2x+7}$

13. $\frac{x}{x^2+2x-15}-\frac{3}{x^2+2x-15}=\frac{x-3}{x^2+2x-15}=\frac{\overset{1}{\cancel{(x-3)}}}{(x+5)\underset{1}{\cancel{(x-3)}}}=\frac{1}{x+5}$

15. $\frac{2x+3}{x^2-x-30}-\frac{x-2}{x^2-x-30}=\frac{2x+3-(x-2)}{x^2-x-30}=\frac{2x+3-x+2}{(x+5)(x-6)}=\frac{\overset{1}{\cancel{(x+5)}}}{\underset{1}{\cancel{(x+5)}}(x-6)}=\frac{1}{x-6}$

17. $\frac{4y+7}{2y^2+7y-4}-\frac{y-5}{2y^2+7y-4}=\frac{4y+7-(y-5)}{2y^2+7y-4}=\frac{4y+7-y+5}{(y+4)(2y-1)}=\frac{3y+12}{(y+4)(2y-1)}=\frac{3\overset{1}{\cancel{(y+4)}}}{\underset{1}{\cancel{(y+4)}}(2y-1)}=\frac{3}{2y-1}$

19. $$\frac{2x^2+3x}{x^2-9x+20}+\frac{2x^2-3}{x^2-9x+20}-\frac{4x^2+2x+1}{x^2-9x+20}=\frac{2x^2+3x+(2x^2-3)-(4x^2+2x+1)}{x^2-9x+20}$$
$$=\frac{2x^2+3x+2x^2-3-4x^2-2x-1}{x^2-9x+20}$$
$$=\frac{\overset{1}{\cancel{(x-4)}}}{\underset{1}{\cancel{(x-4)}}(x-5)}=\frac{1}{x-5}$$

Objective B Exercises

21. For each fraction, multiply the numerator and denominator by the factors whose product with the denominator is the LCM.

23. The LCM of the denominators is xy.
$\frac{4}{x}=\frac{4}{x}\cdot\frac{y}{y}=\frac{4y}{xy}$ $\quad\frac{5}{y}=\frac{5}{y}\cdot\frac{x}{x}=\frac{5x}{xy}$
$\frac{4}{x}+\frac{5}{y}=\frac{4y}{xy}+\frac{5x}{xy}=\frac{4y+5x}{xy}$

25. The LCM of the denominators is $2x$.
$\frac{12}{x}=\frac{12}{x}\cdot\frac{2}{2}=\frac{24}{2x}$ $\quad\frac{5}{2x}=\frac{5}{2x}$
$\frac{12}{x}-\frac{5}{2x}=\frac{24}{2x}-\frac{5}{2x}=\frac{24-5}{2x}=\frac{19}{2x}$

27. The LCM of the denominators is $12x$.

$$\frac{1}{2x}=\frac{1}{2x}\cdot\frac{6}{6}=\frac{6}{12x}$$
$$\frac{5}{4x}=\frac{5}{4x}\cdot\frac{3}{3}=\frac{15}{12x}$$
$$\frac{7}{6x}=\frac{7}{6x}\cdot\frac{2}{2}=\frac{14}{12x}$$
$$\frac{1}{2x}-\frac{5}{4x}+\frac{7}{6x}=\frac{6}{12x}-\frac{15}{12x}+\frac{14}{12x}=\frac{6-15+14}{12x}=\frac{5}{12x}$$

29. The LCM of the denominators is $6x^2$.

$$\frac{5}{3x}=\frac{5}{3x}\cdot\frac{2x}{2x}=\frac{10x}{6x^2}$$
$$\frac{2}{x^2}=\frac{2}{x^2}\cdot\frac{6}{6}=\frac{12}{6x^2}$$
$$\frac{3}{2x}=\frac{3}{2x}\cdot\frac{3x}{3x}=\frac{9x}{6x^2}$$
$$\frac{5}{3x}-\frac{2}{x^2}+\frac{3}{2x}=\frac{10x}{6x^2}-\frac{12}{6x^2}+\frac{9x}{6x^2}=\frac{10x-12+9x}{6x^2}=\frac{19x-12}{6x^2}$$

31. The LCM of the denominators is $20xy$.

$$\frac{2}{x}=\frac{2}{x}\cdot\frac{20y}{20y}=\frac{40y}{20xy}$$
$$\frac{3}{2y}=\frac{3}{2y}\cdot\frac{10x}{10x}=\frac{30x}{20xy}$$
$$\frac{3}{5x}=\frac{3}{5x}\cdot\frac{4y}{4y}=\frac{12y}{20xy}$$
$$\frac{1}{4y}=\frac{1}{4y}\cdot\frac{5x}{5x}=\frac{5x}{20xy}$$
$$\frac{2}{x}-\frac{3}{2y}+\frac{3}{5x}-\frac{1}{4y}=\frac{40y}{20xy}-\frac{30x}{20xy}+\frac{12y}{20xy}-\frac{5x}{20xy}=\frac{40y-30x+12y-5x}{20xy}=\frac{52y-35x}{20xy}$$

33. The LCM of the denominators is $15x$.

$$\frac{2x+1}{3x}=\frac{2x+1}{3x}\cdot\frac{5}{5}=\frac{10x+5}{15x}$$
$$\frac{x-1}{5x}=\frac{x-1}{5x}\cdot\frac{3}{3}=\frac{3x-3}{15x}$$
$$\frac{2x+1}{3x}+\frac{x-1}{5x}=\frac{10x+5}{15x}+\frac{3x-3}{15x}=\frac{10x+5+(3x-3)}{15x}=\frac{13x+2}{15x}$$

35. The LCM of the denominators is $24x$.

$$\frac{x-3}{6x}=\frac{x-3}{6x}\cdot\frac{4}{4}=\frac{4x-12}{24x}$$
$$\frac{x+4}{8x}=\frac{x+4}{8x}\cdot\frac{3}{3}=\frac{3x+12}{24x}$$
$$\frac{x-3}{6x}+\frac{x+4}{8x}=\frac{4x-12}{24x}+\frac{3x+12}{24x}=\frac{4x-12+(3x+12)}{24x}=\frac{7x}{24x}=\frac{7}{24}$$

37. The LCM of the denominators is $45x$.

$$\frac{2x+9}{9x}=\frac{2x+9}{9x}\cdot\frac{5}{5}=\frac{10x+45}{45x}$$
$$\frac{x-5}{5x}=\frac{x-5}{5x}\cdot\frac{9}{9}=\frac{9x-45}{45x}$$
$$\frac{2x+9}{9x}-\frac{x-5}{5x}=\frac{10x+45}{45x}-\frac{9x-45}{45x}=\frac{10x+45-(9x-45)}{45x}=\frac{10x+45-9x+45}{45x}=\frac{x+90}{45x}$$

39. The LCM of the denominators is $2x^2$.

$$\frac{x+4}{2x}=\frac{x+4}{2x}\cdot\frac{x}{x}=\frac{x^2+4x}{2x^2}$$
$$\frac{x-1}{x^2}=\frac{x-1}{x^2}\cdot\frac{2}{2}=\frac{2x-2}{2x^2}$$
$$\frac{x+4}{2x}-\frac{x-1}{x^2}=\frac{x^2+4x}{2x^2}-\frac{2x-2}{2x^2}=\frac{x^2+4x-(2x-2)}{2x^2}=\frac{x^2+4x-2x+2}{2x^2}=\frac{x^2+2x+2}{2x^2}$$

41. The LCM of the denominators is $4x^2$.

$$\frac{x-10}{4x^2}=\frac{x-10}{4x^2}$$
$$\frac{x+1}{2x}=\frac{x+1}{2x}\cdot\frac{2x}{2x}=\frac{2x^2+2x}{4x^2}$$
$$\frac{x-10}{4x^2}+\frac{x+1}{2x}=\frac{x-10}{4x^2}+\frac{2x^2+2x}{4x^2}=\frac{x-10+(2x^2+2x)}{4x^2}=\frac{2x^2+3x-10}{4x^2}$$

43. The LCM of the denominators is $x+4$.

$$\frac{4}{x+4}=\frac{4}{x+4}$$
$$x=\frac{x}{1}\cdot\frac{x+4}{x+4}=\frac{x^2+4x}{x+4}$$
$$\frac{4}{x+4}-x=\frac{4}{x+4}-\frac{x^2+4x}{x+4}$$
$$=\frac{4-(x^2+4x)}{x+4}$$
$$=\frac{4-x^2-4x}{x+4}=\frac{-x^2-4x+4}{x+4}$$

45. The LCM of the denominators is $x+1$.

$$5=\frac{5}{1}\cdot\frac{x+1}{x+1}=\frac{5x+5}{x+1}$$
$$\frac{x-2}{x+1}=\frac{x-2}{x+1}$$
$$5-\frac{x-2}{x+1}=\frac{5x+5}{x+1}-\frac{x-2}{x+1}=\frac{5x+5-(x-2)}{x+1}$$
$$=\frac{5x+5-x+2}{x+1}=\frac{4x+7}{x+1}$$

47. The LCM of the denominators is $24x^2$.

$$\frac{x+3}{6x}=\frac{x+3}{6x}\cdot\frac{4x}{4x}=\frac{4x^2+12x}{24x^2}$$
$$\frac{x-3}{8x^2}=\frac{x-3}{8x^2}\cdot\frac{3}{3}=\frac{3x-9}{24x^2}$$
$$\frac{x+3}{6x}-\frac{x-3}{8x^2}=\frac{4x^2+12x}{24x^2}-\frac{3x-9}{24x^2}$$
$$=\frac{4x^2+12x-(3x-9)}{24x^2}$$
$$=\frac{4x^2+12x-3x+9}{24x^2}$$
$$=\frac{4x^2+9x+9}{24x^2}$$

49. The LCM of the denominators is xy^2.

$$\frac{3x-1}{xy^2}=\frac{3x-1}{xy^2}$$
$$\frac{2x+3}{xy}=\frac{2x+3}{xy}\cdot\frac{y}{y}=\frac{2xy+3y}{xy^2}$$
$$\frac{3x-1}{xy^2}-\frac{2x+3}{xy}=\frac{3x-1}{xy^2}-\frac{2xy+3y}{xy^2}$$
$$=\frac{3x-1-(2xy+3y)}{xy^2}$$
$$=\frac{3x-1-2xy-3y}{xy^2}$$

51. The LCM of the denominators is $24x^2y^2$.

$$\frac{5x+7}{6xy^2}=\frac{5x+7}{6xy^2}\cdot\frac{4x}{4x}=\frac{20x^2+28x}{24x^2y^2}$$
$$\frac{4x-3}{8x^2y}=\frac{4x-3}{8x^2y}\cdot\frac{3y}{3y}=\frac{12xy-9y}{24x^2y^2}$$
$$\frac{5x+7}{6xy^2}-\frac{4x-3}{8x^2y}=\frac{20x^2+28x}{24x^2y^2}-\frac{12xy-9y}{24x^2y^2}$$
$$=\frac{20x^2+28x-(12xy-9y)}{24x^2y^2}$$
$$=\frac{20x^2+28x-12xy+9y}{24x^2y^2}$$

53. The LCM of the denominators is $18xy^2$.

$$\frac{3x-1}{6y^2}=\frac{3x-1}{6y^2}\cdot\frac{3x}{3x}=\frac{9x^2-3x}{18xy^2}$$
$$\frac{x+5}{9xy}=\frac{x+5}{9xy}\cdot\frac{2y}{2y}=\frac{2xy+10y}{18xy^2}$$
$$\frac{3x-1}{6y^2}-\frac{x+5}{9xy}=\frac{9x^2-3x}{18xy^2}-\frac{2xy+10y}{18xy^2}$$
$$=\frac{9x^2-3x-(2xy+10y)}{18xy^2}$$
$$=\frac{9x^2-3x-2xy-10y}{18xy^2}$$

55. The LCM is $(x-3)(x-4)$.

$$\frac{2}{x-3}=\frac{2}{x-3}\cdot\frac{x-4}{x-4}=\frac{2x-8}{(x-3)(x-4)}$$
$$\frac{5}{x-4}=\frac{5}{x-4}\cdot\frac{x-3}{x-3}=\frac{5x-15}{(x-3)(x-4)}$$
$$\frac{2}{x-3}+\frac{5}{x-4}=\frac{2x-8}{(x-3)(x-4)}+\frac{5x-15}{(x-3)(x-4)}$$
$$=\frac{2x-8+(5x-15)}{(x-3)(x-4)}$$
$$=\frac{7x-23}{(x-3)(x-4)}$$

57. The LCM is $(y+6)(y-3)$.

$$\frac{3}{y+6}=\frac{3}{y+6}\cdot\frac{y-3}{y-3}=\frac{3y-9}{(y+6)(y-3)}$$
$$\frac{4}{y-3}=\frac{4}{y-3}\cdot\frac{y+6}{y+6}=\frac{4y+24}{(y+6)(y-3)}$$
$$\frac{3}{y+6}-\frac{4}{y-3}=\frac{3y-9}{(y+6)(y-3)}-\frac{4y+24}{(y+6)(y-3)}$$
$$=\frac{3y-9-(4y+24)}{(y+6)(y-3)}$$
$$=\frac{3y-9-4y-24}{(y+6)(y-3)}=\frac{-y-33}{(y+6)(y-3)}$$

59. The LCM is $(x-4)(x+6)$.

$$\frac{3x}{x-4} = \frac{3x}{x-4} \cdot \frac{x+6}{x+6} = \frac{3x^2+18x}{(x-4)(x+6)}$$

$$\frac{2}{x+6} = \frac{2}{x+6} \cdot \frac{x-4}{x-4} = \frac{2x-8}{(x-4)(x+6)}$$

$$\begin{aligned}\frac{3x}{x-4} + \frac{2}{x+6} &= \frac{3x^2+18x}{(x-4)(x+6)} + \frac{2x-8}{(x-4)(x+6)} \\ &= \frac{3x^2+18x+(2x-8)}{(x-4)(x+6)} \\ &= \frac{3x^2+20x-8}{(x-4)(x+6)}\end{aligned}$$

61. The LCM is $(x+5)(2x+3)$.

$$\frac{6x}{x+5} = \frac{6x}{x+5} \cdot \frac{2x+3}{2x+3} = \frac{12x^2+18x}{(x+5)(2x+3)}$$

$$\frac{3}{2x+3} = \frac{3}{2x+3} \cdot \frac{x+5}{x+5} = \frac{3x+15}{(x+5)(2x+3)}$$

$$\begin{aligned}\frac{6x}{x+5} - \frac{3}{2x+3} &= \frac{12x^2+18x}{(x+5)(2x+3)} - \frac{3x+15}{(x+5)(2x+3)} \\ &= \frac{12x^2+18x-(3x+15)}{(x+5)(2x+3)} \\ &= \frac{12x^2+18x-3x-15}{(x+5)(2x+3)} \\ &= \frac{12x^2+15x-15}{(x+5)(2x+3)} \\ &= \frac{3(4x^2+5x-5)}{(x+5)(2x+3)}\end{aligned}$$

63. The LCM is $x-6$.

$$\frac{4x}{6-x} = \frac{4x}{-(x-6)} \cdot \frac{-1}{-1} = \frac{-4x}{x-6}$$

$$\frac{5}{x-6} = \frac{5}{x-6}$$

$$\frac{4x}{6-x} + \frac{5}{x-6} = \frac{-4x}{x-6} + \frac{5}{x-6} = \frac{-4x+5}{x-6}$$

65. The LCM is $(y+4)(y-4)$.

$$\frac{y}{y^2-16} = \frac{y}{(y+4)(y-4)}$$

$$\frac{1}{y-4} = \frac{1}{y-4} \cdot \frac{y+4}{y+4} = \frac{y+4}{(y+4)(y-4)}$$

$$\begin{aligned}\frac{y}{y^2-16} + \frac{1}{y-4} &= \frac{y}{(y+4)(y-4)} + \frac{y+4}{(y+4)(y-4)} \\ &= \frac{y+(y+4)}{(y+4)(y-4)} = \frac{2y+4}{(y+4)(y-4)} \\ &= \frac{2(y+2)}{(y+4)(y-4)}\end{aligned}$$

67. The LCM is $(x+1)^2$.

$$\frac{(x-1)^2}{(x+1)^2} = \frac{(x-1)^2}{(x+1)^2};\ 1 = \frac{1}{1} \cdot \frac{(x+1)^2}{(x+1)^2} = \frac{(x+1)^2}{(x+1)^2}$$

$$\begin{aligned}\frac{(x-1)^2}{(x+1)^2} - 1 = \frac{(x-1)^2}{(x+1)^2} - \frac{(x+1)^2}{(x+1)^2} &= \frac{(x^2-2x+1)-(x^2+2x+1)}{(x+1)^2} \\ &= \frac{x^2-2x+1-x^2-2x-1}{(x+1)^2} \\ &= \frac{-4x}{(x+1)^2} = -\frac{4x}{(x+1)^2}\end{aligned}$$

69. The LCM is $(1+x)(1-x)$.

$$\frac{x}{1-x^2} = \frac{x}{(1+x)(1-x)};\ 1 = \frac{1}{1} \cdot \frac{(1+x)(1-x)}{(1+x)(1-x)} = \frac{1-x^2}{(1+x)(1-x)};\ \frac{x}{1+x} = \frac{x}{1+x} \cdot \frac{1-x}{1-x} = \frac{x-x^2}{(1+x)(1-x)}$$

$$\begin{aligned}\frac{x}{1-x^2} - 1 + \frac{x}{1+x} &= \frac{x}{(1+x)(1-x)} - \frac{1-x^2}{(1+x)(1-x)} + \frac{x-x^2}{(1+x)(1-x)} \\ &= \frac{x-(1-x^2)+(x-x^2)}{(1+x)(1-x)} \\ &= \frac{x-1+x^2+x-x^2}{(1+x)(1-x)} = \frac{2x-1}{(1+x)(1-x)}\end{aligned}$$

71. The LCM is $(x-5)(x-5)$.

$$\frac{3x-1}{x^2-10x+25} = \frac{3x-1}{(x-5)(x-5)};\ \frac{3}{x-5} = \frac{3}{x-5} \cdot \frac{x-5}{x-5} = \frac{3x-15}{(x-5)(x-5)}$$

$$\begin{aligned}\frac{3x-1}{x^2-10x+25} - \frac{3}{x-5} &= \frac{3x-1}{(x-5)(x-5)} - \frac{3x-15}{(x-5)(x-5)} \\ &= \frac{3x-1-(3x-15)}{(x-5)(x-5)} = \frac{3x-1-3x+15}{(x-5)(x-5)} = \frac{14}{(x-5)(x-5)} = \frac{14}{(x-5)^2}\end{aligned}$$

73. The LCM is $(x+6)(x-7)$.

$$\frac{x+4}{x^2-x-42} = \frac{x+4}{(x+6)(x-7)};\ \frac{3}{7-x} = \frac{3}{-(x-7)} \cdot \frac{x+6}{x+6} = \frac{-3(x+6)}{(x+6)(x-7)} = \frac{-3x-18}{(x+6)(x-7)}$$

$$\begin{aligned}\frac{x+4}{x^2-x-42} + \frac{3}{7-x} &= \frac{x+4}{(x+6)(x-7)} + \frac{-3x-18}{(x+6)(x-7)} \\ &= \frac{x+4+(-3x-18)}{(x+6)(x-7)} = \frac{-2x-14}{(x+6)(x-7)} = \frac{-2(x+7)}{(x+6)(x-7)}\end{aligned}$$

75. The LCM is $(x+1)(x-6)$.

$$\frac{1}{x+1} = \frac{1}{x+1} \cdot \frac{x-6}{x-6} = \frac{x-6}{(x+1)(x-6)};\ \frac{x}{x-6} = \frac{x}{x-6} \cdot \frac{x+1}{x+1} = \frac{x^2+x}{(x+1)(x-6)};\ \frac{5x-2}{x^2-5x-6} = \frac{5x-2}{(x+1)(x-6)}$$

$$\begin{aligned}\frac{1}{x+1} + \frac{x}{x-6} - \frac{5x-2}{x^2-5x-6} &= \frac{x-6}{(x+1)(x-6)} + \frac{x^2+x}{(x+1)(x-6)} - \frac{5x-2}{(x+1)(x-6)} \\ &= \frac{x-6+(x^2+x)-(5x-2)}{(x+1)(x-6)} = \frac{x-6+x^2+x-5x+2}{(x+1)(x-6)} \\ &= \frac{x^2-3x-4}{(x+1)(x-6)} = \frac{\overset{1}{\cancel{(x+1)}}(x-4)}{\underset{1}{\cancel{(x+1)}}(x-6)} = \frac{x-4}{x-6}\end{aligned}$$

77. The LCM is $(x-1)(x-3)$.

$$\frac{3x+1}{x-1}=\frac{3x+1}{x-1}\cdot\frac{x-3}{x-3}=\frac{3x^2-8x-3}{(x-1)(x-3)};\ \frac{x-1}{x-3}=\frac{x-1}{x-3}\cdot\frac{x-1}{x-1}=\frac{x^2-2x+1}{(x-1)(x-3)};\ \frac{x+1}{x^2-4x+3}=\frac{x+1}{(x-1)(x-3)}$$

$$\begin{aligned}\frac{3x+1}{x-1}-\frac{x-1}{x-3}+\frac{x+1}{x^2-4x+3}&=\frac{3x^2-8x-3}{(x-1)(x-3)}-\frac{x^2-2x+1}{(x-1)(x-3)}+\frac{x+1}{(x-1)(x-3)}\\&=\frac{3x^2-8x-3-(x^2-2x+1)+(x+1)}{(x-1)(x-3)}\\&=\frac{3x^2-8x-3-x^2+2x-1+x+1}{(x-1)(x-3)}\\&=\frac{2x^2-5x-3}{(x-1)(x-3)}=\frac{\overset{1}{\cancel{(x-3)}}(2x+1)}{(x-1)\underset{1}{\cancel{(x-3)}}}=\frac{2x+1}{x-1}\end{aligned}$$

79. The LCM is $(x-3)(x+7)$.

$$\frac{2x+9}{3-x}=\frac{2x+9}{-(x-3)}=\frac{-(2x+9)}{x-3}\cdot\frac{x+7}{x+7}=\frac{-(2x^2+23x+63)}{(x-3)(x+7)}=\frac{-2x^2-23x-63}{(x-3)(x+7)};$$
$$\frac{x+5}{x+7}=\frac{x+5}{x+7}\cdot\frac{x-3}{x-3}=\frac{x^2+2x-15}{(x-3)(x+7)};\ \frac{2x^2+3x-3}{x^2+4x-21}=\frac{2x^2+3x-3}{(x-3)(x+7)}$$

$$\begin{aligned}\frac{2x+9}{3-x}+\frac{x+5}{x+7}-\frac{2x^2+3x-3}{x^2+4x-21}&=\frac{-2x^2-23x-63}{(x-3)(x+7)}+\frac{x^2+2x-15}{(x-3)(x+7)}-\frac{2x^2+3x-3}{x^2+4x-21}\\&=\frac{-2x^2-23x-63+x^2+2x-15-(2x^2+3x-3)}{(x-3)(x+7)}\\&=\frac{-2x^2-23x-63+x^2+2x-15-2x^2-3x+3}{(x-3)(x+7)}\\&=\frac{-3x^2-24x-75}{(x-3)(x+7)}=\frac{-3(x^2+8x+25)}{(x-3)(x+7)}\end{aligned}$$

Applying the Concepts

81. a. Strategy x = number of miles per gallon your car gets.

- Determine the number of gallons used per year.
- Multiply the number of gallons by price per gallon.

Solution No. of gallons = miles traveled per year ÷ miles per gallon

$$=12{,}000\div x=\frac{12{,}000}{x}$$

Amount spent on gasoline per year $=\dfrac{12{,}000}{x}\cdot 1.70$

$$=\frac{20{,}400}{x}\text{ dollars}$$

b. Strategy
- Determine how much you would spend each year if you increase gas mileage by 5 mi/gal.
- Subtract the expression found above from the expression found in part a.

Solution $x + 5 =$ increase in miles per gallon

No. of gallons $= 12{,}000 \div (x+5) = \dfrac{12{,}000}{x+5}$

Amt. spent on gasoline per yr. $= \dfrac{12{,}000}{x+5} \cdot 1.70$

$= \dfrac{20{,}400}{x+5}$

$$\frac{20{,}400}{x} - \frac{20{,}400}{x+5} = \frac{20{,}400 \cdot (x+5)}{x(x+5)} - \frac{20{,}400 \cdot x}{x(x+5)}$$
$$= \frac{20{,}400x + 102{,}000 - 20{,}400x}{x(x+5)}$$
$$= \frac{102{,}000}{x(x+5)} \text{ dollars}$$

c. Strategy Evaluate the expression found in part b using 25 for x.

Solution $\dfrac{102{,}000}{x(x+5)}$; $x = 25$

$$\frac{102{,}000}{25(25+5)} = \frac{102{,}000}{25(30)} = \frac{102{,}000}{750} = 136$$

You would save \$136 in one year.

Section 6.4

Objective A Exercises

1. The LCM of x and x^2 is x^2.

$$\frac{1+\frac{3}{x}}{1-\frac{9}{x^2}} = \frac{1+\frac{3}{x}}{1-\frac{9}{x^2}} \cdot \frac{x^2}{x^2} = \frac{1 \cdot x^2 + \frac{3}{x} \cdot x^2}{1 \cdot x^2 - \frac{9}{x^2} \cdot x^2}$$
$$= \frac{x^2+3x}{x^2-9} = \frac{x(\overset{1}{\cancel{x+3}})}{(\underset{1}{\cancel{x+3}})(x-3)} = \frac{x}{x-3}$$

3. The LCM is $x + 4$.

$$\frac{2-\frac{8}{x+4}}{3-\frac{12}{x+4}} = \frac{2-\frac{8}{x+4}}{3-\frac{12}{x+4}} \cdot \frac{x+4}{x+4}$$
$$= \frac{2(x+4) - \frac{8}{x+4}(x+4)}{3(x+4) - \frac{12}{x+4}(x+4)}$$
$$= \frac{2(x+4)-8}{3(x+4)-12} = \frac{2x+8-8}{3x+12-12} = \frac{2x}{3x} = \frac{2}{3}$$

5. The LCM is $y - 2$.

$$\frac{1+\frac{5}{y-2}}{1-\frac{2}{y-2}} = \frac{1+\frac{5}{y-2}}{1-\frac{2}{y-2}} \cdot \frac{y-2}{y-2}$$
$$= \frac{1(y-2) + \frac{5}{y-2}(y-2)}{1(y-2) - \frac{2}{y-2}(y-2)}$$
$$= \frac{(y-2)+5}{(y-2)-2} = \frac{y-2+5}{y-2-2} = \frac{y+3}{y-4}$$

7. The LCM is $x + 7$.

$$\frac{4-\frac{2}{x+7}}{5+\frac{1}{x+7}} = \frac{4-\frac{2}{x+7}}{5+\frac{1}{x+7}} \cdot \frac{x+7}{x+7}$$
$$= \frac{4(x+7) - \frac{2}{x+7}(x+7)}{5(x+7) + \frac{1}{x+7}(x+7)}$$
$$= \frac{4(x+7)-2}{5(x+7)+1} = \frac{4x+28-2}{5x+35+1}$$
$$= \frac{4x+26}{5x+36} = \frac{2(2x+13)}{5x+36}$$

9. The LCM of x and x^2 is x^2.

$$\frac{1-\frac{1}{x}-\frac{6}{x^2}}{1-\frac{9}{x^2}}=\frac{1-\frac{1}{x}-\frac{6}{x^2}}{1-\frac{9}{x^2}}\cdot\frac{x^2}{x^2}$$
$$=\frac{1\cdot x^2-\frac{1}{x}\cdot x^2-\frac{6}{x^2}\cdot x^2}{1\cdot x^2-\frac{9}{x^2}\cdot x^2}$$
$$=\frac{x^2-x-6}{x^2-9}=\frac{(x+2)\overset{1}{\cancel{(x-3)}}}{(x+3)\underset{1}{\cancel{(x-3)}}}=\frac{x+2}{x+3}$$

11. The LCM of x and x^2 is x^2.

$$\frac{1-\frac{5}{x}-\frac{6}{x^2}}{1+\frac{6}{x}+\frac{5}{x^2}}=\frac{1-\frac{5}{x}-\frac{6}{x^2}}{1+\frac{6}{x}+\frac{5}{x^2}}\cdot\frac{x^2}{x^2}$$
$$=\frac{1\cdot x^2-\frac{5}{x}\cdot x^2-\frac{6}{x^2}\cdot x^2}{1\cdot x^2+\frac{6}{x}\cdot x^2+\frac{5}{x^2}\cdot x^2}$$
$$=\frac{x^2-5x-6}{x^2+6x+5}=\frac{\overset{1}{\cancel{(x+1)}}(x-6)}{\underset{1}{\cancel{(x+1)}}(x+5)}$$
$$=\frac{x-6}{x+5}$$

13. The LCM of x and x^2 is x^2.

$$\frac{1-\frac{6}{x}+\frac{8}{x^2}}{\frac{4}{x^2}+\frac{3}{x}-1}=\frac{1-\frac{6}{x}+\frac{8}{x^2}}{\frac{4}{x^2}+\frac{3}{x}-1}\cdot\frac{x^2}{x^2}$$
$$=\frac{1\cdot x^2-\frac{6}{x}\cdot x^2+\frac{8}{x^2}\cdot x^2}{\frac{4}{x^2}\cdot x^2+\frac{3}{x}\cdot x^2-1\cdot x^2}$$
$$=\frac{x^2-6x+8}{4+3x-x^2}=\frac{(x-2)\overset{-1}{\cancel{(x-4)}}}{(1+x)\underset{1}{\cancel{(4-x)}}}$$
$$=\frac{-x+2}{x+1}$$

15. The LCM is $x+3$.

$$\frac{x-\frac{4}{x+3}}{1+\frac{1}{x+3}}=\frac{x-\frac{4}{x+3}}{1+\frac{1}{x+3}}\cdot\frac{x+3}{x+3}$$
$$=\frac{x(x+3)-\frac{4}{x+3}(x+3)}{1(x+3)+\frac{1}{x+3}(x+3)}$$
$$=\frac{x^2+3x-4}{x+3+1}=\frac{\overset{1}{\cancel{(x+4)}}(x-1)}{\underset{1}{\cancel{(x+4)}}}=x-1$$

17. The LCM is $2x+1$.

$$\frac{1-\frac{x}{2x+1}}{x-\frac{1}{2x+1}}=\frac{1-\frac{x}{2x+1}}{x-\frac{1}{2x+1}}\cdot\frac{2x+1}{2x+1}$$
$$=\frac{1(2x+1)-\frac{x}{2x+1}(2x+1)}{x(2x+1)-\frac{1}{2x+1}(2x+1)}$$
$$=\frac{2x+1-x}{2x^2+x-1}=\frac{\overset{1}{\cancel{(x+1)}}}{\underset{1}{\cancel{(x+1)}}(2x-1)}$$
$$=\frac{1}{2x-1}$$

19. The LCM is $x+4$.

$$\frac{x-5+\frac{14}{x+4}}{x+3-\frac{2}{x+4}}=\frac{x-5+\frac{14}{x+4}}{x+3-\frac{2}{x+4}}\cdot\frac{x+4}{x+4}$$
$$=\frac{(x-5)(x+4)+\frac{14}{x+4}(x+4)}{(x+3)(x+4)-\frac{2}{x+4}(x+4)}$$
$$=\frac{x^2-x-20+14}{x^2+7x+12-2}=\frac{x^2-x-6}{x^2+7x+10}$$
$$=\frac{\overset{1}{\cancel{(x+2)}}(x-3)}{\underset{1}{\cancel{(x+2)}}(x+5)}=\frac{x-3}{x+5}$$

21. The LCM is $x - 6$.

$$\frac{x+3-\frac{10}{x-6}}{x+2-\frac{20}{x-6}} = \frac{x+3-\frac{10}{x-6}}{x+2-\frac{20}{x-6}} \cdot \frac{x-6}{x-6} = \frac{(x+3)(x-6)-\frac{10}{x-6}(x-6)}{(x+2)(x-6)-\frac{20}{x-6}(x-6)}$$

$$= \frac{x^2-3x-18-10}{x^2-4x-12-20} = \frac{x^2-3x-28}{x^2-4x-32} = \frac{\overset{1}{\cancel{(x+4)}}(x-7)}{\underset{1}{\cancel{(x+4)}}(x-8)} = \frac{x-7}{x-8}$$

23. The LCM is $2y + 3$.

$$\frac{y-6+\frac{22}{2y+3}}{y-5+\frac{11}{2y+3}} = \frac{y-6+\frac{22}{2y+3}}{y-5+\frac{11}{2y+3}} \cdot \frac{2y+3}{2y+3} = \frac{(y-6)(2y+3)+\frac{22}{2y+3}(2y+3)}{(y-5)(2y+3)+\frac{11}{2y+3}(2y+3)}$$

$$= \frac{2y^2-9y-18+22}{2y^2-7y-15+11} = \frac{2y^2-9y+4}{2y^2-7y-4} = \frac{(2y-1)\overset{1}{\cancel{(y-4)}}}{(2y+1)\underset{1}{\cancel{(y-4)}}} = \frac{2y-1}{2y+1}$$

25. The LCM is $2x - 3$.

$$\frac{x-\frac{2}{2x-3}}{2x-1-\frac{8}{2x-3}} = \frac{x-\frac{2}{2x-3}}{2x-1-\frac{8}{2x-3}} \cdot \frac{2x-3}{2x-3} = \frac{x(2x-3)-\frac{2}{2x-3}(2x-3)}{(2x-1)(2x-3)-\frac{8}{2x-3}(2x-3)}$$

$$= \frac{2x^2-3x-2}{4x^2-8x+3-8} = \frac{2x^2-3x-2}{4x^2-8x-5} = \frac{\overset{1}{\cancel{(2x+1)}}(x-2)}{\underset{1}{\cancel{(2x+1)}}(2x-5)} = \frac{x-2}{2x-5}$$

27. The LCM is x and $x - 1$ is $x(x - 1)$.

$$\frac{\frac{1}{x}-\frac{2}{x-1}}{\frac{3}{x}+\frac{1}{x-1}} = \frac{\frac{1}{x}-\frac{2}{x-1}}{\frac{3}{x}+\frac{1}{x-1}} \cdot \frac{x(x-1)}{x(x-1)} = \frac{\frac{1}{x}(x)(x-1)-\frac{2}{x-1}(x)(x-1)}{\frac{3}{x}(x)(x-1)+\frac{1}{x-1}(x)(x-1)}$$

$$= \frac{x-1-2x}{3(x-1)+x} = \frac{-x-1}{3x-3+x} = \frac{-x-1}{4x-3}$$

29. The LCM of x and $2x - 1$ is $x(2x - 1)$.

$$\frac{\frac{3}{2x-1}-\frac{1}{x}}{\frac{4}{x}+\frac{2}{2x-1}} = \frac{\frac{3}{2x-1}-\frac{1}{x}}{\frac{4}{x}+\frac{2}{2x-1}} \cdot \frac{x(2x-1)}{x(2x-1)} = \frac{\frac{3}{2x-1}(x)(2x-1)-\frac{1}{x}(x)(2x-1)}{\frac{4}{x}(x)(2x-1)+\frac{2}{2x-1}(x)(2x-1)}$$

$$= \frac{3x-(2x-1)}{4(2x-1)+2x} = \frac{3x-2x+1}{8x-4+2x} = \frac{x+1}{10x-4} = \frac{x+1}{2(5x-2)}$$

Applying the Concepts

31. $$1+\frac{1}{1+\frac{1}{2}} = 1+\frac{1}{1+\frac{1}{2}}\cdot\frac{2}{2} = 1+\frac{2}{2+1} = 1+\frac{2}{3} = \frac{3}{3}+\frac{2}{3} = \frac{5}{3}$$

33. $$1-\frac{1}{1-\frac{1}{x}} = 1-\frac{1}{1-\frac{1}{x}}\cdot\frac{x}{x} = 1-\frac{x}{x-1} = \frac{x-1}{x-1}-\frac{x}{x-1} = \frac{x-1-x}{x-1} = -\frac{1}{x-1}$$

35. $$\left(\frac{y}{4}-\frac{4}{y}\right) \div \left(\frac{4}{y}-3+\frac{y}{2}\right) = \frac{\frac{y}{4}-\frac{4}{y}}{\frac{4}{y}-3+\frac{y}{2}} \cdot \frac{4y}{4y}$$

$$= \frac{y^2-16}{16-12y+2y^2}$$

$$= \frac{(y+4)\overset{1}{\cancel{(y-4)}}}{2\underset{1}{\cancel{(y-4)}}(y-2)} = \frac{y+4}{2(y-2)}$$

37. $\dfrac{1+x^{-1}}{1-x^{-1}} = \dfrac{1+\frac{1}{x}}{1-\frac{1}{x}} \cdot \dfrac{x}{x} = \dfrac{x+1}{x-1}$

39. $\dfrac{x^{-1}}{y^{-1}} + \dfrac{x}{y} = \dfrac{\frac{1}{x}}{\frac{1}{y}} + \dfrac{x}{y} = \dfrac{\frac{1}{x}}{\frac{1}{y}} \cdot \dfrac{xy}{xy} + \dfrac{x}{y} = \dfrac{y}{x} + \dfrac{x}{y}$

$= \dfrac{y^2}{xy} + \dfrac{x^2}{xy} = \dfrac{y^2+x^2}{xy}$

Section 6.5

Objective A Exercises

1. 2 cannot be a solution of the equation $\dfrac{6x}{x+1} - \dfrac{x}{x-2} = 4$ because the denominator of $\dfrac{x}{x-2}$ is 0 when $x = 2$.

3. $\dfrac{2x}{3} - \dfrac{5}{2} = -\dfrac{1}{2}$ The LCM is 6.

$$\frac{6}{1}\left(\frac{2x}{3} - \frac{5}{2}\right) = \frac{6}{1}\left(-\frac{1}{2}\right)$$
$$\frac{6}{1}\cdot\frac{2x}{3} - \frac{6}{1}\cdot\frac{5}{2} = -3$$
$$4x - 15 = -3$$
$$4x = 12$$
$$x = 3$$

3 checks as a solution. The solution is 3.

5. $\dfrac{x}{3} - \dfrac{1}{4} = \dfrac{x}{4} - \dfrac{1}{6}$ The LCM is 12.

$$\frac{12}{1}\left(\frac{x}{3} - \frac{1}{4}\right) = \frac{12}{1}\left(\frac{x}{4} - \frac{1}{6}\right)$$
$$\frac{12}{1}\cdot\frac{x}{3} - \frac{12}{1}\cdot\frac{1}{4} = \frac{12}{1}\cdot\frac{x}{4} - \frac{12}{1}\cdot\frac{1}{6}$$
$$4x - 3 = 3x - 2$$
$$x - 3 = -2$$
$$x = 1$$

1 checks as a solution. The solution is 1.

7. $\dfrac{2x-5}{8} + \dfrac{1}{4} = \dfrac{x}{8} + \dfrac{3}{4}$ The LCM is 8.

$$\frac{8}{1}\left(\frac{2x-5}{8} + \frac{1}{4}\right) = \frac{8}{1}\left(\frac{x}{8} + \frac{3}{4}\right)$$
$$\frac{8}{1}\cdot\frac{2x-5}{8} + \frac{8}{1}\cdot\frac{1}{4} = \frac{8}{1}\cdot\frac{x}{8} + \frac{8}{1}\cdot\frac{3}{4}$$
$$2x - 5 + 2 = x + 6$$
$$2x - 3 = x + 6$$
$$x - 3 = 6$$
$$x = 9$$

9 checks as a solution. The solution is 9.

9. $\dfrac{6}{2a+1} = 2$ The LCM is $2a+1$.

$$\frac{2a+1}{1}\cdot\frac{6}{2a+1} = \frac{2a+1}{1}\cdot 2$$
$$6 = (2a+1)2$$
$$6 = 4a + 2$$
$$4 = 4a$$
$$1 = a$$

1 checks as a solution. The solution is 1.

11. $\dfrac{9}{2x-5} = -2$ The LCM is $2x-5$.

$$\frac{2x-5}{1}\cdot\frac{9}{2x-5} = \frac{2x-5}{1}(-2)$$
$$9 = (2x-5)(-2)$$
$$9 = -4x + 10$$
$$-1 = -4x$$
$$\frac{1}{4} = x$$

$\dfrac{1}{4}$ checks as the solution. The solution is $\dfrac{1}{4}$.

13. $2 + \dfrac{5}{x} = 7$ The LCM is x.

$$\frac{x}{1}\left(2 + \frac{5}{x}\right) = \frac{x}{1}\cdot 7$$
$$2x + \frac{x}{1}\cdot\frac{5}{x} = 7x$$
$$2x + 5 = 7x$$
$$5 = 5x$$
$$1 = x$$

1 checks as a solution. The solution is 1.

15. $1 - \dfrac{9}{x} = 4$ The LCM is x.

$$\frac{x}{1}\left(1 - \frac{9}{x}\right) = \frac{x}{1}\cdot 4$$
$$x - \frac{x}{1}\cdot\frac{9}{x} = 4x$$
$$x - 9 = 4x$$
$$-9 = 3x$$
$$-3 = x$$

-3 checks as a solution. The solution is -3.

17. $\dfrac{2}{y} + 5 = 9$ The LCM is y.

$$\frac{y}{1}\left(\frac{2}{y} + 5\right) = \frac{y}{1}\cdot 9$$
$$\frac{y}{1}\cdot\frac{2}{y} + \frac{y}{1}\cdot 5 = 9y$$
$$2 + 5y = 9y$$
$$2 = 4y$$
$$\frac{1}{2} = y$$

$\dfrac{1}{2}$ checks as a solution. The solution is $\dfrac{1}{2}$.

19. The LCM is $x(x-2)$.

$$\frac{3}{x-2}=\frac{4}{x}$$
$$\frac{x(x-2)}{1}\cdot\frac{3}{x-2}=\frac{x(x-2)}{1}\cdot\frac{4}{x}$$
$$3x=(x-2)4$$
$$3x=4x-8$$
$$-x=-8$$
$$x=8$$

8 checks as a solution. The solution is 8.

21. The LCM is $(3x-1)(4x+1)$.

$$\frac{2}{3x-1}=\frac{3}{4x+1}$$
$$\frac{(3x-1)(4x+1)}{1}\cdot\frac{2}{3x-1}=\frac{(3x-1)(4x+1)}{1}\cdot\frac{3}{4x+1}$$
$$(4x+1)2=(3x-1)3$$
$$8x+2=9x-3$$
$$-x+2=-3$$
$$-x=-5$$
$$x=5$$

5 checks as a solution. The solution is 5.

23. The LCM is $(2x+5)(x-1)$.

$$\frac{-3}{2x+5}=\frac{2}{x-1}$$
$$\frac{(2x+5)(x-1)}{1}\cdot\frac{-3}{2x+5}=\frac{(2x+5)(x-1)}{1}\cdot\frac{2}{x-1}$$
$$(x-1)(-3)=(2x+5)2$$
$$-3x+3=4x+10$$
$$-7x+3=10$$
$$-7x=7$$
$$x=-1$$

-1 checks as a solution. The solution is -1.

25. The LCM is $x-4$.

$$\frac{4x}{x-4}+5=\frac{5x}{x-4}$$
$$\frac{x-4}{1}\left(\frac{4x}{x-4}+5\right)=\frac{x-4}{1}\cdot\frac{5x}{x-4}$$
$$\frac{x-4}{1}\cdot\frac{4x}{x-4}+\frac{x-4}{1}\cdot5=5x$$
$$4x+(x-4)5=5x$$
$$4x+5x-20=5x$$
$$9x-20=5x$$
$$-20=-4x$$
$$5=x$$

5 checks as a solution. The solution is 5.

27. The LCM is $a-3$.

$$2+\frac{3}{a-3}=\frac{a}{a-3}$$
$$\frac{a-3}{1}\left(2+\frac{3}{a-3}\right)=\frac{a-3}{1}\cdot\frac{a}{a-3}$$
$$\frac{a-3}{1}\cdot2+\frac{a-3}{1}\cdot\frac{3}{a-3}=a$$
$$(a-3)2+3=a$$
$$2a-6+3=a$$
$$2a-3=a$$
$$a-3=0$$
$$a=3$$

3 does not check as a solution. The equation has no solution.

29.

$$\frac{x}{x-1} = \frac{8}{x+2}$$

The LCM is $(x-1)(x+2)$.

$$\frac{(x-1)(x+2)}{1} \cdot \frac{x}{x-1} = \frac{(x-1)(x+2)}{1} \cdot \frac{8}{x+2}$$
$$(x+2)x = (x-1)8$$
$$x^2 + 2x = 8x - 8$$
$$x^2 - 6x + 8 = 0$$
$$(x-4)(x-2) = 0$$

$$x-4=0 \quad x-2=0$$
$$x=4 \quad x=2$$

Both 4 and 2 check as solutions. The solutions are 4 and 2.

31.

$$\frac{2x}{x+4} = \frac{3}{x-1}$$

The LCM is $(x+4)(x-1)$.

$$\frac{(x+4)(x-1)}{1} \cdot \frac{2x}{x+4} = \frac{(x+4)(x-1)}{1} \cdot \frac{3}{x-1}$$
$$(x-1)2x = (x+4)3$$
$$2x^2 - 2x = 3x + 12$$
$$2x^2 - 5x - 12 = 0$$
$$(2x+3)(x-4) = 0$$

$$2x+3=0 \quad x-4=0$$
$$2x=-3 \quad x=4$$
$$x=-\frac{3}{2}$$

Both $-\frac{3}{2}$ and 4 check as solutions. The solutions are $-\frac{3}{2}$ and 4.

33.

$$x + \frac{6}{x-2} = \frac{3x}{x-2}$$

The LCM is $x-2$.

$$\frac{x-2}{1}\left(x + \frac{6}{x-2}\right) = \frac{x-2}{1} \cdot \frac{3x}{x-2}$$
$$\frac{x-2}{1} \cdot x + \frac{x-2}{1} \cdot \frac{6}{x-2} = 3x$$
$$x^2 - 2x + 6 = 3x$$
$$x^2 - 5x + 6 = 0$$
$$(x-3)(x-2) = 0$$

$$x-3=0 \quad x-2=0$$
$$x=3 \quad x=2$$

2 does not check as a solution. 3 checks as a solution. The solution is 3.

35.

$$\frac{8}{y} = \frac{2}{y-2} + 1$$

The LCM is $y(y-2)$.

$$\frac{y(y-2)}{1} \cdot \frac{8}{y} = \frac{y(y-2)}{1}\left(\frac{2}{y-2} + 1\right)$$
$$(y-2)8 = \frac{y(y-2)}{1} \cdot \frac{2}{y-2} + \frac{y(y-2)}{1} \cdot 1$$
$$8y - 16 = 2y + y^2 - 2y$$
$$8y - 16 = y^2$$
$$0 = y^2 - 8y + 16$$
$$0 = (y-4)(y-4)$$

$$y-4=0 \quad y-4=0$$
$$y=4 \quad y=4$$

4 checks as a solution. The solution is 4.

Applying the Concepts

37.
$$\frac{3}{5}y-\frac{1}{3}(1-y)=\frac{2y-5}{15} \quad \text{The LCM is 15.}$$
$$\frac{15}{1}\cdot\frac{3}{5}y-\frac{15}{1}\cdot\frac{1}{3}(1-y)=\frac{15}{1}\cdot\frac{2y-5}{15}$$
$$9y-5(1-y)=2y-5$$
$$9y-5+5y=2y-5$$
$$14y-5=2y-5$$
$$12y=0$$
$$y=0$$
0 checks as a solution. The solution is 0.

39.
$$\frac{b+2}{5}=\frac{1}{4}b-\frac{3}{10}(b-1) \quad \text{The LCM is 20.}$$
$$\frac{20}{1}\cdot\frac{b+2}{5}=\frac{20}{1}\cdot\frac{1}{4}b-\frac{20}{1}\cdot\frac{3}{10}(b-1)$$
$$4(b+2)=5b-6(b-1)$$
$$4b+8=5b-6b+6$$
$$4b+8=-b+6$$
$$5b=-2$$
$$b=-\frac{2}{5}$$
$-\frac{2}{5}$ checks as a solution. The solution is $-\frac{2}{5}$.

41.
$$\frac{x+1}{x^2+x-2}=\frac{x+2}{x^2-1}+\frac{3}{x+2}$$
$$\frac{x+1}{(x+2)(x-1)}=\frac{x+2}{(x+1)(x-1)}+\frac{3}{x+2} \quad \text{The LCM is } (x+2)(x+1)(x-1).$$
$$\frac{(x+2)(x+1)(x-1)}{1}\cdot\frac{x+1}{(x+2)(x-1)}=\frac{(x+2)(x+1)(x-1)}{1}\cdot\frac{x+2}{(x+1)(x-1)}+\frac{(x+2)(x+1)(x-1)}{1}\cdot\frac{3}{x+2}$$
$$(x+1)(x+1)=(x+2)(x+2)+3(x+1)(x-1)$$
$$x^2+2x+1=x^2+4x+4+3x^2-3$$
$$-3x^2-2x=0$$
$$3x^2+2x=0$$
$$x(3x+2)=0$$

$$x=0 \qquad 3x+2=0$$
$$3x=-2$$
$$x=-\frac{2}{3}$$
Both 0 and $-\frac{2}{3}$ check as solutions. The solutions are 0 and $-\frac{2}{3}$.

Section 6.6

Objective A Exercises

1. A proportion is the equality of two ratios or rates.

3.
$$\frac{x}{12}=\frac{3}{4}$$
$$12\cdot\frac{x}{12}=12\cdot\frac{3}{4}$$
$$x=9$$
The solution is 9.

5. $$\frac{4}{9}=\frac{x}{27}$$
$$27\cdot\frac{4}{9}=27\cdot\frac{x}{27}$$
$$12=x$$
The solution is 12.

7. $$\frac{x+3}{12}=\frac{5}{6}$$
$$12\cdot\frac{x+3}{12}=12\cdot\frac{5}{6}$$
$$x+3=10$$
$$x=7$$
The solution is 7.

9. $$\frac{18}{x+4}=\frac{9}{5}$$
$$5(x+4)\cdot\frac{18}{x+4}=5(x+4)\cdot\frac{9}{5}$$
$$90=(x+4)9$$
$$90=9x+36$$
$$54=9x$$
$$6=x$$
The solution is 6.

11. $$\frac{2}{x}=\frac{4}{x+1}$$
$$x(x+1)\cdot\frac{2}{x}=x(x+1)\cdot\frac{4}{x+1}$$
$$(x+1)2=4x$$
$$2x+2=4x$$
$$2=2x$$
$$1=x$$
The solution is 1.

13. $$\frac{x+3}{4}=\frac{x}{8}$$
$$8\cdot\frac{x+3}{4}=8\cdot\frac{x}{8}$$
$$2(x+3)=x$$
$$2x+6=x$$
$$x+6=0$$
$$x=-6$$
The solution is −6.

15. $$\frac{2}{x-1}=\frac{6}{2x+1}$$
$$(x-1)(2x+1)\cdot\frac{2}{x-1}=(x-1)(2x+1)\cdot\frac{6}{2x+1}$$
$$(2x+1)2=(x-1)6$$
$$4x+2=6x-6$$
$$-2x+2=-6$$
$$-2x=-8$$
$$x=4$$
The solution is 4.

17. $$\frac{2x}{7}=\frac{x-2}{14}$$
$$14\cdot\frac{2x}{7}=14\cdot\frac{x-2}{14}$$
$$4x=x-2$$
$$3x=-2$$
$$x=-\frac{2}{3}$$
The solution is $-\frac{2}{3}$.

Objective B Exercises

19. **Strategy** Write and solve a proportion using x to represent the number of voters who favored the amendment.

Solution
$$\frac{4}{7}=\frac{x}{35{,}000}$$
$$35{,}000\cdot\frac{4}{7}=\frac{x}{35{,}000}\cdot 35{,}000$$
$$20{,}000=x$$
20,000 voters voted in favor of the amendment.

21. **Strategy** Write and solve a proportion using d to represent the distance between the two cities.

Solution
$$\frac{\frac{3}{8}}{25}=\frac{2\frac{5}{8}}{d}$$
$$25d\cdot\frac{\frac{3}{8}}{25}=25d\cdot\frac{\frac{21}{8}}{d}$$
$$\frac{3d}{8}=\frac{25\cdot 21}{8}$$
$$3d=25\cdot 21$$
$$3d=525$$
$$d=175$$
The distance between the two cities is 175 mi.

23. Strategy Write and solve a proportion using s to represent the additional sales tax. The total sales tax is $780 + s$.

Solution

$$\frac{12{,}000}{780} = \frac{13{,}500}{780 + s}$$
$$780(780 + s) \cdot \frac{12{,}000}{780} = 780(780 + s) \cdot \frac{13{,}500}{780 + s}$$
$$(780 + s) \cdot 12{,}000 = 780 \cdot 13{,}500$$
$$9{,}360{,}000 + 12{,}000s = 10{,}530{,}000$$
$$12{,}000s = 1{,}170{,}000$$
$$s = 97.50$$

The sales tax will be $97.50 higher.

25. Strategy Write and solve a proportion using x to represent the height of the person.

Solution

$$\frac{1.25}{x} = \frac{1}{54}$$
$$54x \cdot \frac{1.25}{x} = 54x \cdot \frac{1}{54}$$
$$67.5 = x$$

The person is 67.5 in. tall.

27. Strategy Write and solve a proportion using x to represent the number of elk in the preserve.

Solution

$$\frac{2}{15} = \frac{10}{x}$$
$$2x = 150$$
$$x = 75$$

There are approximately 75 elk in the preserve.

Objective C Exercises

29. Strategy Triangle ABC is similar to triangle DEF. Write and solve a proportion, using x to represent the length of AC.

Solution

$$\frac{AB}{DE} = \frac{AC}{DF}$$
$$\frac{4}{9} = \frac{x}{15}$$
$$45 \cdot \frac{4}{9} = 45 \cdot \frac{x}{15}$$
$$20 = 3x$$
$$6.66 \approx x$$

The length of AC is approximately 6.7 cm.

31. Strategy Triangle ABC is similar to triangle DEF. Write and solve a proportion, using x to represent the height of triangle ABC.

Solution

$$\frac{CB}{FE} = \frac{\text{height of triangle } ABC}{\text{height of triangle } DEF}$$
$$\frac{5}{12} = \frac{x}{7}$$
$$84 \cdot \frac{5}{12} = 84 \cdot \frac{x}{7}$$
$$35 = 12x$$
$$2.9 \approx x$$

The height of triangle ABC is approximately 2.9 m.

33. Strategy To find the perimeter of triangle DEF:

- Use a proportion to find the length of DF.
- Add the measures of the three sides of triangle DEF.

Solution

$$\frac{AB}{DE} = \frac{AC}{DF}$$
$$\frac{4}{6} = \frac{5}{x}$$
$$6x \cdot \frac{4}{6} = 6x \cdot \frac{5}{x}$$
$$4x = 30$$
$$x = 7.5$$
$$\text{Perimeter} = DE + EF + DE$$
$$= 6 + 9 + 7.5$$
$$= 22.5$$

The perimeter of triangle DEF is 22.5 ft.

35. Strategy To find the area of triangle ABC:
- Use a proportion to find the height of triangle ABC. Let h represent the height of triangle ABC.
- Use the formula for the area of a triangle to find the area. The base of triangle ABC is 12 m.

Solution

$$\frac{AB}{DE} = \frac{\text{height of triangle } ABC}{\text{height of triangle } DEF}$$
$$\frac{12}{18} = \frac{h}{12}$$
$$36 \cdot \frac{12}{18} = 36 \cdot \frac{h}{12}$$
$$24 = 3h$$
$$8 = h$$
$$A = \frac{1}{2}bh$$
$$= \frac{1}{2}(12)(8)$$
$$= 48$$

The area of the triangle is $48\,\text{m}^2$.

37. Strategy $\angle CAE = \angle CBD$ and $\angle C = \angle C$, thus triangle AEC is similar to triangle BDC because two angles of triangle BDC are equal to two angles of triangle AEC. Write and solve a proportion to find the length of BC.

Solution

$$\frac{AE}{BD} = \frac{AC}{BC}$$
$$\frac{8}{5} = \frac{10}{BC}$$
$$5BC \cdot \frac{8}{5} = 5BC \cdot \frac{10}{BC}$$
$$8BC = 50$$
$$BC = 6.25$$

The length of BC is 6.25 cm.

39. Strategy Triangle ABC is similar to triangle DBE. Use a proportion to find AB. $DA = AB - DB$

Solution

$$\frac{AC}{DE} = \frac{AB}{DB}$$
$$\frac{10}{6} = \frac{15}{DB}$$
$$6DB \cdot \frac{10}{6} = 6DB \cdot \frac{15}{DB}$$
$$10DB = 90$$
$$DB = 9$$
$$DA = AB - DB$$
$$DA = 15 - 9$$
$$DA = 6$$

The length of DA is 6 in.

41. Strategy Triangle MNO is similar to triangle PQO. Solve a proportion to find the length of OP. Let x represent the length of OP and $39 - x$ represent the length of OM.

Solution

$$\frac{NO}{OQ} = \frac{OM}{OP}$$
$$\frac{24}{12} = \frac{39 - x}{x}$$
$$12x \cdot \frac{24}{12} = 12x \cdot \frac{39 - x}{x}$$
$$24x = 468 - 12x$$
$$36x = 468$$
$$x = 13$$

The length of OP is 13 cm.

43. Strategy Triangle ABO is similar to triangle DCO. Write and solve a proportion to find the distance DC.

Solution

$$\frac{AB}{DC} = \frac{BO}{OC}$$
$$\frac{14}{DC} = \frac{8}{20}$$
$$20DC \cdot \frac{14}{DC} = 20DC \cdot \frac{8}{20}$$
$$280 = 8DC$$
$$35 = DC$$

The distance across the river is 35 m.

Applying the Concepts

45. Strategy Write and solve a proportion using x to represent the first person's share of the winnings.

Solution

$$\frac{25}{90} = \frac{x}{4.5}$$
$$90\left(\frac{25}{90}\right) = 90\left(\frac{x}{4.5}\right)$$
$$25 = 20x$$
$$1.25 = x$$

The first person won \$1.25 million.

47. Strategy Write and solve a proportion using x to represent the number of foul shots made by the player.

Solution

$$\frac{5}{1} = \frac{x}{42}$$
$$42(5) = 42\left(\frac{x}{42}\right)$$
$$210 = x$$

The player made 210 foul shots.

Section 6.7

Objective A Exercises

1.
$$\begin{aligned} 3x + y &= 10 \\ 3x + (-3x) + y &= -3x + 10 \\ y &= -3x + 10 \end{aligned}$$

3.
$$\begin{aligned} 4x - y &= 3 \\ 4x + (-4x) - y &= -4x + 3 \\ -y &= -4x + 3 \\ (-1)(-y) &= (-1)(-4x + 3) \\ y &= 4x - 3 \end{aligned}$$

5.
$$\begin{aligned} 3x + 2y &= 6 \\ 3x + (-3x) + 2y &= -3x + 6 \\ 2y &= -3x + 6 \\ \frac{1}{2} \cdot 2y &= \frac{1}{2}(-3x + 6) \\ y &= -\frac{3}{2}x + 3 \end{aligned}$$

7.
$$\begin{aligned} 2x - 5y &= 10 \\ 2x + (-2x) - 5y &= -2x + 10 \\ -5y &= -2x + 10 \\ \left(-\frac{1}{5}\right)(-5y) &= -\frac{1}{5}(-2x + 10) \\ y &= \frac{2}{5}x - 2 \end{aligned}$$

9.
$$\begin{aligned} 2x + 7y &= 14 \\ 2x + (-2x) + 7y &= -2x + 14 \\ 7y &= -2x + 14 \\ \frac{1}{7} \cdot 7y &= \frac{1}{7}(-2x + 14) \\ y &= -\frac{2}{7}x + 2 \end{aligned}$$

11.
$$\begin{aligned} x + 3y &= 6 \\ x + (-x) + 3y &= -x + 6 \\ 3y &= -x + 6 \\ \frac{1}{3} \cdot 3y &= \frac{1}{3}(-x + 6) \\ y &= -\frac{1}{3}x + 2 \end{aligned}$$

13.
$$\begin{aligned} y - 2 &= 3(x + 2) \\ y - 2 &= 3x + 6 \\ y - 2 + 2 &= 3x + 6 + 2 \\ y &= 3x + 8 \end{aligned}$$

15.
$$\begin{aligned} y - 1 &= -\frac{2}{3}(x + 6) \\ y - 1 &= -\frac{2}{3}x - 4 \\ y - 1 + 1 &= -\frac{2}{3}x - 4 + 1 \\ y &= -\frac{2}{3}x - 3 \end{aligned}$$

17.
$$\begin{aligned} x + 6y &= 10 \\ x + 6y + (-6y) &= -6y + 10 \\ x &= -6y + 10 \end{aligned}$$

19.
$$\begin{aligned} 2x - y &= 6 \\ 2x - y + y &= y + 6 \\ 2x &= y + 6 \\ \frac{1}{2} \cdot 2x &= \frac{1}{2}(y + 6) \\ x &= \frac{1}{2}y + 3 \end{aligned}$$

21.
$$\begin{aligned} 4x + 3y &= 12 \\ 4x + 3y + (-3y) &= -3y + 12 \\ 4x &= -3y + 12 \\ \frac{1}{4} \cdot 4x &= \frac{1}{4}(-3y + 12) \\ x &= -\frac{3}{4}y + 3 \end{aligned}$$

23.
$$\begin{aligned} x - 4y - 3 &= 0 \\ x - 4y - 3 + 3 &= 3 \\ x - 4y &= 3 \\ x - 4y + 4y &= 4y + 3 \\ x &= 4y + 3 \end{aligned}$$

25.
$$\begin{aligned} d &= rt \\ \frac{1}{r} \cdot d &= \frac{1}{r} \cdot rt \\ \frac{d}{r} &= t \end{aligned}$$

27.
$$\begin{aligned} PV &= nRT \\ \frac{1}{nR} \cdot PV &= \frac{1}{nR} \cdot nRT \\ \frac{PV}{nR} &= T \end{aligned}$$

29.
$$\begin{aligned} P &= 2l + 2w \\ P + (-2w) &= 2l + 2w + (-2w) \\ P - 2w &= 2l \\ \frac{1}{2}(P - 2w) &= \frac{1}{2} \cdot 2l \\ \frac{P - 2w}{2} &= l \end{aligned}$$

31.
$$\begin{aligned} A &= \frac{1}{2}h(b_1 + b_2) \\ 2 \cdot A &= 2 \cdot \frac{1}{2}h(b_1 + b_2) \\ 2A &= h(b_1 + b_2) \\ 2A &= hb_1 + hb_2 \\ 2A + (-hb_2) &= hb_1 + hb_2 + (-hb_2) \\ 2A - hb_2 &= hb_1 \\ \frac{1}{h}(2A - hb_2) &= \frac{1}{h}(hb_1) \\ \frac{2A - hb_2}{h} &= b_1 \end{aligned}$$

33.
$$\begin{aligned} V &= \frac{1}{3}Ah \\ 3 \cdot V &= 3 \cdot \frac{1}{3}Ah \\ 3V &= Ah \\ \frac{1}{A} \cdot 3V &= \frac{1}{A} \cdot Ah \\ \frac{3V}{A} &= h \end{aligned}$$

35.
$$R = \frac{C-S}{t}$$
$$t \cdot R = t \cdot \frac{C-S}{t}$$
$$Rt = C - S$$
$$Rt + (-C) = C + (-C) - S$$
$$Rt - C = -S$$
$$-1(Rt - C) = -1(-S)$$
$$C - Rt = S$$

37.
$$A = P + Prt$$
$$A = P(1 + rt)$$
$$\frac{1}{1+rt} \cdot A = \frac{1}{1+rt} \cdot P(1+rt)$$
$$\frac{A}{1+rt} = P$$

39.
$$A = Sw + w$$
$$A = w(S+1)$$
$$\frac{1}{S+1} \cdot A = \frac{1}{S+1} \cdot w(S+1)$$
$$\frac{A}{S+1} = w$$

Applying the Concepts

41. a.
$$B = \frac{F}{S-V}$$
$$(S-V) \cdot B = (S-V) \cdot \frac{F}{S-V}$$
$$BS - BV = F$$
$$BS = F + BV$$
$$S = \frac{F+BV}{B}$$

b. $S = \frac{20{,}000 + (200)(80)}{200} = 180$

The required selling price is \$180.

c. $S = \frac{15{,}000 + (600)(50)}{600} = 75$

The required selling price is \$75.

Section 6.8

Objective A Exercises

1. The rate of work is the amount of a task that is completed per unit of time.

3. **Strategy** • Time for both sprinklers to fill the fountain: t

	Rate	Time	Part
First sprinkler	$\frac{1}{3}$	t	$\frac{t}{3}$
Second sprinkler	$\frac{1}{6}$	t	$\frac{t}{6}$

• The sum of the parts completed by each sprinkler must equal 1.

Solution
$$\frac{t}{3} + \frac{t}{6} = 1$$
$$6\left(\frac{t}{3} + \frac{t}{6}\right) = 6 \cdot 1$$
$$2t + t = 6$$
$$3t = 6$$
$$t = 2$$

It will take 2 h to fill the fountain with both sprinklers working.

5. **Strategy** • Time to remove the earth with both skiploaders working together: t

	Rate	Time	Part
First skiploader	$\frac{1}{12}$	t	$\frac{t}{12}$
Larger skiploader	$\frac{1}{4}$	t	$\frac{t}{4}$

• The sum of the task completed by each skiploader must equal 1.

Solution
$$\frac{t}{12} + \frac{t}{4} = 1$$
$$12\left(\frac{t}{12} + \frac{t}{4}\right) = 12 \cdot 1$$
$$t + 3t = 12$$
$$4t = 12$$
$$t = 3$$

With both skiploaders working together, it would take 3 h to remove the earth.

7. **Strategy** • Time for the computers working together: t

	Rate	Time	Part
First computer	$\frac{1}{75}$	t	$\frac{t}{75}$
Second computer	$\frac{1}{50}$	t	$\frac{t}{50}$

• The sum of the parts of the task completed by each computer must equal 1.

Solution
$$\frac{t}{75} + \frac{t}{50} = 1$$
$$150\left(\frac{t}{75} + \frac{t}{50}\right) = 150 \cdot 1$$
$$2t + 3t = 150$$
$$5t = 150$$
$$t = 30$$

With both computers working, it would take 30 h to solve the problem.

9. Strategy • Time to cool the room with both air conditioners working: t

	Rate	Time	Part
Small air conditioner	$\frac{1}{75}$	t	$\frac{t}{75}$
Large air conditioner	$\frac{1}{50}$	t	$\frac{t}{50}$

• The sum of the parts completed by each air conditioner must equal 1.

Solution

$$\frac{t}{75}+\frac{t}{50}=1$$
$$150\left(\frac{t}{75}+\frac{t}{50}\right)=150\cdot 1$$
$$2t+3t=150$$
$$5t=150$$
$$t=30$$

It would take 30 min to cool the room with both air conditioners working.

11. Strategy • Time for the second pipeline to fill the tank: t

	Rate	Time	Part
First pipeline	$\frac{1}{45}$	30	$\frac{30}{45}$
Second pipeline	$\frac{1}{t}$	30	$\frac{30}{t}$

• The sum of the parts completed by each pipeline must equal 1.

Solution

$$\frac{30}{45}+\frac{30}{t}=1$$
$$45t\left(\frac{30}{45}+\frac{30}{t}\right)=45t\cdot 1$$
$$30t+1350=45t$$
$$1350=15t$$
$$90=t$$

It would take the second pipeline 90 min to fill the tank.

13. Strategy • Time for the apprentice working alone to complete the wall: t

	Rate	Time	Part
Mason	$\frac{1}{10}$	6	$\frac{6}{10}$
Apprentice	$\frac{1}{t}$	6	$\frac{6}{t}$

• The sum of the parts completed by each mason must equal 1.

Solution

$$\frac{6}{10}+\frac{6}{t}=1$$
$$10t\left(\frac{6}{10}+\frac{6}{t}\right)=10t\cdot 1$$
$$6t+60=10t$$
$$60=4t$$
$$15=t$$

It would take the apprentice 15 h to construct the wall.

15. Strategy • Time for the second technician to complete the task: t

	Rate	Time	Part
First technician	$\frac{1}{4}$	2	$\frac{2}{4}$
Second technician	$\frac{1}{6}$	t	$\frac{t}{6}$

• The sum of the parts of the task completed by each technician must equal 1.

Solution

$$\frac{2}{4}+\frac{t}{6}=1$$
$$12\left(\frac{2}{4}+\frac{t}{6}\right)=12\cdot 1$$
$$6+2t=12$$
$$2t=6$$
$$t=3$$

It will take the second technician 3 h to complete the wiring.

17. Strategy • Time for one of the welders to complete the job: t

	Rate	Time	Part
First welder	$\frac{1}{t}$	10	$\frac{10}{t}$
Second welder	$\frac{1}{t}$	30	$\frac{30}{t}$

• The sum of the parts of the task completed by each welder must equal 1.

Solution

$$\frac{10}{t}+\frac{30}{t}=1$$
$$t\left(\frac{10}{t}+\frac{30}{t}\right)=t\cdot 1$$
$$10+30=t$$
$$40=t$$

It would have taken one of the welders 40 h to complete the welds.

19. Strategy • Time for one machine to complete the task: t

	Rate	Time	Part
First machine	$\frac{1}{t}$	7	$\frac{7}{t}$
Second machine	$\frac{1}{t}$	21	$\frac{21}{t}$

• The sum of the parts of the task completed by each machine must equal 1.

Solution

$$\frac{7}{t}+\frac{21}{t}=1$$
$$t\left(\frac{7}{t}+\frac{21}{t}\right)=t\cdot 1$$
$$7+21=t$$
$$28=t$$

It would have taken one machine 28 h to fill the boxes.

Objective B Exercises

21. Strategy
- Find the rate of the jogger, r.
- Multiply the rate by $2h$.

Solution

$$r = \frac{24mi}{3h} = \frac{8mi}{h}$$
$$\frac{8mi}{h} \cdot 2h = 16mi$$

The jogger ran 16 mi in 2 h.

23. Strategy
- Rate of the technician through the congested traffic: r

	Distance	Rate	Time
Congested traffic	10	r	$\frac{10}{r}$
Expressway	20	$r+20$	$\frac{20}{r+10}$

- The total time for the trip was 1 hour.

Solution

$$\frac{10}{r} + \frac{20}{r+20} = 1$$
$$r(r+20)\left(\frac{10}{r} + \frac{20}{r+20}\right) = r(r+20)$$
$$(r+20)10 + 20r = r^2 + 20r$$
$$10r + 200 + 20r = r^2 + 20r$$
$$r^2 - 10r - 200 = 0$$
$$(r-20)(r+10) = 0$$

$r - 20 = 0 \qquad r + 10 = 0$
$r = 20 \qquad r = -10$

The solution -10 is not possible because the rate cannot be negative. The rate of travel in the congested area was 20 mph.

25. Strategy
- Rate of jogger: r
  Rate of cyclist: $r + 12$

	Distance	Rate	Time
Jogger	8	r	$\frac{8}{r}$
Cyclist	20	$r+12$	$\frac{20}{r+12}$

- The time of the jogger equals the time of the cyclist.

Solution

$$\frac{8}{r} = \frac{20}{r+12}$$
$$8(r+12) = 20r$$
$$8r + 96 = 20r$$
$$96 = 12r$$
$$8 = r$$

$20 = r + 12$

The rate of the jogger is 8 mph.
The rate of the cyclist is 20 mph.

27. Strategy
- Rate of the helicopter: r
  Rate of the jet: $4r$

	Distance	Rate	Time
Helicopter	180	r	$\frac{180}{r}$
Jet	1080	$4r$	$\frac{1080}{4r}$

- The total time for the trip was 5 h.

Solution

$$\frac{180}{r} + \frac{1080}{4r} = 5$$
$$4r\left(\frac{180}{r} + \frac{1080}{4r}\right) = 4r \cdot 5$$
$$720 + 1080 = 20r$$
$$1800 = 20r$$
$$90 = r$$

$4r = 90 \cdot 4 = 360$

The rate of the jet is 360 mph.

29. Strategy
- Walking rate: r
  Running rate: $r + 3$

	Distance	Rate	Time
Walking	$\frac{2}{3}r$	r	$\frac{2}{3}$
Running	$\frac{1}{3}(r+3)$	$r+3$	$\frac{1}{3}$

Note: Convert minutes to hours.
- Total distance walked: 5 mi.

Solution

$$\frac{2}{3}r + \frac{1}{3}(r+3) = 5$$
$$\frac{2}{3}r + \frac{1}{3}r + 1 = 5$$
$$r = 4$$

Camille's walking rate is 4 mph.

31. Strategy
- Rate of the cyclist: r
  Rate of the car: $r + 36$

	Distance	Rate	Time
Cyclist	96	r	$\frac{96}{r}$
Car	384	$r+36$	$\frac{384}{r+36}$

- The time of the cyclist equals the time of the car.

Solution

$$\frac{96}{r} = \frac{384}{r+36}$$
$$96(r+36) = 384r$$
$$96r + 3456 = 384r$$
$$3456 = 288r$$
$$12 = r$$
$$48 = r + 36$$

The rate of the car is 48 mph.

33. Strategy • Rate of the wind: r

	Distance	Rate	Time
With the wind	600	$180+r$	$\frac{600}{180+r}$
Against the wind	480	$180-r$	$\frac{480}{180-r}$

• The time with the wind equals the time against the wind.

Solution

$$\frac{600}{180+r} = \frac{480}{180-r}$$
$$(180+r)(180-r)\frac{600}{180+r} = (180+r)(180-r)\frac{480}{180-r}$$
$$(180-r)(600) = (180+r)(480)$$
$$108{,}000 - 600r = 86{,}400 + 480r$$
$$21{,}600 = 1080r$$
$$20 = r$$

The rate of the wind is 20 mph.

35. Strategy • Rate of the gulf current: r

	Distance	Rate	Time
With gulf current	170	$28+r$	$\frac{170}{28+r}$
Against gulf current	110	$28-r$	$\frac{110}{28-r}$

• The time with the gulf current equals the time against the gulf current.

Solution

$$\frac{170}{28+r} = \frac{110}{28-r}$$
$$170(28-r) = 110(28+r)$$
$$4760 - 170r = 3080 + 110r$$
$$1680 = 280r$$
$$6 = r$$

The rate of the gulf current is 6 mph.

37. Strategy • Constant rate of trucker: r

	Distance	Rate	Time
Constant rate	330	r	$\frac{330}{r}$
Reduced rate	30	$r-25$	$\frac{30}{r-25}$

• The total time for the trip is 7 h.

Solution

$$\frac{330}{r} + \frac{30}{r-25} = 7$$
$$r(r-25)\left(\frac{330}{r} + \frac{30}{r-25}\right) = r(r-25)\cdot 7$$
$$330(r-25) + 30r = 7r(r-25)$$
$$330r - 8250 + 30r = 7r^2 - 175r$$
$$360r - 8250 = 7r^2 - 175r$$
$$0 = 7r^2 - 535r + 8250$$
$$0 = (7r-150)(r-55)$$

$0 = 7r - 150$ $\quad$ $0 = r - 55$

$150 = 7r$ $\quad$ $55 = r$

$\frac{150}{7} = r$

$21\frac{3}{7} = r$

The solution $21\frac{3}{7}$ mph is not possible because the rate would be negative when the trucker had to reduce the speed by 25 mph. The rate of the trucker for the first 330 mi was 55 mph.

Applying the Concepts

39. Strategy • Usual speed: r
Speed in bad weather: $r-10$

	Distance	Rate	Time
Usual	150	r	$\frac{150}{r}$
Bad weather	150	$r-10$	$\frac{150}{r-10}$

• The time during bad weather is $\frac{1}{2}$ h more than the usual time.

Solution

$$\frac{150}{r-10}=\frac{150}{r}+\frac{1}{2}$$
$$2r(r-10)\frac{150}{r-10}=2r(r-10)\left(\frac{150}{r}+\frac{1}{2}\right)$$
$$300r=300(r-10)+r(r-10)$$
$$300r=300r-3000+r^2-10r$$
$$0=r^2-10r-3000$$
$$0=(r-60)(r+50)$$

$r-60=0 \qquad r+50=0$
$r=60 \qquad r=-50$

The solution -50 is not possible because rate cannot be negative. The bus usually travels 60 mph.

Chapter 6 Review Exercises

1. $$\frac{6a^2b^7}{25x^3y}\div\frac{12a^3b^4}{5x^2y^2}=\frac{6a^2b^7}{25x^3y}\cdot\frac{5x^2y^2}{12a^3b^4}=\frac{\overset{1}{\cancel{6}}\cdot\overset{1}{\cancel{5}}a^2b^7x^2y^2}{\underset{5}{\cancel{25}}\cdot\underset{2}{\cancel{12}}a^3b^4x^3y}$$
$$=\frac{b^3y}{10ax}$$

2. $$\frac{x+7}{15x}+\frac{x-2}{20x}$$
$$=\frac{4(x+7)}{60x}+\frac{3(x-2)}{60x}=\frac{4x+28+3x-6}{60x}$$
$$=\frac{7x+22}{60x}$$

3. $$\frac{3x^3+9x^2}{6xy^2-18y^2}\cdot\frac{4xy^3-12y^3}{5x^2+15x}=\frac{3x^2\overset{1}{\cancel{(x+3)}}}{6y^2\underset{1}{\cancel{(x-3)}}}\cdot\frac{4y^3\overset{1}{\cancel{(x-3)}}}{5x\underset{1}{\cancel{(x+3)}}}$$
$$=\frac{12x^2y^3}{30xy^2}$$
$$=\frac{2xy}{5}$$

4. $$\frac{2x(x-y)}{x^2y(x+y)}\div\frac{3(x-y)}{x^2y^2}=\frac{2x\overset{1}{\cancel{(x-y)}}}{x^2y(x+y)}\cdot\frac{x^2y^2}{3\underset{1}{\cancel{(x-y)}}}$$
$$=\frac{2x\cdot x^2y^2}{3x^2y(x+y)}$$
$$=\frac{2xy}{3(x+y)}$$

5. $$\frac{x-\frac{16}{5x-2}}{3x-4-\frac{88}{5x-2}}$$
$$\frac{(5x-2)\left(x-\frac{16}{5x-2}\right)}{(5x-2)\left(3x-4-\frac{88}{5x-2}\right)}=\frac{x(5x-2)-16}{(3x-4)(5x-2)-88}$$
$$=\frac{5x^2-2x-16}{15x^2-26x+8-88}$$
$$=\frac{5x^2-2x-16}{15x^2-26x-80}$$
$$=\frac{(5x+8)(x-2)}{(5x+8)(3x-10)}$$
$$=\frac{x-2}{3x-10}$$

6. $$\frac{x^2+x-30}{15+2x-x^2}$$
$$\frac{x^2+x-30}{-1(x^2-2x-15)}=\frac{(x+6)(x-5)}{-(x-5)(x+3)}$$
$$=-\frac{x+6}{x+3}$$

7. $$\frac{16x^5y^3}{24xy^{10}}=\frac{2x^4}{3y^7}$$

8. $$\frac{20}{x+2}=\frac{5}{16}$$
$$16(x+2)\left(\frac{20}{x+2}\right)=16(x+2)\left(\frac{5}{16}\right)$$
$$16(20)=(x+2)5$$
$$320=5x+10$$
$$310=5x$$
$$62=x$$

9. $$\frac{10-23y+12y^2}{6y^2-y-5}\div\frac{4y^2-13y+10}{18y^2+3y-10}$$
$$=\frac{10-23y+12y^2}{6y^2-y-5}\cdot\frac{18y^2+3y-10}{4y^2-13y+10}$$
$$=\frac{(2-3y)(5-4y)}{(6y+5)(y-1)}\cdot\frac{(3y-2)(6y+5)}{(4y-5)(y-2)}$$
$$=\frac{(2-3y)\overset{-1}{\cancel{(5-4y)}}}{\cancel{(6y+5)}(y-1)}\cdot\frac{(3y-2)\overset{1}{\cancel{(6y+5)}}}{\cancel{(4y-5)}(y-2)}$$
$$=\frac{-(2-3y)(3y-2)}{(y-1)(y-2)}=\frac{(3y-2)(3y-2)}{(y-1)(y-2)}$$
$$=\frac{(3y-2)^2}{(y-1)(y-2)}$$

10. $$3ax-x=5$$
$$x(3a-1)=5$$
$$\frac{x(3a-1)}{3a-1}=\frac{5}{3a-1}$$
$$x=\frac{5}{3a-1}$$

11.
$$\begin{aligned}\frac{2}{x}+\frac{3}{4}&=1\\ 4x\left(\frac{2}{x}+\frac{3}{4}\right)&=4x\cdot 1\\ 8+3x&=4x\\ 8+3x-3x&=4x-3x\\ 8&=x\end{aligned}$$

12.
$$\begin{aligned}\frac{x}{y}+\frac{3}{x}&=\frac{x}{x}\cdot\frac{x}{y}+\frac{y}{y}\cdot\frac{3}{x}\\ &=\frac{x^2}{xy}+\frac{3y}{xy}\\ &=\frac{x^2+3y}{xy}\end{aligned}$$

13.
$$\begin{aligned}5x+4y&=20\\ 5x-5x+4y&=20-5x\\ 4y&=20-5x\\ y&=\frac{20-5x}{4}\\ y&=-\frac{5}{4}x+5\end{aligned}$$

14.
$$\frac{8ab^2}{15x^3y}\cdot\frac{5xy^4}{16a^2b}$$
$$=\frac{\overset{1}{\cancel{8}}\cdot\overset{1}{\cancel{5}}ab^2xy^4}{\underset{3}{\cancel{15}}\cdot\underset{2}{\cancel{16}}\,a^2bx^3y}=\frac{by^3}{6ax^2}$$

15.
$$\frac{x^2\left(1-\frac{1}{x}\right)}{x^2\left(1-\frac{8x-7}{x^2}\right)}=\frac{x^2-x}{x^2-8x+7}=\frac{x\overset{1}{\cancel{(x-1)}}}{(x-7)\underset{1}{\cancel{(x-1)}}}=\frac{x}{x-7}$$

16.
$$\frac{x}{12x^2+16x-3}\qquad\frac{4x^2}{6x^2+7x-3}$$
$$(6x-1)(2x+3)\qquad(2x+3)(3x-1)$$
$$\text{LCM}=(6x-1)(2x+3)(3x-1)$$
$$\frac{x(3x-1)}{(6x-1)(2x+3)(3x-1)},\quad\frac{4x^2(6x-1)}{(6x-1)(2x+3)(3x-1)}$$
$$\frac{3x^2-x}{(6x-1)(2x+3)(3x-1)},\quad\frac{24x^3-4x^2}{(6x-1)(2x+3)(3x-1)}$$

17.
$$\begin{aligned}T&=2(ab+bc+ca)\\ T&=2ab+2bc+2ca\\ T-2bc&=2ab+2ca\\ T-2bc&=a(2b+2c)\\ \frac{T-2bc}{2b+2c}&=a\end{aligned}$$

18.
$$\begin{aligned}\frac{5}{7}+\frac{x}{2}&=2-\frac{x}{7}\\ 14\left(\frac{5}{7}+\frac{x}{2}\right)&=14\left(2-\frac{x}{7}\right)\\ 10+7x&=28-2x\\ 9x&=18\\ x&=2\end{aligned}$$

19.
$$\begin{aligned}\frac{2+\frac{1}{x}}{3-\frac{2}{x}}&=\frac{2+\frac{1}{x}}{3-\frac{2}{x}}\cdot\frac{x}{x}\\ &=\frac{2x+1}{3x-2}\end{aligned}$$

20.
$$\begin{aligned}\frac{2x}{x-5}-\frac{x+1}{x-2}&=\frac{2x}{x-5}\cdot\frac{x-2}{x-2}-\frac{x+1}{x-2}\cdot\frac{x-5}{x-5}\\ &=\frac{2x^2-4x}{(x-5)(x-2)}-\frac{x^2-4x-5}{(x-5)(x-2)}\\ &=\frac{2x^2-4x-(x^2-4x-5)}{(x-5)(x-2)}\\ &=\frac{2x^2-4x-x^2+4x+5}{(x-5)(x-2)}\\ &=\frac{x^2+5}{(x-5)(x-2)}\end{aligned}$$

21.
$$\begin{aligned}i&=\frac{100m}{c}\\ ic&=100m\\ c&=\frac{100m}{i}\end{aligned}$$

22.
$$\begin{aligned}\frac{x+8}{x+4}&=1+\frac{5}{x+4}\\ (x+4)\left(\frac{x+8}{x+4}\right)&=(x+4)\left(1+\frac{5}{x+4}\right)\\ x+8&=x+4+5\\ x+8&=x+9\\ 8&\neq 9\end{aligned}$$
The equation has no solution.

23.
$$\begin{aligned}&\frac{20x^2-45x}{6x^3+4x^2}\div\frac{40x^3-90x^2}{12x^2+8x}\\ &=\frac{20x^2-45x}{6x^3+4x^2}\cdot\frac{12x^2+8x}{40x^3-90x^2}\\ &=\frac{5x(4x-9)}{2x^2(3x+2)}\cdot\frac{4x(3x+2)}{10x^2(4x-9)}\\ &=\frac{5x\overset{1}{\cancel{(4x-9)}}}{2x^2\underset{1}{\cancel{(3x+2)}}}\cdot\frac{4x\overset{1}{\cancel{(3x+2)}}}{10x^2\underset{1}{\cancel{(4x-9)}}}\\ &=\frac{1}{x^2}\end{aligned}$$

24.
$$\frac{2y}{5y-7}+\frac{3}{7-5y}=\frac{2y}{5y-7}-\frac{3}{5y-7}=\frac{2y-3}{5y-7}$$

25.
$$\begin{aligned}\frac{5x+3}{2x^2+5x-3}-\frac{3x+4}{2x^2+5x-3}&=\frac{5x+3-(3x+4)}{2x^2+5x-3}\\ &=\frac{5x+3-3x-4}{2x^2+5x-3}\\ &=\frac{2x-1}{2x^2+5x-3}\\ &=\frac{2x-1}{(2x-1)(x+3)}\\ &=\frac{1}{x+3}\end{aligned}$$

26. $10x^2 - 11x + 3 \qquad 20x^2 - 17x + 3$

$(5x-3)(2x-1) \qquad (5x-3)(4x-1)$

$\text{LCM} = (5x-3)(2x-1)(4x-1)$

27.
$$\begin{aligned} 4x + 9y &= 18 \\ 9y &= -4x + 18 \\ \frac{9y}{9} &= \frac{-4x}{9} + \frac{18}{9} \\ y &= -\frac{4}{9}x + 2 \end{aligned}$$

28.
$$\begin{aligned} \frac{2x^2-5x-3}{3x^2-7x-6}\cdot\frac{3x^2+8x+4}{x^2+4x+4} &= \frac{(2x+1)\cancel{(x-3)}}{\cancel{(3x+2)}\cancel{(x-3)}}\cdot\frac{\cancel{(3x+2)}\cancel{(x+2)}}{\cancel{(x+2)}(x+2)} \\ &= \frac{2x+1}{x+2} \end{aligned}$$

29.
$$\begin{aligned} \frac{20}{2x+3} &= \frac{17x}{2x+3} - 5 \\ (2x+3)\left(\frac{20}{2x+3}\right) &= \left(\frac{17x}{2x+3} - 5\right) \\ 20 &= 17x - 5(2x+3) \\ 20 &= 17x - 10x - 15 \\ 20 &= 7x - 15 \\ 35 &= 7x \\ 5 &= x \end{aligned}$$

30.
$$\begin{aligned} \frac{x-1}{x+2} + \frac{3x-2}{5-x} + \frac{5x^2+15x-11}{x^2-3x-10} &= \frac{x-1}{x+2} - \frac{3x-2}{x-5} + \frac{5x^2+15x-11}{(x-5)(x+2)} \\ &= \frac{(x-1)(x-5) - (3x-2)(x+2) + 5x^2 + 15x - 11}{(x+2)(x-5)} \\ &= \frac{x^2 - 6x + 5 - (3x^2 + 4x - 4) + 5x^2 + 15x - 11}{(x+2)(x-5)} \\ &= \frac{x^2 - 6x + 5 - 3x^2 - 4x + 4 + 5x^2 + 15x - 11}{(x+2)(x-5)} \\ &= \frac{3x^2+5x-2}{(x+2)(x-5)} = \frac{(3x-1)\cancel{(x+2)}}{\cancel{(x+2)}(x-5)} = \frac{3x-1}{x-5} \end{aligned}$$

31.
$$\begin{aligned} \frac{6}{x-7} &= \frac{8}{x-6} \\ (x-7)(x-6)\left(\frac{6}{x-7}\right) &= (x-7)(x-6)\left(\frac{8}{x-6}\right) \\ (x-6)6 &= (x-7)8 \\ 6x - 36 &= 8x - 56 \\ -2x &= -20 \\ x &= 10 \end{aligned}$$

32.
$$\begin{aligned} \frac{3}{20} &= \frac{x}{80} \\ 80\left(\frac{3}{20}\right) &= 80\left(\frac{x}{80}\right) \\ 12 &= x \end{aligned}$$

33. Strategy Triangle NMO is similar to triangle QPO. Solve a proportion to find the length of QO. Let x represent the length of QO and $25 - x$ represent the length of NO.

Solution

$$\frac{MO}{PO} = \frac{NO}{QO}$$
$$\frac{6}{9} = \frac{25 - x}{x}$$
$$9x \cdot \frac{6}{9} = 9x \cdot \frac{25 - x}{x}$$
$$6x = 9(25 - x)$$
$$6x = 225 - 9x$$
$$15x = 225$$
$$x = 15$$

The length of QO is 15 cm.

34. Strategy To find the area of triangle DEF:
- Solve proportions to find the height (h) of triangle DEF and the length of DE.
- Use the height and the length of the base to find the area of triangle DEF.

Solution

$$\frac{AB}{DF} = \frac{8}{h} \qquad \frac{AB}{DF} = \frac{AC}{DE}$$
$$\frac{9}{12} = \frac{8}{h} \qquad \frac{9}{12} = \frac{12}{DE}$$
$$h = 10\frac{2}{3} \qquad DE = 16$$

The height is $10\frac{2}{3}$ in. The base is 16 in.

$$A = \frac{1}{2}bh$$
$$= \frac{1}{2} \cdot 16 \cdot \frac{32}{3}$$
$$= \frac{256}{3}$$

The area of triangle DEF is $\frac{256}{3}$ in^2.

35. Strategy
- Time to fill the pool using both hoses: t

	Rate	Time	Part
First hose	$\frac{1}{15}$	t	$\frac{t}{15}$
Second hose	$\frac{1}{10}$	t	$\frac{t}{10}$

- The sum of the parts of the task completed by each hose must equal 1.

Solution

$$\frac{t}{15} + \frac{t}{10} = 1$$
$$30\left(\frac{t}{15} + \frac{t}{10}\right) = 30 \cdot 1$$
$$2t + 3t = 30$$
$$5t = 30$$
$$t = 6$$

It would take 6 h to fill the pool.

36. Strategy
- Rate of the bus: r

	Distance	Rate	Time
Bus	245	r	$\frac{245}{r}$
Car	315	$r + 10$	$\frac{315}{r+10}$

- The time of the bus equals the time of the car.

Solution

$$\frac{245}{r} = \frac{315}{r + 10}$$
$$315r = 245(r + 10)$$
$$315r = 245r + 2450$$
$$70r = 2450$$
$$r = 35$$
$$r + 10 = 45$$

The rate of the car is 45 mph.

37. Strategy
- Rate of the wind: r

	Distance	Rate	Time
With wind	2100	$400 + r$	$\frac{2100}{400+r}$
Against wind	1900	$400 - r$	$\frac{1900}{400-r}$

- The time with the wind equals the time against the wind.

Solution

$$\frac{2100}{400 + r} = \frac{1900}{400 - r}$$
$$2100(400 - r) = 1900(400 + r)$$
$$840{,}000 - 2100r = 760{,}000 + 1900r$$
$$80{,}000 = 4000r$$
$$20 = r$$

The rate of the wind is 20 mph.

38. Unknown runs: x

Write and solve a proportion.

$$\frac{15}{100} = \frac{x}{9}$$
$$900\left(\frac{15}{100}\right) = 900\left(\frac{x}{9}\right)$$
$$135 = 100x$$
$$1.35 = x$$

The ERA is 1.35.

Chapter 6 Test

1. $\frac{x}{x + 3} - \frac{2x - 5}{x^2 + x - 6}$

LCM $= (x + 3)(x - 2)$

$$\frac{x}{x + 3} = \frac{x(x - 2)}{(x + 3)(x - 2)} = \frac{x^2 - 2x}{(x + 3)(x - 2)}$$
$$\frac{x}{x + 3} - \frac{2x - 5}{x^2 + x - 6}$$
$$= \frac{x^2 - 2x}{(x + 3)(x - 2)} - \frac{2x - 5}{(x + 3)(x - 2)}$$
$$= \frac{x^2 - 2x - 2x + 5}{(x + 3)(x - 2)}$$
$$= \frac{x^2 - 4x + 5}{(x + 3)(x - 2)}$$

2.
$$\frac{3}{x+4}=\frac{5}{x+6}$$
$$(x+4)(x+6)\left(\frac{3}{x+4}\right)=(x+4)(x+6)\left(\frac{5}{x+6}\right)$$
$$(x+6)3=(x+4)5$$
$$3x+18=5x+20$$
$$-2x=2$$
$$x=-1$$

3.
$$\frac{x^2+2x-3}{x^2+6x+9}\cdot\frac{2x^2-11x+5}{2x^2+3x-5}$$
$$=\frac{(x+3)(x-1)}{(x+3)(x+3)}\cdot\frac{(2x-1)(x-5)}{(2x+5)(x-1)}$$
$$=\frac{(2x-1)(x-5)}{(x+3)(2x+5)}$$

4.
$$\frac{16x^5y}{24x^2y^4}=\frac{2x^3}{3y^3}$$

5.
$$d=s+rt$$
$$d-s=rt$$
$$\frac{d-s}{r}=t$$

6.
$$\frac{6}{x}-2=1$$
$$\frac{6}{x}=\frac{3}{1}$$
$$x\left(\frac{6}{x}\right)=x\left(\frac{3}{1}\right)$$
$$6=3x$$
$$2=x$$

7.
$$\frac{x^2+4x-5}{1-x^2}=\frac{(x+5)(x-1)}{(1+x)(1-x)}$$
$$=-\frac{x+5}{x+1}$$

8.
$6x-3 \qquad 2x^2+x-1$

$3(2x-1) \qquad (2x-1)(x+1)$

$\text{LCM}=3(2x-1)(x+1)$

9.
$$\frac{2}{2x-1}-\frac{3}{3x+1}$$
$$=\frac{2(3x+1)-3(2x-1)}{(2x-1)(3x+1)}$$
$$=\frac{6x+2-6x+3}{(2x-1)(3x+1)}=\frac{5}{(2x-1)(3x+1)}$$

10.
$$\frac{x^2+3x+2}{x^2+5x+4}\div\frac{x^2-x-6}{x^2+2x-15}$$
$$=\frac{x^2+3x+2}{x^2+5x+4}\cdot\frac{x^2+2x-15}{x^2-x-6}$$
$$=\frac{(x+2)(x+1)}{(x+4)(x+1)}\cdot\frac{(x+5)(x-3)}{(x-3)(x+2)}$$
$$=\frac{x+5}{x+4}$$

11.
$$\frac{1+\frac{1}{x}-\frac{12}{x^2}}{1+\frac{2}{x}-\frac{8}{x^2}}$$
$$=\frac{x^2\left(1+\frac{1}{x}-\frac{12}{x^2}\right)}{x^2\left(1+\frac{2}{x}-\frac{8}{x^2}\right)}$$
$$=\frac{x^2+x-12}{x^2+2x-8}=\frac{(x+4)(x-3)}{(x+4)(x-2)}$$
$$=\frac{x-3}{x-2}$$

12.
$\dfrac{3}{x^2-2x} \qquad \dfrac{x}{x^2-4}$

$x(x-2) \qquad (x+2)(x-2)$

$\text{LCM}=x(x-2)(x+2)$

$$\frac{3(x+2)}{x(x-2)(x+2)}=\frac{3x+6}{x(x-2)(x+2)}$$
$$\frac{x(x)}{x(x-2)(x+2)}=\frac{x^2}{x(x-2)(x+2)}$$

13.
$$\frac{2x}{x^2+3x-10}-\frac{4}{x^2+3x-10}$$
$$=\frac{2x-4}{x^2+3x-10}$$
$$=\frac{2(x-2)}{(x+5)(x-2)}=\frac{2}{x+5}$$

14.
$$3x-8y=16$$
$$-8y=-3x+16$$
$$\frac{-8y}{-8}=\frac{-3x}{-8}+\frac{16}{-8}$$
$$y=\frac{3}{8}x-2$$

15.
$$\frac{2x}{x+1}-3=\frac{-2}{x+1}$$
$$(x+1)\left(\frac{2x}{x+1}-3\right)=(x+1)\left(\frac{-2}{x+1}\right)$$
$$2x-3(x+1)=-2$$
$$2x-3x-3=-2$$
$$-x=1$$
$$x=-1$$

-1 does not check. The equation has no solution.

16.
$$\frac{x^3y^4}{x^2-4x+4}\cdot\frac{x^2-x-2}{x^6y^4}$$
$$=\frac{x^3y^4}{(x-2)(x-2)}\cdot\frac{(x-2)(x+1)}{x^6y^4}$$
$$=\frac{x+1}{x^3(x-2)}$$

17. Strategy $\angle CAE = \angle CBD$ and $\angle C = \angle C$, thus triangle CAE is similar to triangle CBD. Write and solve a proportion to find the length of CE. Let x represent the length of CD, then
$CE = CD + DE = x + 8$.
$AE = AB + BC = 5 + 3 = 8$.

Solution

$$\frac{AC}{BC} = \frac{CE}{CD}$$
$$\frac{8}{3} = \frac{8+x}{x}$$
$$3x \cdot \frac{8}{3} = 3x \cdot \frac{8+x}{x}$$
$$8x = 24 + 3x$$
$$5x = 24$$
$$x = 4.8$$
$$CE = x + 8 = 4.8 + 8 = 12.8$$

The length of CE is 12.8 ft.

18. Strategy Write and solve a proportion using x to represent the additional salt.

Solution

$$\frac{4}{10} = \frac{x+4}{15}$$
$$30\left(\frac{4}{10}\right) = 30\left(\frac{x+4}{15}\right)$$
$$12 = 2(x+4)$$
$$12 = 2x + 8$$
$$4 = 2x$$
$$2 = x$$

An additional 2 lb of salt are needed.

19. Strategy
- Time to fill the pool with both pipes: t

	Rate	Time	Part
First pipe	$\frac{1}{6}$	t	$\frac{t}{6}$
Second pipe	$\frac{1}{12}$	t	$\frac{t}{12}$

- The sum of the parts of the task completed by each pipe must equal 1.

Solution

$$\frac{t}{6} + \frac{t}{12} = 1$$
$$12\left(\frac{t}{6} + \frac{t}{12}\right) = 12 \cdot 1$$
$$2t + t = 12$$
$$3t = 12$$
$$t = 4$$

It would take 4 h to fill the pool.

20. Strategy
- Rate of the wind: r

	Distance	Rate	Time
With wind	260	$110 + r$	$\frac{260}{110+r}$
Against wind	180	$110 - r$	$\frac{180}{110-r}$

- The time with the wind equals the time against the wind.

Solution

$$\frac{260}{110+r} = \frac{180}{110-r}$$
$$260(110 - r) = 180(110 + r)$$
$$28{,}600 - 260r = 19{,}800 + 180r$$
$$8800 = 440r$$
$$20 = r$$

The rate of the wind is 20 mph.

21. Strategy Write and solve a proportion using x to represent the number of sprinklers needed for a 3600-ft^2 lawn.

Solution

$$\frac{3}{200} = \frac{x}{3600}$$
$$3600\left(\frac{3}{200}\right) = 3600\left(\frac{x}{3600}\right)$$
$$54 = x$$

54 sprinklers are needed for a 3600-ft^2 lawn.

Cumulative Review Exercises

1. $$\left(\frac{2}{3}\right)^2 \div \left(\frac{3}{2} - \frac{2}{3}\right) + \frac{1}{2} = \left(\frac{2}{3}\right)^2 \div \left(\frac{9}{6} - \frac{4}{6}\right) + \frac{1}{2}$$
$$= \frac{4}{9} \div \left(\frac{5}{6}\right) + \frac{1}{2}$$
$$= \frac{4}{\cancel{9}_3} \cdot \frac{\cancel{6}^2}{5} + \frac{1}{2}$$
$$= \frac{8}{15} + \frac{1}{2} = \frac{16}{30} + \frac{15}{30} = \frac{31}{30}$$

2. $-a^2 + (a - b)^2$
$$-(-2)^2 + (-2 - 3)^2 = -4 + (-5)^2$$
$$= -4 + 25 = 21$$

3. $$-2x - (-3y) + 7x - 5y = -2x + 3y + 7x - 5y$$
$$= -2x + 7x + 3y - 5y$$
$$= 5x - 2y$$

4. $$2[3x - 7(x - 3) - 8] = 2[3x - 7x + 21 - 8]$$
$$= 2[-4x + 13]$$
$$= -8x + 26$$

5. $$4 - \frac{2}{3}x = 7$$
$$-\frac{2}{3}x = 3$$
$$-\frac{3}{2}\left(-\frac{2}{3}x\right) = 3\left(-\frac{3}{2}\right)$$
$$x = -\frac{9}{2}$$

6. $$3[x - 2(x - 3)] = 2(3 - 2x)$$
$$3[x - 2x + 6] = 6 - 4x$$
$$3[-x + 6] = 6 - 4x$$
$$-3x + 18 = 6 - 4x$$
$$x = -12$$

7.
$$P \times B = A$$
$$16\frac{2}{3}\% \times 60 = A$$
$$\frac{1}{6} \times 60 = A$$
$$10 = A$$

8. $(a^2b^5)(ab^2) = a^3b^7$

9. $(a - 3b)(a + 4b)$
$= a^2 + 4ab - 3ab - 12b^2$
$= a^2 + ab - 12b^2$

10. $\dfrac{15b^4 - 5b^2 + 10b}{5b}$
$= \dfrac{15b^4}{5b} - \dfrac{5b^2}{5b} + \dfrac{10b}{5b}$
$= 3b^3 - b + 2$

11.
$$\begin{array}{r} x^2 + 2x + 4 \\ x - 2\overline{)x^3 + 0x^2 + 0x - 8} \\ \underline{x^3 - 2x^2} \\ 2x^2 + 0x \\ \underline{2x^2 - 4x} \\ 4x - 8 \\ \underline{4x - 8} \end{array}$$
$(x^3 - 8) \div (x - 2) = x^2 + 2x + 4$

12. $12x^2 - x - 1$
$12 \cdot (-1) = -12$
Factors of -12 whose sum is -1: -4 and 3
$12x^2 - 4x + 3x - 1$
$= 4x(3x - 1) + 1(3x - 1) = (3x - 1)(4x + 1)$

13. $y^2 - 7y + 6$
$1(6) = 6$
Factors of 6 whose sum is -7: -6 and -1
$y^2 - 6y - y + 6$
$= y(y - 6) - 1(y - 6) = (y - 6)(y - 1)$

14. $2a^3 + 7a^2 - 15a$
The GCF is a: $a(2a^2 + 7a - 15)$
$2 \cdot (-15) = -30$
Factors of -30 whose sum is 7: 10 and -3
$a(2a^2 + 10a - 3a - 15)$
$= a[2a(a + 5) - 3(a + 5)] = a(a + 5)(2a - 3)$

15. $4b^2 - 100$
The GCF is 4: $4(b^2 - 25)$
$4[b^2 - (5)^2] = 4(b + 5)(b - 5)$

16. $(x + 3)(2x - 5) = 0$
$x + 3 = 0 \qquad 2x - 5 = 0$
$x = -3 \qquad 2x = 5$
$x = \dfrac{5}{2}$

17. $\dfrac{12x^4y^2}{18xy^7} = \dfrac{2x^3}{3y^5}$

18. $\dfrac{x^2 - 7x + 10}{25 - x^2}$
$= \dfrac{\overset{1}{\cancel{(x-5)}}(x - 2)}{\underset{-1}{\cancel{(5-x)}}(5 + x)} = -\dfrac{x - 2}{x + 5}$

19. $\dfrac{x^2 - x - 56}{x^2 + 8x + 7} \div \dfrac{x^2 - 13x + 40}{x^2 - 4x - 5}$
$= \dfrac{x^2 - x - 56}{x^2 + 8x + 7} \cdot \dfrac{x^2 - 4x - 5}{x^2 - 13x + 40}$
$= \dfrac{\cancel{(x-8)}\cancel{(x+7)}}{\cancel{(x+7)}\cancel{(x+1)}} \cdot \dfrac{\cancel{(x-5)}\cancel{(x+1)}}{\cancel{(x-8)}\cancel{(x-5)}} = 1$

20. $\dfrac{2}{2x - 1} - \dfrac{1}{x + 1}$
$= \dfrac{2(x + 1) - 1(2x - 1)}{(2x - 1)(x + 1)}$
$= \dfrac{2x + 2 - 2x + 1}{(2x - 1)(x + 1)} = \dfrac{3}{(2x - 1)(x + 1)}$

21. $\dfrac{1 - \frac{2}{x} - \frac{15}{x^2}}{1 - \frac{25}{x^2}}$
$= \dfrac{x^2\left(1 - \frac{2}{x} - \frac{15}{x^2}\right)}{x^2\left(1 - \frac{25}{x^2}\right)}$
$= \dfrac{x^2 - 2x - 15}{x^2 - 25}$
$= \dfrac{\cancel{(x-5)}(x + 3)}{\cancel{(x-5)}(x + 5)} = \dfrac{x + 3}{x + 5}$

22.
$$\frac{3x}{x - 3} - 2 = \frac{10}{x - 3}$$
$$(x - 3)\left(\frac{3x}{x - 3} - 2\right) = (x - 3)\left(\frac{10}{x - 3}\right)$$
$$3x - 2(x - 3) = 10$$
$$3x - 2x + 6 = 10$$
$$x + 6 = 10$$
$$x = 4$$

23.
$$\frac{2}{x - 2} = \frac{12}{x + 3}$$
$$(x - 2)(x + 3)\left(\frac{2}{x - 2}\right) = (x - 2)(x + 3)\left(\frac{12}{x + 3}\right)$$
$$(x + 3)2 = (x - 2)12$$
$$2x + 6 = 12x - 24$$
$$-10x = -30$$
$$x = 3$$

24.
$$f = v + at$$
$$f - v = at$$
$$\frac{f - v}{a} = t$$

25. **Strategy** • The unknown number: x

Solution $5x - 13 = -8$
$5x = 5$
$x = 1$

26. Strategy • Percent of silver in the alloy: x

	Amount	Percent	Quantity
40% silver	60	0.40	0.40(60)
Silver alloy	120	x	$120x$
Mixture	180	0.60	0.60(180)

• The sum of the quantities before mixing is equal to the quantity after mixing.

Solution

$$0.40(60) + 120x = 0.60(180)$$
$$24 + 120x = 108$$
$$120x = 84$$
$$x = 0.70$$

The silver alloy is 70% silver.

27. Strategy • Height: x
Base: $2x - 2$
• Use the equation for the area of a triangle.

Solution

$$\frac{1}{2}bh = A$$
$$\frac{1}{2}(2x-2)x = 30$$
$$(x-1)x = 30$$
$$x^2 - x = 30$$
$$x^2 - x - 30 = 0$$
$$(x-6)(x+5) = 0$$
$$x - 6 = 0$$
$$x = 6$$
$$2x - 2 = 12 - 2 = 10$$

The base is 10 in.
The height is 6 in.

28. Strategy Write and solve a proportion using x to represent the cost for a \$5000 policy.

Solution

$$\frac{16}{1000} = \frac{x}{5000}$$
$$\overset{5}{\cancel{5000}} \cdot \frac{16}{\cancel{1000}} = \frac{x}{\cancel{5000}} \cdot \cancel{5000}$$
$$80 = x$$

The cost of a \$5000 policy is \$80.

29. Strategy • Time for both pipes working together: t

	Rate	Time	Part
First pipe	$\frac{1}{9}$	t	$\frac{t}{9}$
Second pipe	$\frac{1}{18}$	t	$\frac{t}{18}$

• The sum of the parts of the task completed by each pipe must equal 1.

Solution

$$\frac{t}{9} + \frac{t}{18} = 1$$
$$18\left(\frac{t}{9} + \frac{t}{18}\right) = 18 \cdot 1$$
$$2t + t = 18$$
$$3t = 18$$
$$t = 6$$

It would take both pipes 6 min to fill the tank.

30. Strategy • Rate of the current: r

	Distance	Rate	Time
With current	14	$5+r$	$\frac{14}{5+r}$
Against current	6	$5-r$	$\frac{6}{5-r}$

• The time with the current equals the time against the current.

Solution

$$\frac{14}{5+r} = \frac{6}{5-r}$$
$$(5+r)(5-r)\left(\frac{14}{5+r}\right) = (5+r)(5-r)\left(\frac{6}{5-r}\right)$$
$$(5-r)14 = (5+r)6$$
$$70 - 14r = 30 + 6r$$
$$-20r = -40$$
$$r = 2$$

The rate of the current is 2 mph.

Chapter 7: Linear Equations in Two Variables

Prep Test

1. $-\dfrac{5-(-7)}{4-8} = -\dfrac{5+7}{-4} = \dfrac{12}{4} = 3$

2. $\dfrac{a-b}{c-d}$; $a = 3, b = -2, c = -3, d = 2$

$\dfrac{3-(-2)}{(-3)-2} = \dfrac{3+2}{-3-2} = \dfrac{5}{-5} = -1$

3. $-3(x-4) = -3x + 12$

4.
$$\begin{aligned} 3x + 6 &= 0 \\ 3x &= -6 \\ x &= -2 \end{aligned}$$

5.
$$\begin{aligned} 4x + 5y &= 0;\ y = 0 \\ 4x + 5(0) &= 20 \\ 4x &= 20 \\ x &= 5 \end{aligned}$$

6.
$$\begin{aligned} 3x - 7y &= 11;\ x = -1 \\ 3(-1) - 7y &= 11 \\ -3 - 7y &= 11 \\ -7y &= 14 \\ y &= -2 \end{aligned}$$

7. $\dfrac{12x - 15}{-3} = -4x + 5$

8.
$$\begin{aligned} \frac{2x+1}{3} &= \frac{3x}{4} \\ 12 \cdot \frac{2x+1}{3} &= 12 \cdot \frac{3x}{4} \\ 8x + 4 &= 9x \\ 4 &= x \end{aligned}$$

9.
$$\begin{aligned} 3x - 5y &= 15 \\ -5y &= -3x + 15 \\ y &= \frac{-3x+15}{-5} \\ y &= \frac{3}{5}x - 3 \end{aligned}$$

10.
$$\begin{aligned} y + 3 &= -\frac{1}{2}(x+4) \\ y + 3 &= -\frac{1}{2}x - 2 \\ y &= -\frac{1}{2}x - 5 \end{aligned}$$

Go Figure

A B C D

$\dfrac{AB}{AC} = \dfrac{1}{4}$ $\dfrac{BC}{CD} = \dfrac{1}{2}$

Since $\dfrac{AB}{AC} = \dfrac{1}{4}$, $AC = 4AB$

Also,
$$\begin{aligned} AC &= AB + BC \\ 4AB &= AB + BC \\ 3AB &= BC \end{aligned}$$

So,
$$\begin{aligned} \frac{BC}{CD} &= \frac{1}{2} \\ \frac{3AB}{CD} &= \frac{1}{2} \\ \frac{AB}{CD} &= \frac{1}{6} \end{aligned}$$

The ratio of AB to CD is $\dfrac{1}{6}$.

Section 7.1

Objective A Exercises

1.

3.

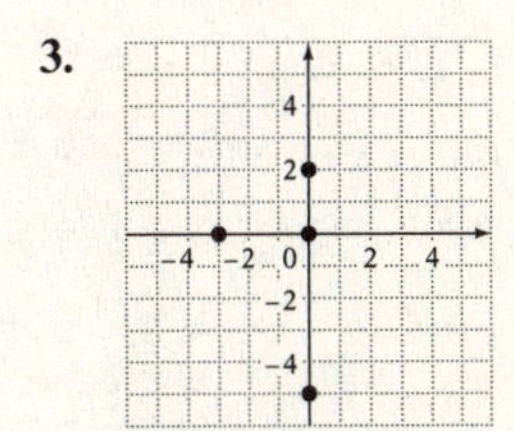

5.

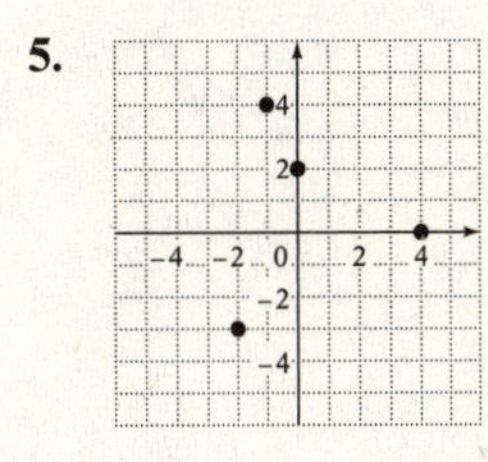

7. $A(2, 3)$
$B(4, 0)$
$C(-4, 1)$
$D(-2, -2)$

9. $A(-2, 5)$
$B(3, 4)$
$C(0, 0)$
$D(-3, -2)$

11. **a.** Abscissa of point A: 2
Abscissa of point C: -4

b. Ordinate of point B: 1
Ordinate of point D: -3

13. Students should explain that the ordered pairs are being plotted in reverse order. In an ordered pair, the first number indicates a movement to the left or right, and the second number indicates a movement up or down.

Objective B Exercises

15.
$$\begin{array}{c|l} y = -x+7 & \\ \hline 4 & -(3)+7 \\ & -3+7 \\ & 4 \end{array}$$
$4 = 4$
Yes, (3, 4) is a solution of $y = -x+7$.

17.
$$\begin{array}{c|l} \multicolumn{2}{c}{y = \frac{1}{2}x - 1} \\ \hline 2 & \frac{1}{2}(-1) - 1 \\ & -\frac{1}{2} - 1 \\ & -\frac{3}{2} \end{array}$$
$2 \neq -\frac{3}{2}$
No, $(-1, 2)$ is not a solution of $y = \frac{1}{2}x - 1$.

19.
$$\begin{array}{r|l} \multicolumn{2}{c}{2x - 5y = 4} \\ \hline 2(4) - 5(1) & 4 \\ 8 - 5 & 4 \\ 3 \neq & 4 \end{array}$$
No, (4, 1) is not a solution of $2x - 5y = 4$.

21.
$$\begin{array}{r|l} \multicolumn{2}{c}{3x - 4y = -4} \\ \hline 3(0) - 4(4) & -4 \\ 0 - 16 & -4 \\ -16 \neq & -4 \end{array}$$
No, (0, 4) is not a solution of $3x - 4y = -4$.

23.

x	$y = 2x$
-2	$2(-2)$
-1	$2(-1)$
0	$2(0)$
2	$2(2)$

y	(x, y)
-4	$(-2, -4)$
-2	$(-1, -2)$
0	(0, 0)
4	(2, 4)

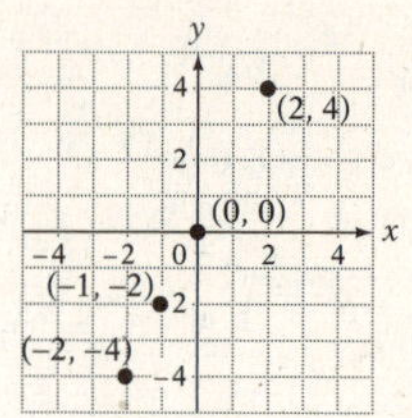

25.

x	$y = \frac{2}{3}x + 1$
-3	$\frac{2}{3}(-3) + 1$
0	$\frac{2}{3}(0) + 1$
3	$\frac{2}{3}(3) + 1$

y	(x, y)
-1	$(-3, -1)$
1	(0, 1)
3	(3, 3)

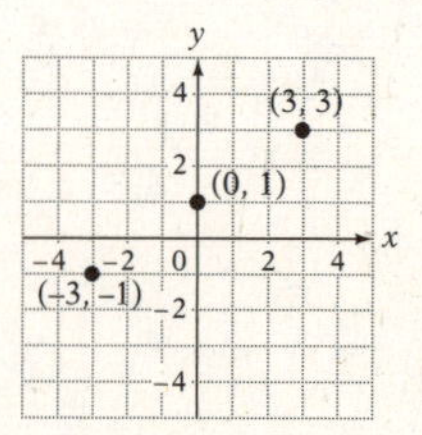

27. Solve $2x + 3y = 6$ for y.
$$2x + 3y = 6$$
$$3y = -2x + 6$$
$$y = -\frac{2}{3}x + 2$$

x	$y = -\frac{2}{3}x + 2$
-3	$-\frac{2}{3}(-3) + 2$
0	$-\frac{2}{3}(0) + 2$
3	$-\frac{2}{3}(3) + 2$

y	(x, y)
4	$(-3, 4)$
2	(0, 2)
0	(3, 0)

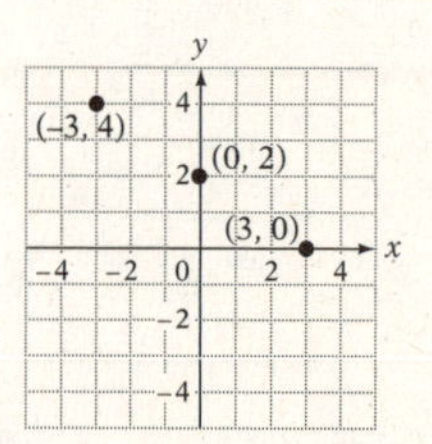

Objective C Exercises

29. {(24, 600), (32, 750), (22, 430), (15, 300), (4.4, 68), (17, 370), (15, 310), (4.4, 55)}. No, the relation is not a function. The two ordered pairs (4.4, 68) and (4.4, 55) have the same first component but different second components.

31. Find the home runs per at-bat for each player by dividing the player's number of home runs by the number of at-bats.

Bonds: $45 \div 390 \approx 0.115$
Pujois: $43 \div 591 \approx 0.073$
Sosa: $40 \div 517 \approx 0.077$
Sheffield: $39 \div 576 \approx 0.068$
Bagwell: $39 \div 605 \approx 0.064$

The relation is {(390, 0.115), (591, 0.073), (517, 0.077), (576, 0.068), (605, 0.064)}.
Yes, the relation is a function since no two ordered pairs have the same first coordinate.

33. The relation is {(−2, 1), (−1, −1), (0, −3), (3, −9)}. No two ordered pairs have the same first coordinate, so, yes, y is a function of x.

35. The relation is {(1, 0), (2, 1), (2, −1), (3, 2), (3, −2), (4, 3), (4, −3)}. No, y is not a function of x. Several ordered pairs have the same first coordinate but different second coordinates.

37. The relation is {(−2, 4), (−1, 1), (0, 0), (1, 1), (2, 4)}. No two ordered pairs have the same first coordinate, so, yes, y is a function of x.

Objective D Exercises

39. $f(x) = 3x - 4$
$f(4) = 3(4) - 4 = 12 - 4 = 8$

41. $f(x) = x^2$
$f(3) = 3^2 = 9$

43. $G(x) = x^2 + x$
$G(-2) = (-2)^2 + (-2) = 4 - 2 = 2$

45. $s(t) = \dfrac{3}{t-1}$

$s(-2) = \dfrac{3}{-2-1} = \dfrac{3}{-3} = -1$

47. $h(x) = 3x^2 - 2x + 1$
$h(3) = 3(3)^2 - 2(3) + 1 = 3(9) - 6 + 1$
$= 27 - 6 + 1 = 22$

49. $f(x) = \dfrac{x}{x+5}$

$f(-3) = \dfrac{-3}{-3+5} = \dfrac{-3}{2} = -\dfrac{3}{2}$

51. $g(x) = x^3 - x^2 + 2x - 7$
$g(0) = 0^3 - 0^2 + 2(0) - 7$
$= 0 - 0 + 0 - 7 = -7$

Applying the Concepts

53. A relation and a function are similar in that both are sets of ordered pairs. A function is a specific type of relation. A function is a relation in which there are no two ordered pairs with the same first coordinate. This can also be stated as follows: A function is a relation in which no two ordered pairs with the same first coordinate have different second coordinates.

55. No, it is not possible to evaluate $f(x) = \dfrac{5}{x-1}$ when $x = 1$. When $x = 1$, the denominator is $1 - 1 = 0$, and division by 0 is not defined.

Section 7.2

Objective A Exercises

1.

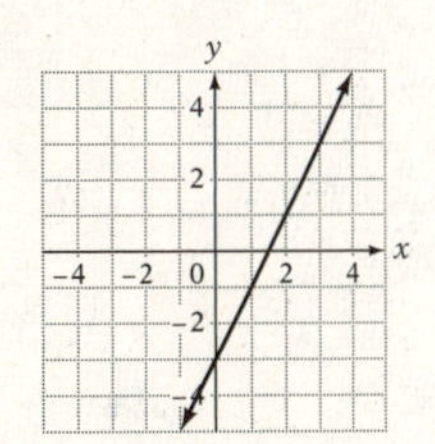

3.

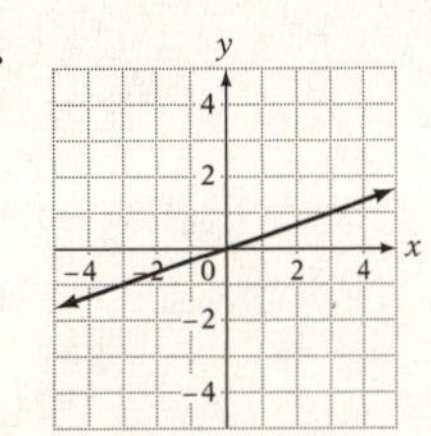

5.

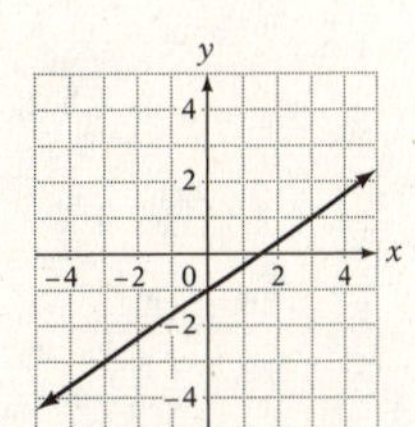

7.

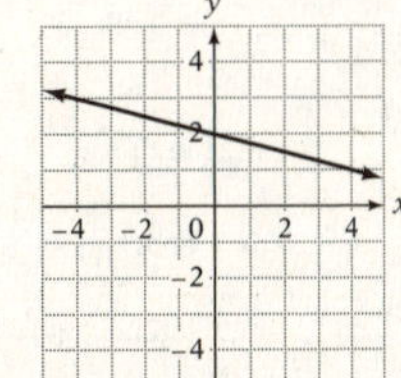

9.

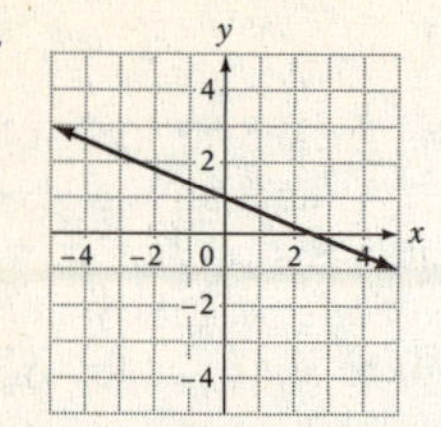

11.

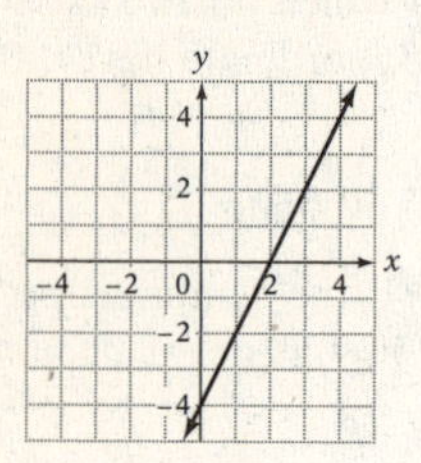

13.

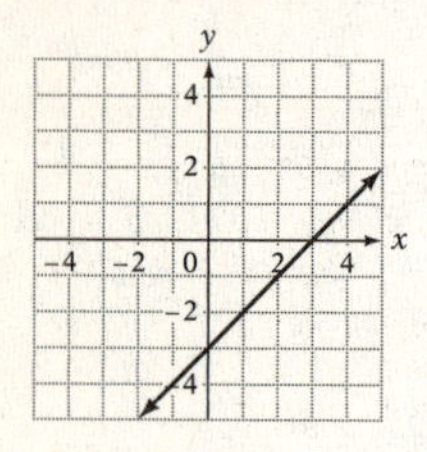

15.

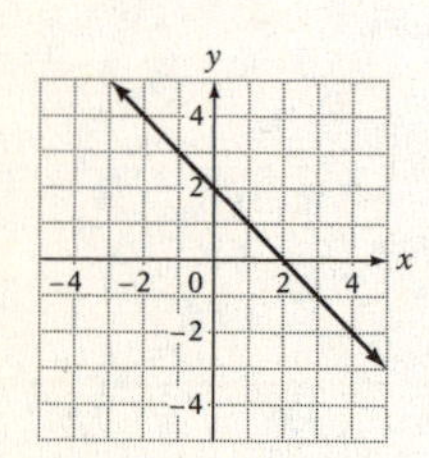

17.

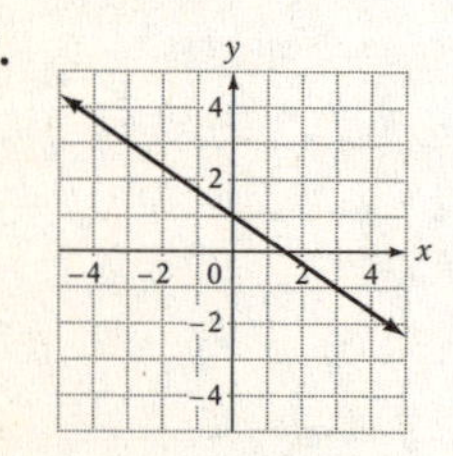

Objective B Exercises

19.

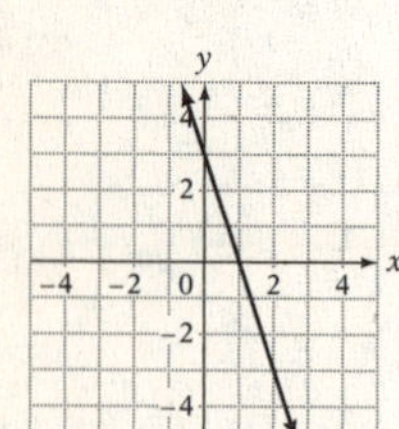

21.

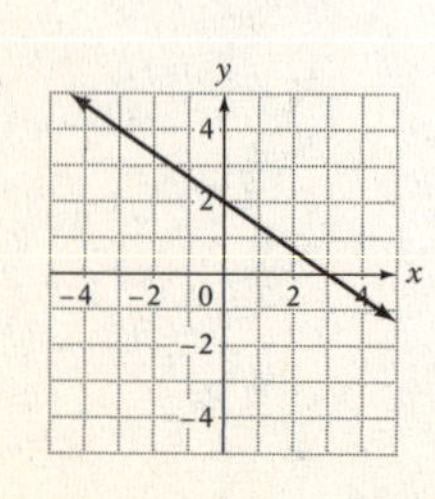

23.

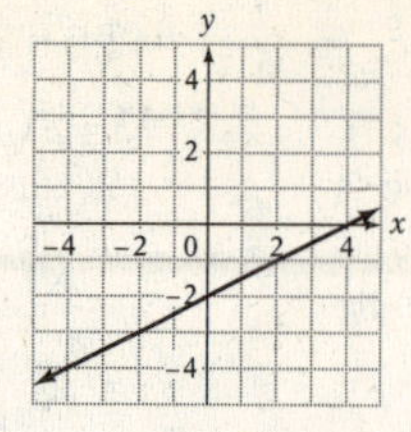

25.

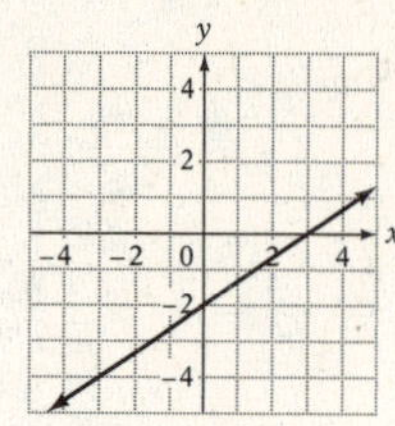

27.

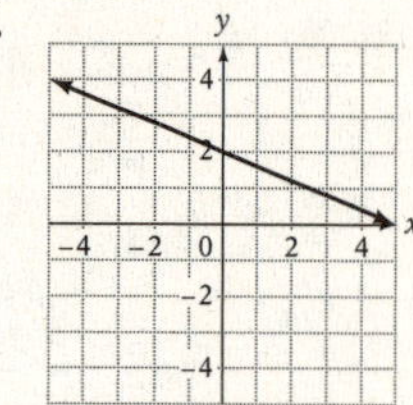

29.

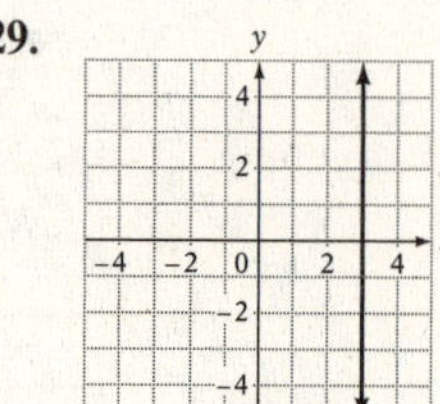

31.

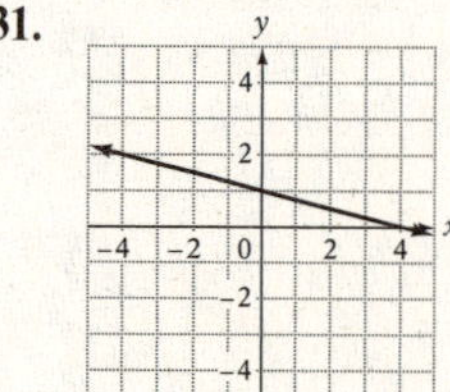

33.

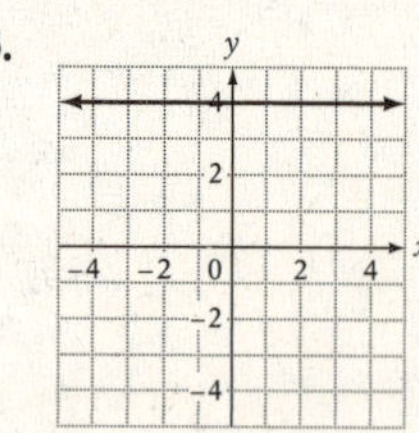

35.

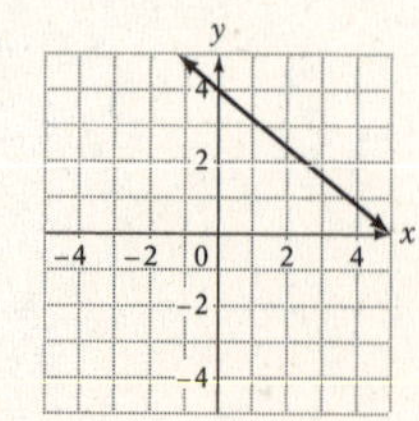

Objective C Exercises

37.

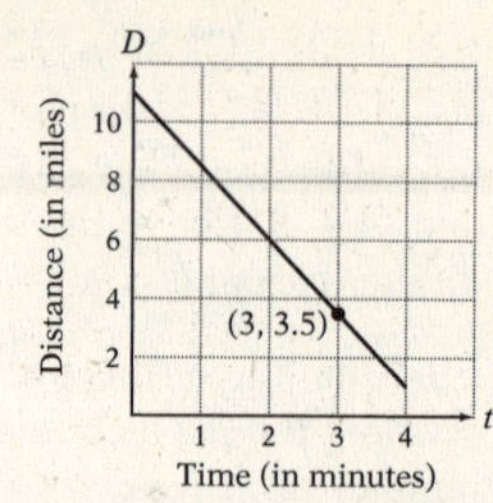

The ordered pair (3, 3.5) means that after flying for 3 min, the helicopter is 3.5 mi away from the victims.

39.

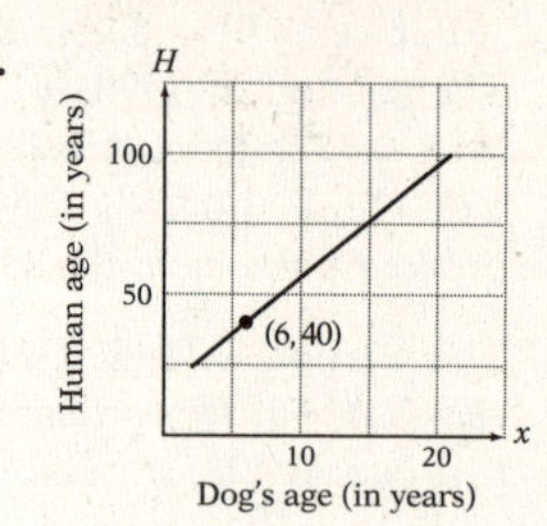

The ordered pair (6, 40) means that a dog 6 years old is equivalent in age to a human 40 years old.

Applying the Concepts

41.

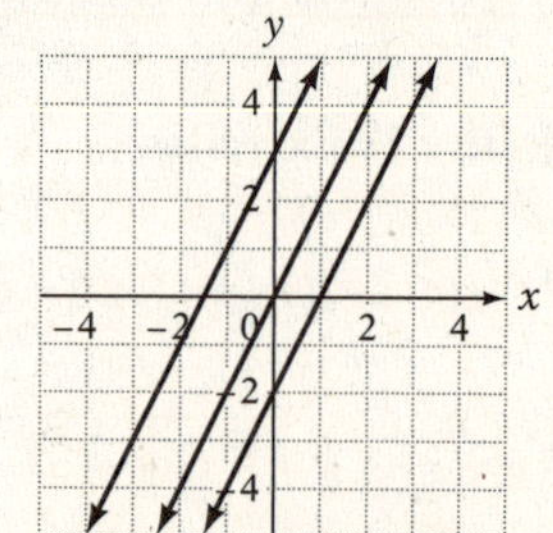

All the graphs are parallel.

43. $y = 3x + 2$

$x = 1 \quad y = 3(1) + 2 = 5$

$x = 2 \quad y = 3(2) + 2 = 8$

The value of y increases. The change in y is 3.

$x = 13 \quad y = 3(13) + 2 = 41$

$x = 14 \quad y = 3(14) + 2 = 44$

The change in y is 3.

45. a. Use $C = 0.99$ when $0 < t \leq 15$.

For $t = 5$, $C = 0.99$.

The cost of a 5-minute call is $.99.

b. Use $C = 0.15(t - 15) + 0.99$ when $t > 15$.
For $t = 20$,
$C = 0.15(20 - 15) + 0.99 = 0.75 + 0.99 = 1.74$.
The cost of a 20 minute call is \$1.74.

Section 7.3

Objective A Exercises

1. x-intercept
$x - y = 3$
$x - 0 = 3$
$x = 3$
$(3, 0)$

y-intercept
$x - y = 3$
$0 - y = 3$
$-y = 3$
$y = -3$
$(0, -3)$

3. x-intercept
$y = 3x - 6$
$0 = 3x - 6$
$6 = 3x$
$2 = x$
$(2, 0)$

y-intercept
$(0, b)$
$b = -6$
$(0, -6)$

5. x-intercept
$x - 5y = 10$
$x - 5(0) = 10$
$x = 10$
$(10, 0)$

y-intercept
$x - 5y = 10$
$0 - 5y = 10$
$-5y = 10$
$y = -2$
$(0, -2)$

7. x-intercept
$y = 3x + 12$
$0 = 3x + 12$
$-12 = 3x$
$-4 = x$
$(-4, 0)$

y-intercept
$(0, b)$
$b = 12$
$(0, 12)$

9. x-intercept
$2x - 3y = 0$
$2x - 3(0) = 0$
$2x = 0$
$x = 0$
$(0, 0)$

y-intercept
$2x - 3y = 0$
$2(0) - 3y = 0$
$-3y = 0$
$y = 0$
$(0, 0)$

11. x-intercept
$y = -\frac{1}{2}x + 3$
$0 = -\frac{1}{2}x + 3$
$\frac{1}{2}x = 3$
$x = 6$
$(6, 0)$

y-intercept
$(0, b)$
$b = 3$
$(0, 3)$

13. x-intercept
$5x + 2y = 10$
$5x + 2(0) = 10$
$5x = 10$
$x = 2$
$(2, 0)$

y-intercept
$5x + 2y = 10$
$5(0) + 2y = 10$
$2y = 10$
$y = 5$
$(0, 5)$

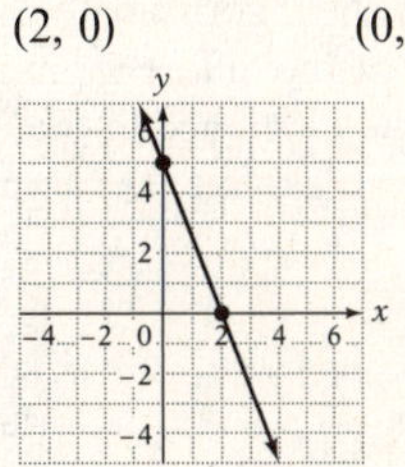

15. x-intercept
$y = \frac{3}{4}x - 3$
$0 = \frac{3}{4}x - 3$
$3 = \frac{3}{4}x$
$4 = x$
$(4, 0)$

y-intercept
$(0, b)$
$b = -3$
$(0, -3)$

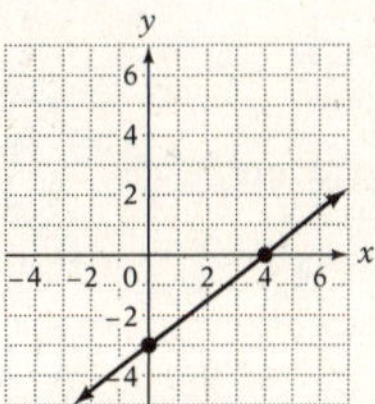

17. x-intercept
$5y - 3x = 15$
$5(0) - 3x = 15$
$-3x = 15$
$x = -5$
$(-5, 0)$

y-intercept
$5y - 3x = 15$
$5y - 3(0) = 15$
$5y = 15$
$y = 3$
$(0, 3)$

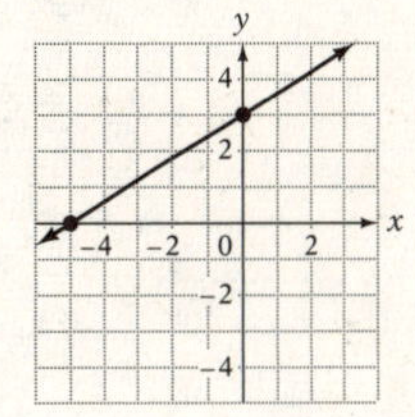

Objective B Exercises

19. The slope of a nonvertical line is calculated by selecting two points on the line, finding the difference between the y-coordinates, finding the difference between the x-coordinates, and then writing the ratio of the difference in the y-coordinates to the difference in the x-coordinates. For a vertical line, the slope is undefined.

21. $P_1(4, 2), P_2(3, 4)$
$m = \frac{y_2 - y_1}{x_2 - x_1} = \frac{4-2}{3-4} = \frac{2}{-1} = -2$
The slope is -2.

23. $P_1(-1, 3), P_2(2, 4)$
$m = \frac{y_2 - y_1}{x_2 - x_1} = \frac{4-3}{2-(-1)} = \frac{1}{3}$
The slope is $\frac{1}{3}$.

25. $P_1(2, 4), P_2(4, -1)$
$m = \frac{y_2 - y_1}{x_2 - x_1} = \frac{-1-4}{4-2} = \frac{-5}{2} = -\frac{5}{2}$
The slope is $-\frac{5}{2}$.

27. $P_1(3, -4), P_2(3, 5)$
$m = \frac{y_2 - y_1}{x_2 - x_1} = \frac{5-(-4)}{3-3} = \frac{9}{0}$
The slope is undefined.

29. $P_1(4, -2), P_2(3, -2)$
$m = \frac{y_2 - y_1}{x_2 - x_1} = \frac{-2-(-2)}{3-4} = \frac{0}{-1} = 0$
The line has zero slope.

31. $P_1(0, -1), P_2(3, -2)$
$m = \frac{y_2 - y_1}{x_2 - x_1} = \frac{-2-(-1)}{3-0} = \frac{-1}{3} = -\frac{1}{3}$
The slope is $-\frac{1}{3}$.

33. $P_1(-3, 4), P_2(2, -5)$
$m = \frac{-5-4}{2-(-3)} = -\frac{9}{5}$
$Q_1(3, 6), Q_2(-2, -3)$
$m = \frac{-3-6}{-2-3} = \frac{9}{5}$
Neither. The slopes are not equal nor is their product -1.

35. $P_1(0, 1), P_2(2, 4)$
$m = \frac{4-1}{2-0} = \frac{3}{2}$
$Q_1(-4, -7), Q_2(2, 5)$
$m = \frac{5-(-7)}{2-(-4)} = \frac{12}{6} = 2$
Neither. The slopes are not equal nor is their product -1.

37. $P_1(-2, 4), P_2(2, 4)$
$m = \frac{4-4}{2-(-2)} = 0$
$Q_1(-3, 6), Q_2(4, 6)$
$m = \frac{6-6}{4-(-3)} = 0$
Parallel. The slopes are equal.

39. $P_1(7, -1), P_2(-4, 6)$
$m = \frac{6-(-1)}{-4-7} = -\frac{7}{11}$
$Q_1(3, 0), Q_2(-5, 3)$
$m = \frac{3-0}{-5-3} = -\frac{3}{8}$
Neither. The slopes are not equal nor is their product -1.

41. From the graph, let $(x_1, y_1) = (1, 55)$ and $(x_2, y_2) = (5, 187)$.
$m = \frac{y_2 - y_1}{x_2 - x_1} = \frac{187-55}{5-1} = \frac{132}{4} = 33$
The sales are given in millions and $x = 0$ represents the year 2002.
$m = 33$. The worldwide sales of camera-phones are increasing by 33 million units per year.

43. From the graph, let $(x_1, y_1) = (15, 30{,}125)$ and $(x_2, y_2) = (30, 27{,}425)$
$m = \frac{y_2 - y_1}{x_2 - x_1} = \frac{27{,}425 - 30{,}125}{30-15} = \frac{-2700}{15} = -180$
$m = -180$. The value of the car is decreasing \$180 for each additional 1000 mi the car is driven.

Objective C Exercises

45. Write the equation in slope-intercept form.
$2x - 3y = 6$
$-3y = -2x + 6$
$y = \frac{2}{3}x - 2$
The slope is $\frac{2}{3}$; the y-intercept is $(0, -2)$.

47. Write the equation in slope-intercept form.
$2x + 5y = 10$
$5y = -2x + 10$
$y = -\frac{2}{5}x + 2$
The slope is $-\frac{2}{5}$; the y-intercept is $(0, 2)$.

49. Write the equation in slope-intercept form.
$x - 4y = 0$
$-4y = -x$
$y = \frac{1}{4}x$
The slope is $\frac{1}{4}$; the y-intercept is $(0, 0)$.

51.

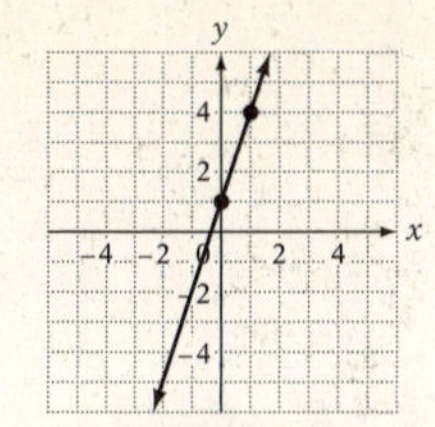

53.

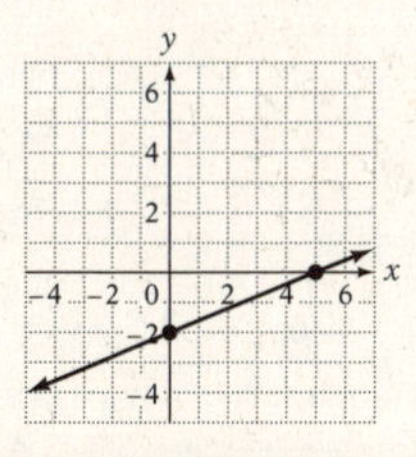

55.

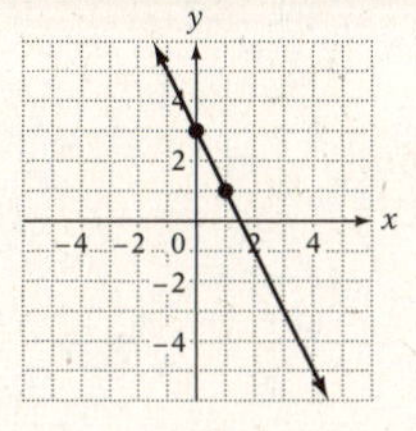

57.

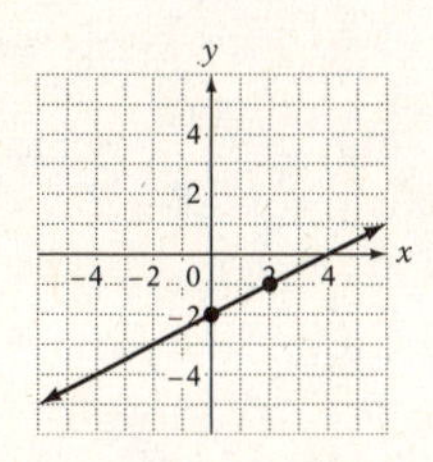

59.

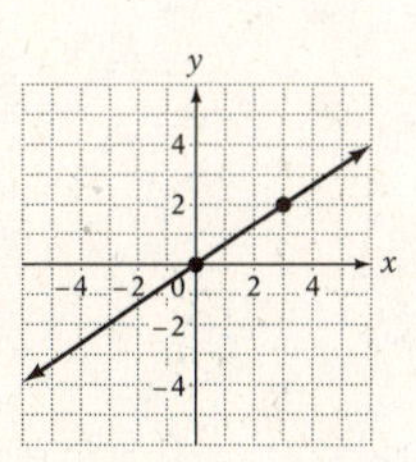

61.

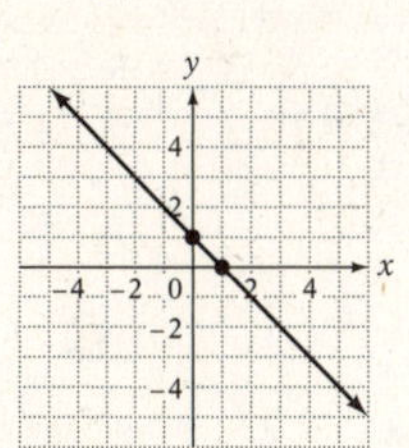

63.

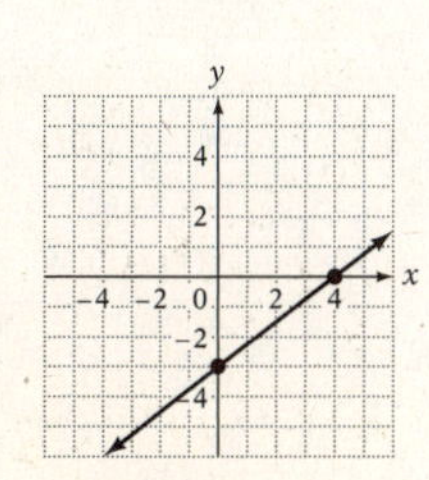

65.

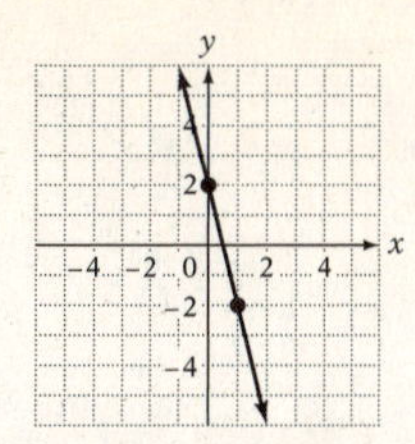

Applying the Concepts

67. Yes. If two lines have the same slope and y-intercept, their equations are identical. Therefore, the graphs of the lines would be the same.

69. The sign indicates a 6% downgrade, which means that the average slope of the road is 0.06. In other words, for every 100 ft in the horizontal direction, the road drops 6 ft in the vertical direction.

Section 7.4

Objective A Exercises

1. The point-slope formula is $y - y_1 = m(x - x_1)$. This formula is used to find the equation of a line when its slope and a point on the line are known.

3. $m = 2,\ b = 2$

$$y = mx + b$$
$$y = 2x + 2$$

5. $m = -3,\ (x_1, y_1) = (-1, 2)$

$$\begin{aligned} y - y_1 &= m(x - x_1) \\ y - 2 &= -3(x - (-1)) \\ y - 2 &= -3(x + 1) \\ y - 2 &= -3x - 3 \\ y &= -3x - 1 \end{aligned}$$

7. $m = \frac{1}{3},\ (x_1, y_1) = (3, 1)$

$$\begin{aligned} y - y_1 &= m(x - x_1) \\ y - 1 &= \frac{1}{3}(x - 3) \\ y - 1 &= \frac{1}{3}x - 1 \\ y &= \frac{1}{3}x \end{aligned}$$

9. $m = \frac{3}{4},\ (x_1, y_1) = (4, -2)$

$$\begin{aligned} y - y_1 &= m(x - x_1) \\ y - (-2) &= \frac{3}{4}(x - 4) \\ y + 2 &= \frac{3}{4}(x - 4) \\ y + 2 &= \frac{3}{4}x - 3 \\ y &= \frac{3}{4}x - 5 \end{aligned}$$

11. $m = -\frac{3}{5}, (x_1, y_1) = (5, -3)$

$$y - y_1 = m(x - x_1)$$
$$y - (-3) = -\frac{3}{5}(x - 5)$$
$$y + 3 = -\frac{3}{5}x + 3$$
$$y = -\frac{3}{5}x$$

13. $m = \frac{1}{4}, (x_1, y_1) = (2, 3)$

$$y - y_1 = m(x - x_1)$$
$$y - 3 = \frac{1}{4}(x - 2)$$
$$y - 3 = \frac{1}{4}x - \frac{1}{2}$$
$$y = \frac{1}{4}x + \frac{5}{2}$$

Objective B Exercises

15. $m = \frac{y_2 - y_1}{x_2 - x_1} = \frac{-7 - (-1)}{-2 - 1} = \frac{-6}{-3} = 2$

$$m = 2, (x_1, y_1) = (1, -1)$$
$$y - y_1 = m(x - x_1)$$
$$y - (-1) = 2(x - 1)$$
$$y + 1 = 2x - 2$$
$$y = 2x - 3$$

The equation of the line is $y = 2x - 3$.

17. $m = \frac{y_2 - y_1}{x_2 - x_1} = \frac{-5 - 1}{1 - (-2)} = \frac{-6}{3} = -2$

$$m = -2, (x_1, y_1) = (-2, 1)$$
$$y - y_1 = m(x - x_1)$$
$$y - 1 = -2[x - (-2)]$$
$$y - 1 = -2[x + 2]$$
$$y - 1 = -2x - 4$$
$$y = -2x - 3$$

The equation of the line is $y = -2x - 3$.

19. $m - \frac{y_2 - y_1}{x_2 - x_1} = \frac{-2 - 0}{-3 - 0} = \frac{2}{3}$

$$m = \frac{2}{3}, (x_1, y_1) = (0, 0)$$
$$y - y_1 = m(x - x_1)$$
$$y - 0 = \frac{2}{3}(x - 0)$$
$$y = \frac{2}{3}x$$

The equation of the line is $y = \frac{2}{3}x$.

21. $m = \frac{y_2 - y_1}{x_2 - x_1} = \frac{0 - 3}{-4 - 2} = \frac{-3}{-6} = \frac{1}{2}$

$$m = \frac{1}{2}, (x_1, y_1) = (2, 3)$$
$$y - y_1 = m(x - x_1)$$
$$y - 3 = \frac{1}{2}(x - 2)$$
$$y - 3 = \frac{1}{2}x - 1$$
$$y = \frac{1}{2}x + 2$$

The equation of the line is $y = \frac{1}{2}x + 2$.

23. $m = \frac{y_2 - y_1}{x_2 - x_1} = \frac{-5 - 1}{4 - (-4)} = \frac{-6}{8} = -\frac{3}{4}$

$$m = -\frac{3}{4}, (x_1, y_1) = (-4, 1)$$
$$y - y_1 = m(x - x_1)$$
$$y - 1 = -\frac{3}{4}[x - (-4)]$$
$$y - 1 = -\frac{3}{4}[x + 4]$$
$$y - 1 = -\frac{3}{4}x - 3$$
$$y = -\frac{3}{4}x - 2$$

The equation of the line is $y = -\frac{3}{4}x - 2$.

25. $m = \frac{y_2 - y_1}{x_2 - x_1} = \frac{4 - 1}{2 - (-2)} = \frac{3}{4}$

$$m = \frac{3}{4}, (x_1, y_1) = (-2, 1)$$
$$y - y_1 = m(x - x_1)$$
$$y - 1 = \frac{3}{4}[x - (-2)]$$
$$y - 1 = \frac{3}{4}[x + 2]$$
$$y - 1 = \frac{3}{4}x + \frac{3}{2}$$
$$y = \frac{3}{4}x + \frac{5}{2}$$

The equation of the line is $y = \frac{3}{4}x + \frac{5}{2}$.

Objective C Exercises

27.

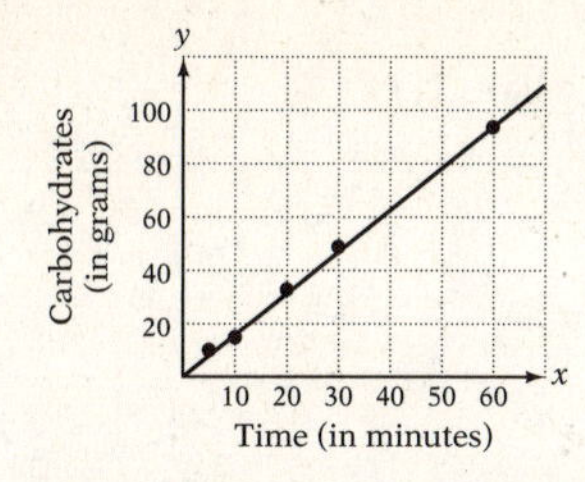

The tennis player is using 1.55 g of carbohydrates per minute.

29.

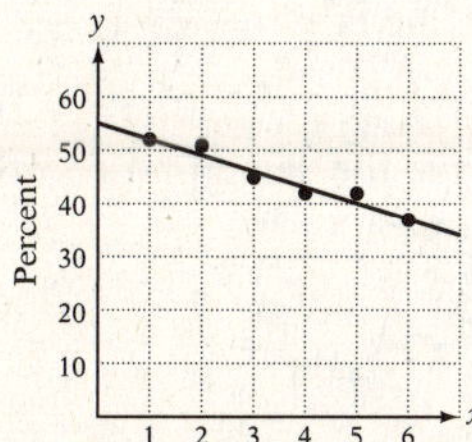

The percent of music purchased in stores is decreasing 3% per year.

Applying the Concepts

31. $m_1 = \dfrac{1-2}{4-(-3)} = \dfrac{-1}{7} = -\dfrac{1}{7}$

$m_2 = \dfrac{0-2}{-1-(-3)} = \dfrac{-2}{2} = -1$

No; because $m_1 \neq m_2$, the third point is not on the line.

33. $m_1 = \dfrac{3-(-5)}{1-(-3)} = \dfrac{8}{4} = 2$

$m_2 = \dfrac{9-(-5)}{4-(-3)} = \dfrac{14}{7} = 2$

Yes; because $m_1 = m_2$, the third point is on the line.

35.
$$\begin{aligned} y &= mx + 1 \\ 4 &= m(-2) + 1 \\ 4 &= -2m + 1 \\ 3 &= -2m \\ -\frac{3}{2} &= m \end{aligned}$$

The slope is $-\dfrac{3}{2}$.

37. $m_1 = \dfrac{-7-(-3)}{6-0} = \dfrac{-4}{6} = -\dfrac{2}{3}$

$$\begin{aligned} m_2 = \frac{n-(-3)}{3-0} &= -\frac{2}{3} \\ \frac{n+3}{3} &= \frac{-2}{3} \\ n+3 &= -2 \\ n &= -5 \end{aligned}$$

The value of n is -5.

39.
$$\begin{aligned} y - y_1 &= \frac{y_2 - y_1}{x_2 - x_1}(x - x_1) \\ y - 3 &= \frac{-1-3}{4-(-2)}[x-(-2)] \\ y - 3 &= \frac{-4}{6}(x+2) \\ y - 3 &= -\frac{2}{3}x - \frac{4}{3} \\ y &= -\frac{2}{3}x + \frac{5}{3} \end{aligned}$$

41. The condition $x_1 \neq x_2$ is placed on the two-point formula because if $x_1 = x_2$, the denominator would equal zero, and division by zero is not defined.

Chapter 7 Review Exercises

1. a.

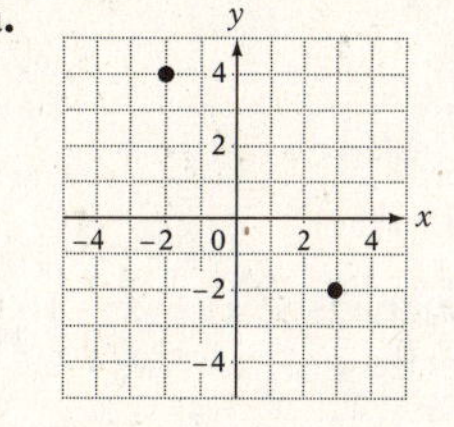

b. Abscissa of point A: -2

c. Ordinate of point B: -4

2.

x	$y = -\frac{1}{2}x - 2$	y	(x, y)
-4	$-\frac{1}{2}(-4) - 2$	0	$(-4, 0)$
-2	$-\frac{1}{2}(-2) - 2$	-1	$(-2, -1)$
0	$-\frac{1}{2}(0) - 2$	-2	$(0, -2)$
2	$-\frac{1}{2}(2) - 2$	-3	$(2, -3)$

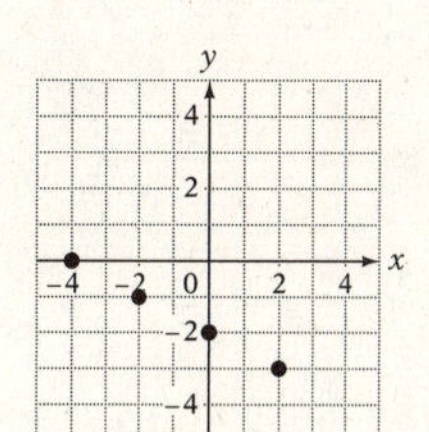

3. $(x_1, y_1) = (-1, 3)$
$(x_2, y_2) = (2, -5)$
$$y - y_1 = \frac{y_2 - y_1}{x_2 - x_1}(x - x_1)$$
$$y - 3 = \frac{-5 - 3}{2 + 1}(x + 1)$$
$$y - 3 = \frac{-8}{3}(x + 1)$$
$$y - 3 = -\frac{8}{3}x - \frac{8}{3}$$
$$y = -\frac{8}{3}x + \frac{1}{3}$$
The equation of the line is $y = -\frac{8}{3}x + \frac{1}{3}$.

4. $m = -\frac{5}{2}$, $(x_1, y_1) = (6, 1)$
$$y - y_1 = m(x - x_1)$$
$$y - 1 = -\frac{5}{2}(x - 6)$$
$$y - 1 = -\frac{5}{2}x + 15$$
$$y = -\frac{5}{2}x + 16$$
The equation of the line is $y = -\frac{5}{2}x + 16$.

5.

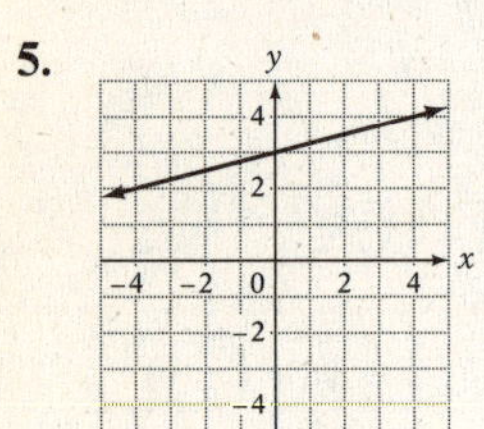

6.

7. $P_1(7, -5)$, $P_2(6, -1)$
$$m = \frac{-1 - (-5)}{6 - 7} = \frac{4}{-1} = -4$$
$Q_1(4, 5)$, $Q_2(2, -3)$
$$m = \frac{-3 - 5}{2 - 4} = \frac{-8}{-2} = 4$$
Neither. The slopes are not equal nor is their product -1.

8. $f(x) = x^2 - 2$
$f(-1) = (-1)^2 - 2 = 1 - 2 = -1$

9. $(x_1, y_1) = (-2, 5)$
$(x_2, y_2) = (4, 1)$
$$y - y_1 = \frac{y_2 - y_1}{x_2 - x_1}(x - x_1)$$
$$y - 5 = \frac{1 - 5}{4 + 2}(x + 2)$$
$$y - 5 = \frac{-4}{6}(x + 2)$$
$$y - 5 = -\frac{2}{3}x - \frac{4}{3}$$
$$y = -\frac{2}{3}x + \frac{11}{3}$$
The equation of the line is $y = -\frac{2}{3}x + \frac{11}{3}$.

10. The relation is $\{(-2, 1), (0, 3), (3, 0), (5, -2)\}$. No two ordered pairs have the same first coordinate, so, yes, y is a function of x.

11. $m = \frac{y_2 - y_1}{x_2 - x_1} = \frac{1 - 8}{-2 - 9} = \frac{-7}{-11} = \frac{7}{11}$
The slope is $\frac{7}{11}$.

12. $3x - 2y = 24$

y-intercept	x-intercept
$-2y = 24$	$3x = 24$
$y = -12$	$x = 8$
$(0, -12)$	$(8, 0)$

13. $m = \frac{y_2 - y_1}{x_2 - x_1} = \frac{-3 - (-3)}{-2 - 4} = \frac{0}{-6} = 0$
The line has zero slope.

14.

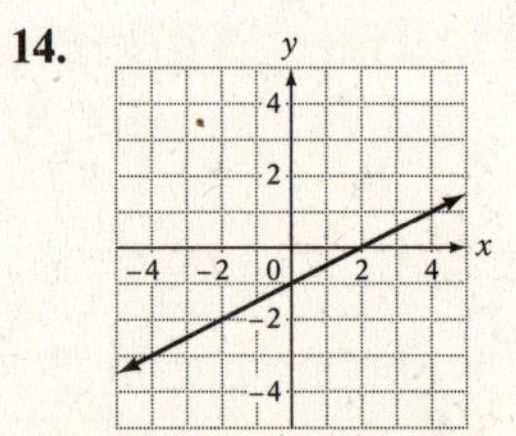

15.

16.

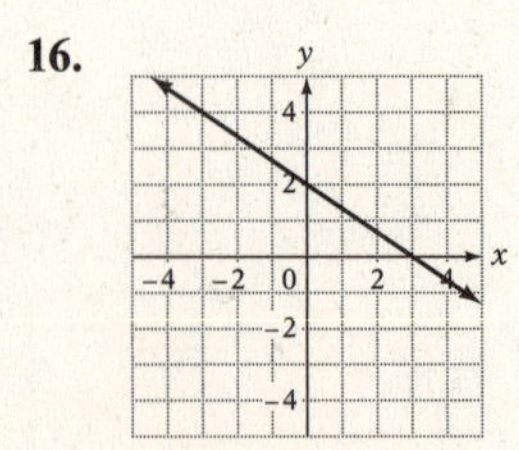

17.

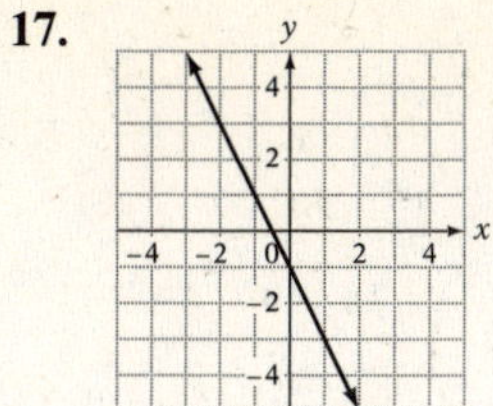

18.

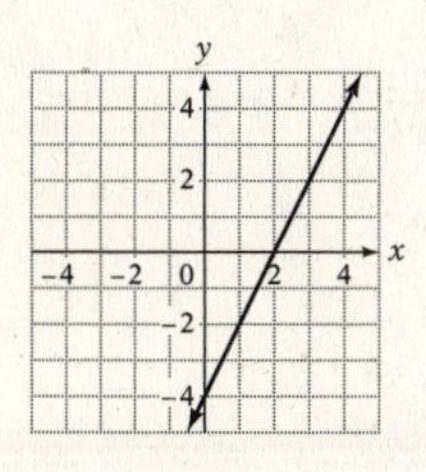

19.

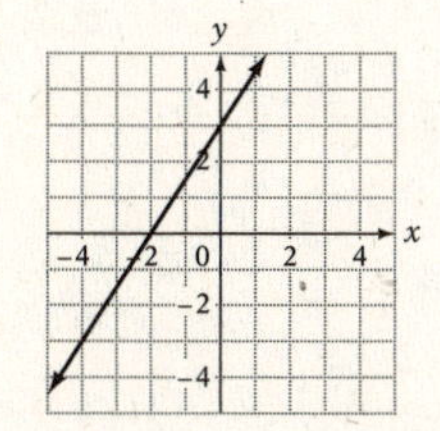

20. {(55, 95), (57, 101), (53, 94), (57, 98), (60, 100), (61, 105), (58, 97), (54, 95)}
No, the relation is not a function. The two ordered pairs (57, 101) and (57, 98) have the same first component but different second components.

21.

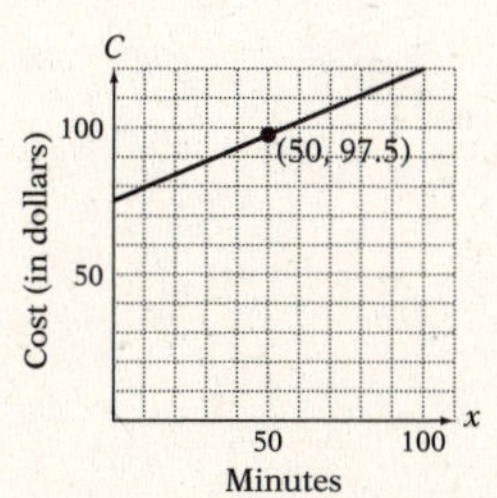

The cost of 50 min of access time for one month is \$97.50.

22.

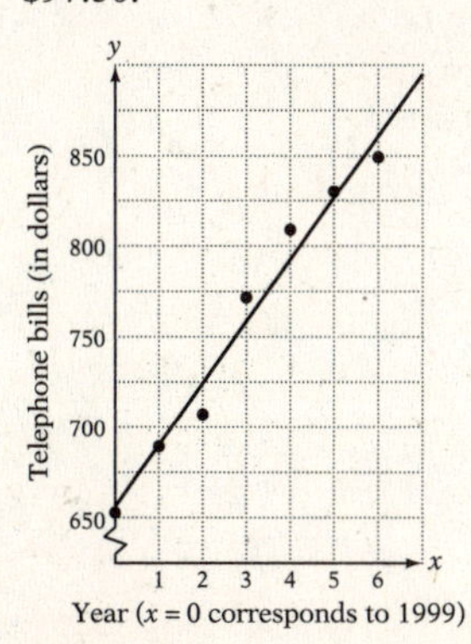

The average annual telephone bill for a family is increasing by \$34 per year.

Chapter 7 Test

1.
$$\begin{aligned} 2x - 3y &= 15 \\ 2(3) - 3y &= 15 \\ 6 - 3y &= 15 \\ -3y &= 9 \\ y &= -3 \end{aligned}$$
The ordered-pair solution is (3, −3).

2.

x	$y = -\frac{3}{2}x + 1$
−2	$-\frac{3}{2}(-2) + 1$
0	$-\frac{3}{2}(0) + 1$
4	$-\frac{3}{2}(4) + 1$

y	(x, y)
4	(−2, 4)
1	(0, 1)
−5	(4, −5)

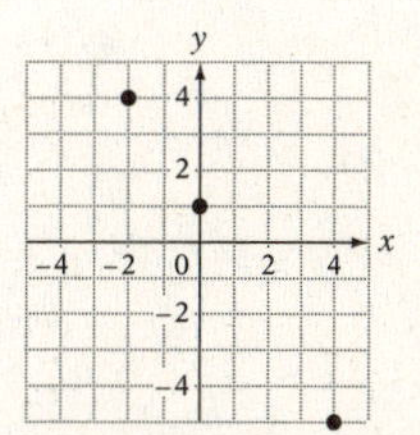

3. The relation is {(−2, −4), (0, −3), (4, −1)}.
No two ordered pairs have the same first coordinate, so, yes, y is a function of x.

4.
$$\begin{aligned} f(t) &= t^2 + t \\ f(2) &= (2)^2 + 2 \\ &= 4 + 2 = 6 \end{aligned}$$

5.
$$\begin{aligned} f(x) &= x^2 - 2x \\ f(-1) &= (-1)^2 - 2(-1) \\ &= 1 + 2 = 3 \end{aligned}$$

6. {(3.5, 25), (4.0, 30), (5.2, 45), (5.0, 38), (4.0, 42), (6.3, 12), (5.4, 34)}
No, the relation is not a function. The two ordered pairs (4.0, 30) and (4.0, 42) have the same first component but different second components.

7.

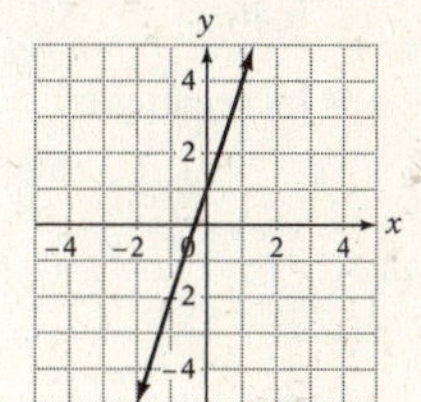

8.

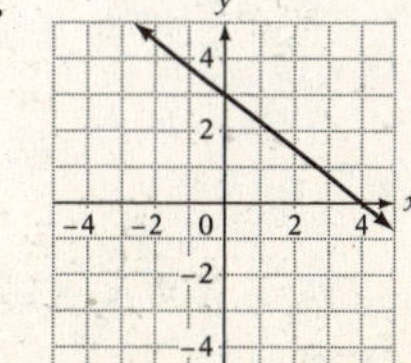

9.

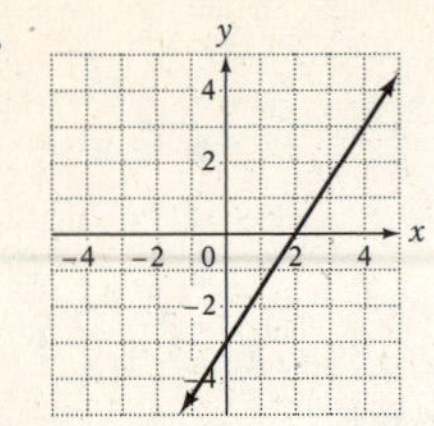

10.

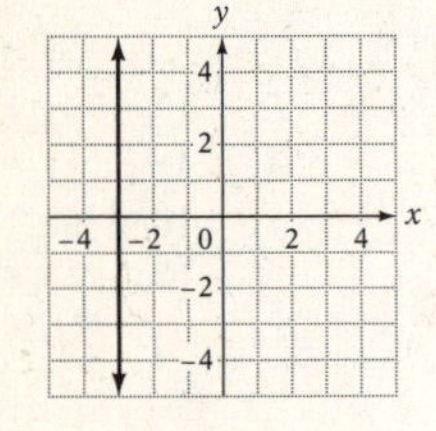

11.

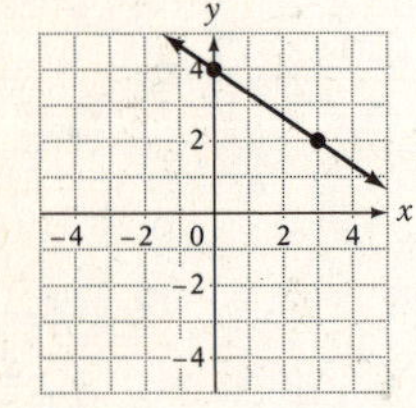

12.

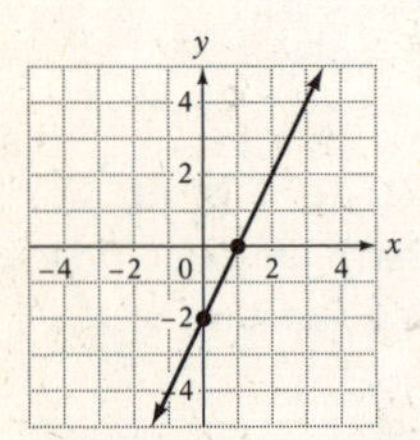

13.

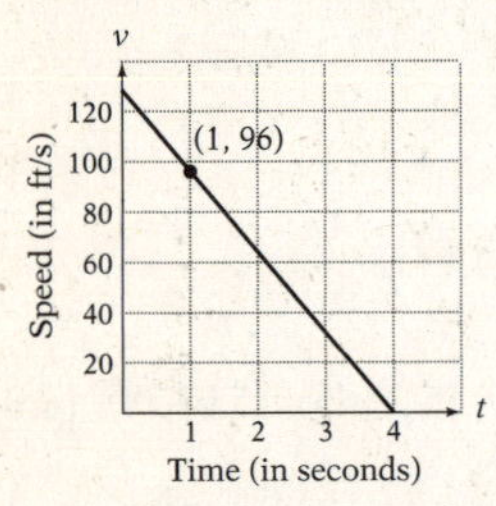

The ordered pair (1, 96) means that after 1 s, the speed of the ball is 96 ft/s.

14. $m = \dfrac{16.33 - 14.49}{5 - 1} = \dfrac{1.84}{4} = 0.46$

The average hourly wage is increasing by \$.46 per year.

15.

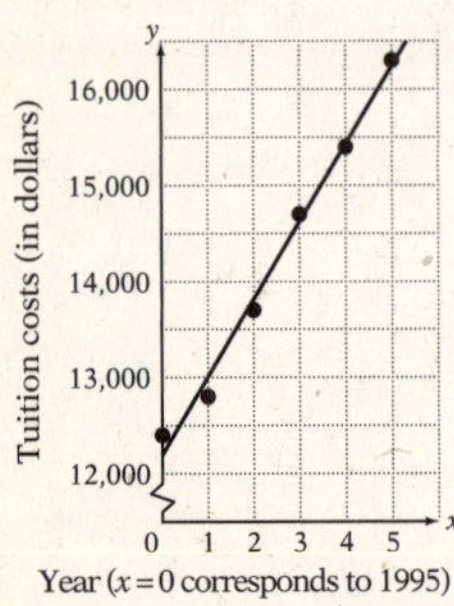

The average annual tuition for a private 4-year college is increasing \$809 per year.

16. $6x - 4y = 12$

y-intercept
$0 - 4y = 12$
$-4y = 12$
$y = -3$
$(0, -3)$

x-intercept
$6x - 0 = 12$
$6x = 12$
$x = 2$
$(2, 0)$

17. $y = \dfrac{1}{2}x + 1$

y-intercept
$y = 0 + 1$
$y = 1$
$(0, 1)$

x-intercept
$0 = \dfrac{1}{2}x + 1$
$2 \cdot -1 = \dfrac{1}{2}x \cdot 2$
$-2 = x$
$(-2, 0)$

18. $m = \dfrac{y_2 - y_1}{x_2 - x_1} = \dfrac{1 - (-3)}{4 - 2} = \dfrac{4}{2} = 2$

The slope is 2.

19. $P_1(2, 5),\ P_2(-1, 1)$

$m = \dfrac{1 - 5}{-1 - 2} = \dfrac{4}{3}$

$Q_1(-2, 3),\ Q_2(4, 11)$

$m = \dfrac{11 - 3}{4 - (-2)} = \dfrac{8}{6} = \dfrac{4}{3}$

The slopes are the same. The lines are parallel.

20. $m = \dfrac{y_2 - y_1}{x_2 - x_1} = \dfrac{7 - 2}{-5 - (-5)} = \dfrac{5}{0}$

The slope is undefined.

21. $2x + 3y = 6$

$3y = -2x + 6$

$y = -\dfrac{2}{3}x + 2$

$m = -\dfrac{2}{3}$

22. $m = 3,\ (x_1, y_1) = (0, -1) = b$

$y = mx + b$

$y = 3x - 1$

23. $m = \dfrac{2}{3},\ (x_1, y_1) = (-3, 1)$

$y - y_1 = m(x - x_1)$

$y - 1 = \dfrac{2}{3}(x + 3)$

$y - 1 = \dfrac{2}{3}x + 2$

$y = \dfrac{2}{3}x + 3$

24. $(x_1, y_1) = (5, -4)$
$(x_2, y_2) = (-3, 1)$

$$y - y_1 = \frac{y_2 - y_1}{x_2 - x_1}(x - x_1)$$
$$y + 4 = \frac{1+4}{-3-5}(x - 5)$$
$$y + 4 = \frac{5}{-8}(x - 5)$$
$$y + 4 = -\frac{5}{8}x + \frac{25}{8}$$
$$y = -\frac{5}{8}x - \frac{7}{8}$$

25. $(x_1, y_1) = (-2, 0)$
$(x_2, y_2) = (5, -2)$

$$y - y_1 = \frac{y_2 - y_1}{x_2 - x_1}(x - x_1)$$
$$y - 0 = \frac{-2-0}{5+2}(x + 2)$$
$$y = \frac{-2}{7}(x + 2)$$
$$y = -\frac{2}{7}x - \frac{4}{7}$$

Cumulative Review Exercises

1. $12 - 18 \div 3 \cdot (-2)^2 = 12 - 6 \cdot (4)$
$= 12 - 24$
$= -12$

2. $$\frac{a-b}{a^2-c} = \frac{-2-(3)}{(-2)^2-(-4)} = \frac{-5}{4+4} = -\frac{5}{8}$$

3. $$f(x) = \frac{2}{x-1}$$
$$f(-2) = \frac{2}{-2-1} = \frac{2}{-3} = -\frac{2}{3}$$

4. $$2x - \frac{2}{3} = \frac{7}{3}$$
$$2x = \frac{9}{3}$$
$$2x = 3$$
$$x = \frac{3}{2}$$

5. $3x - 2[x - 3(2 - 3x)] = x - 7$
$3x - 2[x - 6 + 9x] = x - 7$
$3x - 2[10x - 6] = x - 7$
$3x - 20x + 12 = x - 7$
$-17x + 12 = x - 7$
$-18x = -19$
$$x = \frac{19}{18}$$

6. $$6\frac{2}{3}\% = \frac{6\frac{2}{3}}{100} = \frac{\frac{20}{3}}{100}$$
$$= \frac{\overset{1}{\cancel{20}}}{3} \times \frac{1}{\underset{5}{\cancel{100}}} = \frac{1}{15}$$

7. $(-2x^2y)^3(2xy^2)^2$
$= (-8x^6y^3)(4x^2y^4)$
$= -32x^8y^7$

8. $$\frac{-15x^7}{5x^5} = -3x^2$$

9.
$$\begin{array}{r} x+3 \\ x-7\overline{)x^2-4x-21} \\ \underline{x^2-7x} \\ 3x-21 \\ \underline{3x-21} \end{array}$$
$(x^2 - 4x - 21) \div (x - 7) = x + 3$

10. $5x^2 + 15x + 10$
$= 5(x^2 + 3x + 2)$
$= 5(x + 2)(x + 1)$

11. $x(a + 2) + y(a + 2) = (a + 2)(x + y)$

12. $x(x - 2) = 8$
$x^2 - 2x - 8 = 0$
$(x - 4)(x + 2) = 0$
$x - 4 = 0 \quad x + 2 = 0$
$x = 4 \quad x = -2$
The solutions are 4 and -2.

13. $$\frac{x^5y^3}{x^2-x-6} \cdot \frac{x^2-9}{x^2y^4}$$
$$= \frac{x^5y^3}{\cancel{(x-3)}(x+2)} \cdot \frac{(x+3)\cancel{(x-3)}}{x^2y^4}$$
$$= \frac{x^3(x+3)}{y(x+2)}$$

14. $$\frac{3x}{x^2+5x-24} - \frac{9}{x^2+5x-24}$$
$$= \frac{3x-9}{x^2+5x-24} = \frac{3\cancel{(x-3)}}{(x+8)\cancel{(x-3)}}$$
$$= \frac{3}{x+8}$$

15. $$3 - \frac{1}{x} = \frac{5}{x}$$
$$x\left(\frac{3}{1} - \frac{1}{x}\right) = x \cdot \frac{5}{x}$$
$3x - 1 = 5$
$3x = 6$
$x = 2$

16. $4x - 5y = 15$
$-5y = -4x + 15$
$x = \frac{4}{5}x - 3$

17. $y = 2x - 1$
$y = 2(-2) - 1$
$y = -4 - 1$
$y = -5$
The ordered-pair solution is $(-2, -5)$.

18. $m = \frac{y_2 - y_1}{x_2 - x_1}$
$= \frac{3 - 3}{-2 - 2} = \frac{0}{-4} = 0$
The line has zero slope.

19. $m = \frac{1}{2}$, $(x_1, y_1) = (2, -1)$
$y - y_1 = m(x - x_1)$
$y + 1 = \frac{1}{2}(x - 2)$
$y + 1 = \frac{1}{2}x - 1$
$y = \frac{1}{2}x - 2$
The equation of the line is $y = \frac{1}{2}x - 2$.

20. $m = -3$, $(x_1, y_1) = (0, 2) = b$
$y = mx + b$
$y = -3x + 2$
The equation of the line is $y = -3x + 2$.

21. $m = 2$, $(x_1, y_1) = (-1, 0)$
$y - y_1 = m(x - x_1)$
$y = 2(x + 1)$
$y = 2x + 2$
The equation of the line is $y = 2x + 2$.

22. $m = \frac{2}{3}$, $(x_1, y_1) = (6, 1)$
$y - y_1 = m(x - x_1)$
$y - 1 = \frac{2}{3}(x - 6)$
$y - 1 = \frac{2}{3}x - 4$
$y = \frac{2}{3}x - 3$
The equation of the line is $y = \frac{2}{3}x - 3$.

23. $S = R - rR$
$S = 89 - 0.30(89)$
$S = 89 - 26.70$
$S = 62.30$
The sale price is \$62.30.

24. Strategy Measure of second angle: x
Measure of first angle: $x + 3$
Measure of third angle: $2x + 5$
Use the equation $A + B + C = 180°$.

Solution $x + (x + 3) + (2x + 5) = 180$
$4x + 8 = 180$
$4x = 172$
$x = 43$
$x + 3 = 46$
$2x + 5 = 91$
The measures of the angles are 43°, 46°, and 91°.

25. Strategy Write and solve a proportion using x to represent the value of the home.

Solution $\frac{625}{50{,}000} = \frac{1375}{x}$
$625x = 68{,}750{,}000$
$x = 110{,}000$
The value of the home is \$110,000.

26. Strategy • Time working together: t

	Rate	Time	Part
Electrician	$\frac{1}{6}$	t	$\frac{t}{6}$
Apprentice	$\frac{1}{10}$	t	$\frac{t}{10}$

• The sum of the parts of the task completed by each person must equal 1.

Solution $\frac{t}{6} + \frac{t}{10} = 1$
$30\left(\frac{t}{6} + \frac{t}{10}\right) = 30 \cdot 1$
$5t + 3t = 30$
$8t = 30$
$t = 3\frac{3}{4}$

It would take $3\frac{3}{4}$ h for both, working together, to wire the garage.

27.

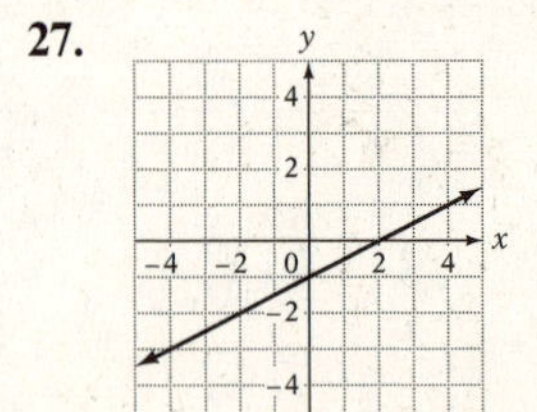

28.

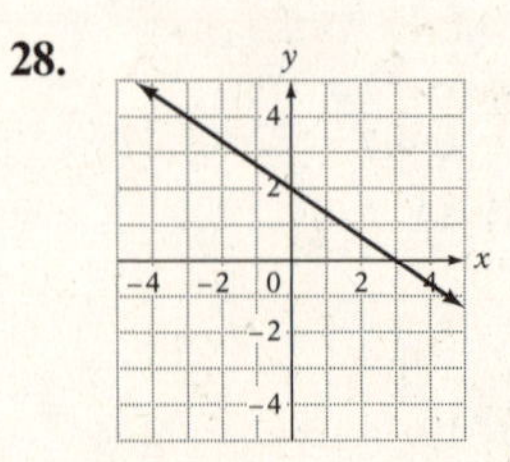

Chapter 8: Systems of Linear Equations

Prep Test

1. $3x - 4y = 24$
$-4y = -3x + 24$
$y = \frac{3}{4}x - 6$

2. $50 + 0.07x = 0.05(x + 1400)$
$50 + 0.07x = 0.05x + 70$
$0.02x = 20$
$x = 1000$

3. $-3(2x - 7y) + 3(2x + 4y) = -6x + 21y + 6x + 12y$
$= 33y$

4. $4x + 2(3x - 50) = 4x + 6x - 10$
$= 10x - 10$

5. $3(-4) - 5(2) = -22$
$-12 - 10 = -22$
$-22 = -22$
Yes, $(-4, 2)$ is a solution of $3x - 5y = -22$.

6. $3x - 4y = 12$
x-intercept: Let $y = 0$
$3x - 4(0) = 12$
$3x = 12$
$x = 4$
x-intercept is $(4, 0)$.
y-intercept: Let $x = 0$
$3(0) - 4y = 12$
$-4y = 12$
$y = -3$
y-intercept is $(0, -3)$.

7. **Strategy** To determine if the graphs of the equations are parallel, write each equation in slope-intercept form and then compare slopes.

Solution $3x + y = 6$
$y = -3x + 6$
$y = -3x - 4$
The slopes are equal and the y-intercepts are different, so, yes, the lines are parallel.

8.

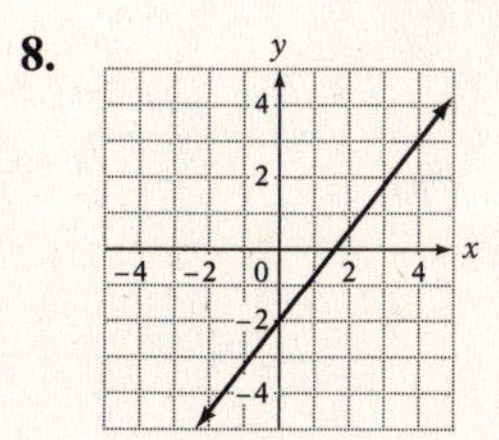

9. **Strategy**
- Amount of 55% acetic acid: x
- Amount of 80% acetic acid: 20 ml

	Amount	Percent	Quantity
55% acetic acid	x	0.55	$0.55x$
80% acetic acid	20	0.80	0.80(20)
75% acetic acid	$x + 20$	0.75	$0.75(x + 20)$

- The sum of the quantities before mixing is equal to the quantity after mixing.

Solution $0.55x + 0.80(20) = 0.75(x + 20)$
$0.55x + 16 = 0.75x + 15$
$0.20x = 1$
$x = 5$
5 ml of 55% acetic acid are needed.

10. **Strategy**
- Time second hiker spends hiking: t
- Time first hiker spends hiking: $t + 0.5$

	Distance	Rate	Time
Hiker 1	d	3 mph	$t + 0.5$
Hiker 2	d	4 mph	t

- The hikers travel the same distance.

Solution $3(t + 0.5) = 4t$
$3t + 1.5 = 4t$
$t = 1.5$
It will take 1.5 h after the second hiker starts for the hikers to be side-by-side.

Go Figure

Assuming Carla and James run and walk at the same rate, Carla will arrive home first. This is because James stops running when he is halfway home. In order for Carla to run half the time, she must run farther than halfway home.

Section 8.1

Objective A Exercises

1. **I.** The lines are parallel, so the system is **(c)** inconsistent.
II. The lines intersect, so the system is **(a)** independent.
III. The lines are the same, so the system is **(b)** dependent.

3. **Strategy** To find the solution of the system, identify the point where the graphs intersect.

Solution The point $(2, -1)$ is the solution of the system of equations.

5. **Strategy** To find the solution of the system, identify the point where the graphs intersect.

Solution The lines are the same. So, the ordered-pair solutions of $y = -\frac{3}{2}x + 1$ are solutions of the system of equations.

7. **Strategy** To find the solution of the system, identify the point where the graphs intersect.

Solution The lines do not intersect. So, there is no solution.

9. **Strategy** To find the solution of the system, identify the point where the graphs intersect.

Solution The point $(-2, 4)$ is the solution of the system of equations.

11.
$$\begin{array}{r|l} 3x + 4y = & 18 \\ \hline 3(2) + 4(3) & 18 \\ 6 + 12 & 18 \\ 18 = 18 & \end{array} \qquad \begin{array}{r|l} 2x - y = & 1 \\ \hline 2(2) - 3 & 1 \\ 4 - 3 & 1 \\ 1 = 1 & \end{array}$$

Yes, $(2, 3)$ is a solution of the system of equations.

13.
$$\begin{array}{r|l} 5x - 2y = & 14 \\ \hline 5(4) - 2(3) & 14 \\ 20 - 6 & 14 \\ 14 = 14 & \end{array} \qquad \begin{array}{r|l} x + y = & 8 \\ \hline 4 + 3 & 8 \\ 7 \neq 8 & \end{array}$$

No, $(4, 3)$ is not a solution of the system of equations.

15.
$$\begin{array}{l|l} y = & 2x - 7 \\ \hline -3 & 2(2) - 7 \\ -3 & 4 - 7 \\ -3 = -3 & \end{array} \qquad \begin{array}{r|l} 3x - y = & 9 \\ \hline 3(2) - (-3) & 9 \\ 6 + 3 & 9 \\ 9 = 9 & \end{array}$$

Yes, $(2, -3)$ is a solution of the system of equations.

17.
$$\begin{array}{l|l} y = & x \\ \hline 0 & (0) \\ 0 = 0 & \end{array} \qquad \begin{array}{r|l} 3x + 4y = & 0 \\ \hline 3(0) + 4(0) & 0 \\ 0 + 0 = 0 & \\ 0 = 0 & \end{array}$$

Yes, $(0, 0)$ is a solution of the system of equations.

19. (4, 1)

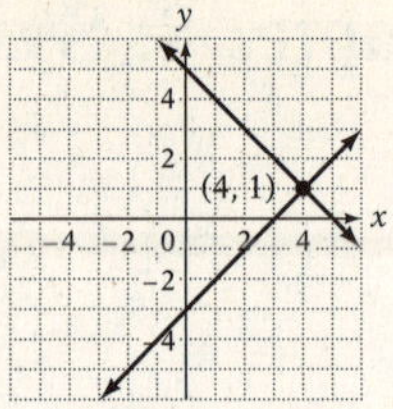

21. (4, 1)

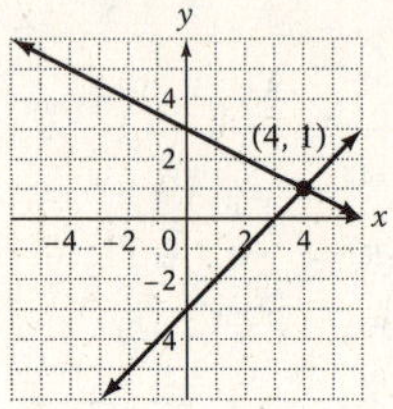

23. (4, 3)

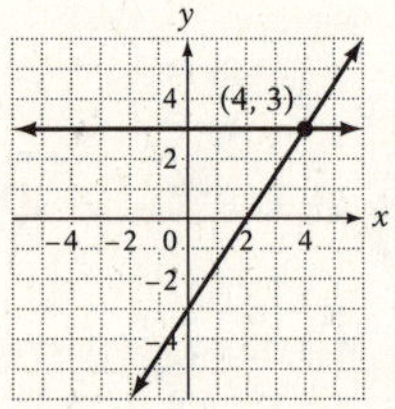

25. (3, −2)

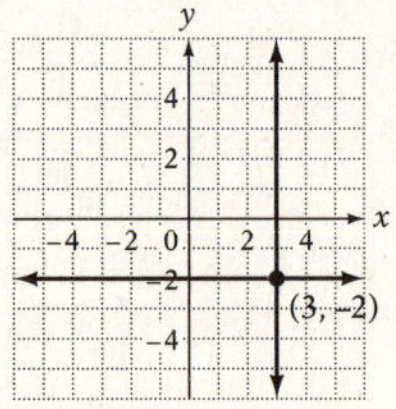

27. (2, −2)

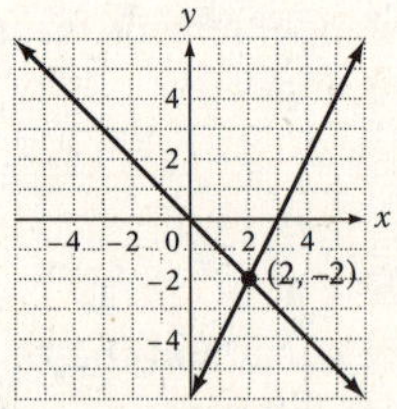

29. No solution

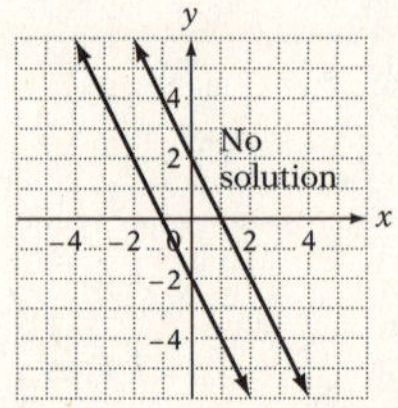

31.

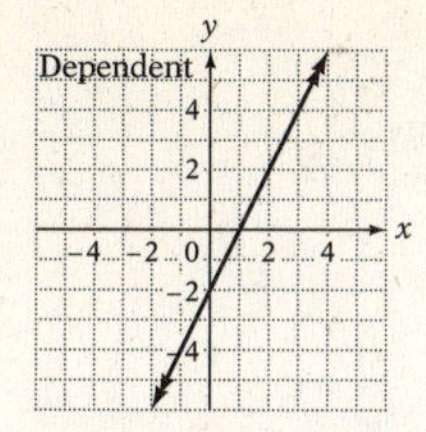

The ordered-pair solutions of $y = 2x - 2$

33.

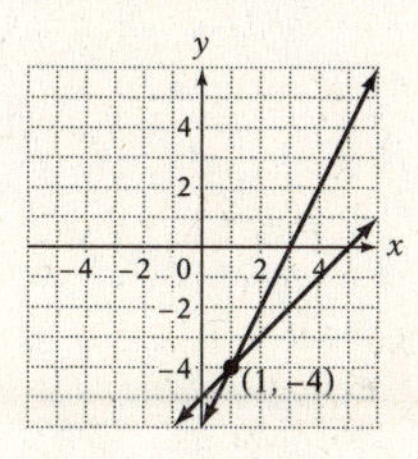

35.

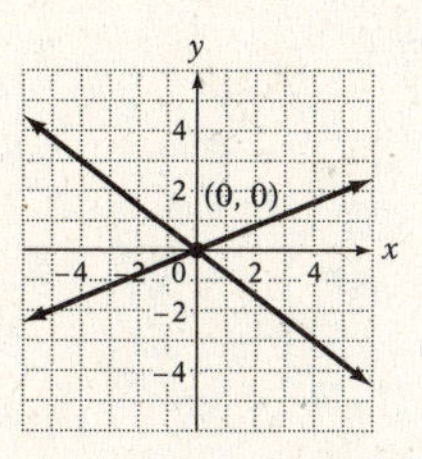

37.

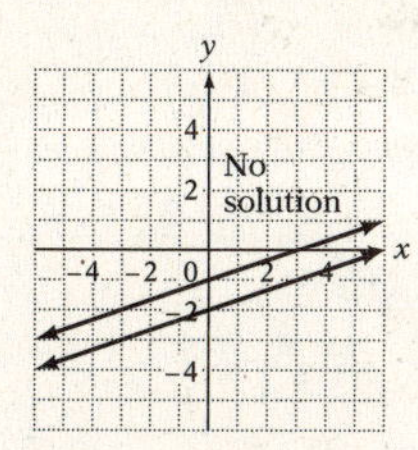

Applying the Concepts

39. **a.** Sometimes true

b. Always true

c. Never true

d. Always true

41. If the system of equations is inconsistent, the graphs at the equations are parallel and therefore do not intersect.

43. Answers will vary. An example is:
$2x + 6y = 1$ has no solution.
$x + 3y = 2$

Section 8.2

Objective A Exercises

1. The following should be included:
(i) Solve, if necessary, one of the equations for x or y (or whatever variables are used in the system).
(ii) Assuming Equation (1) was solved for y, replace y in Equation (2) by the expression to which it is equal.
(iii) Solve for x in Equation (2).
(iv) Use the value of x to determine y and write the answer as an ordered pair.
(v) Check the result.

3. (1) $2x + 3y = 7$
(2) $x = 2$
Substitute in Equation (1).
$2x + 3y = 7$
$2(2) + 3y = 7$
$4 + 3y = 7$
$3y = 3$
$y = 1$
The solution is (2, 1).

5. (1) $y = x - 3$
(2) $x + y = 5$
Substitute in Equation (2).
$x + y = 5$
$x + (x - 3) = 5$
$2x - 3 = 5$
$2x = 8$
$x = 4$
Substitute in Equation (1).
$y = x - 3$
$y = 4 - 3$
$y = 1$
The solution is (4, 1).

7. (1) $x = y - 2$
(2) $x + 3y = 2$
Substitute in Equation (2).
$x + 3y = 2$
$(y - 2) + 3y = 2$
$4y - 2 = 2$
$4y = 4$
$y = 1$
Substitute in Equation (1).
$x = y - 2$
$x = 1 - 2$
$x = -1$
The solution is (−1, 1).

9. (1) $y = 4 - 3x$
(2) $3x + y = 5$
Substitute in Equation (2).
$3x + y = 5$
$3x + (4 - 3x) = 5$
$0 + 4 = 5$
$4 = 5$
The system is inconsistent and has no solution.

11. (1) $x = 3y + 3$
(2) $2x - 6y = 12$
Substitute in Equation (2).
$2x - 6y = 12$
$2(3y + 3) - 6y = 12$
$6y + 6 - 6y = 12$
$0 + 6 = 12$
$6 = 12$
The system is inconsistent and has no solution.

13. (1) $3x + 5y = -6$
(2) $x = 5y + 3$
Substitute in Equation (1).
$3x + 5y = -6$
$3(5y + 3) + 5y = -6$
$15y + 9 + 5y = -6$
$20y + 9 = -6$
$20y = -15$
$y = -\frac{3}{4}$
Substitute in Equation (2).
$x = 5y + 3$
$x = 5\left(-\frac{3}{4}\right) + 3$
$x = -\frac{15}{4} + 3$
$x = -\frac{15}{4} + \frac{12}{4}$
$x = -\frac{3}{4}$
The solution is $\left(-\frac{3}{4}, -\frac{3}{4}\right)$.

15. (1) $3x + y = 4$
(2) $4x - 3y = 1$
Substitute in Equation (1) for y.
$3x + y = 4$
$y = -3x + 4$
Substitute in Equation (2).
$4x - 3y = 1$
$4x - 3(-3x + 4) = 1$
$4x + 9x - 12 = 1$
$13x - 12 = 1$
$13x = 13$
$x = 1$
Substitute in Equation (1).
$3x + y = 4$
$3(1) + y = 4$
$3 + y = 4$
$y = 1$
The solution is (1, 1).

17. (1) $3x - y = 6$
(2) $x + 3y = 2$
Solve Equation (2) for x.
$x + 3y = 2$
$x = -3y + 2$
Substitute in Equation (1).
$3x - y = 6$
$3(-3y + 2) - y = 6$
$-9y + 6 - y = 6$
$-10y + 6 = 6$
$-10y = 0$
$y = 0$
Substitute in Equation (2).
$x + 3y = 2$
$x + 3(0) = 2$
$x + 0 = 2$
$x = 2$
The solution is (2, 0).

19. (1) $3x - y = 5$
(2) $2x + 5y = -8$
Solve Equation (1) for y.
$3x - y = 5$
$-y = -3x + 5$
$y = 3x - 5$
Substitute in Equation (2).
$2x + 5y = -8$
$2x + 5(3x - 5) = -8$
$2x + 15x - 25 = -8$
$17x - 25 = -8$
$17x = 17$
$x = 1$
Substitute in Equation (1).
$3x - y = 5$
$3(1) - y = 5$
$3 - y = 5$
$-y = 2$
$y = -2$
The solution is (1, −2).

21. (1) $4x + 3y = 0$
(2) $2x - y = 0$
Solve Equation (2) for y.
$2x - y = 0$
$-y = -2x$
$y = 2x$
Substitute in Equation (1).
$4x + 3y = 0$
$4x + 3(2x) = 0$
$4x + 6x = 0$
$10x = 0$
$x = 0$
Substitute in Equation (2).
$2x - y = 0$
$2(0) - y = 0$
$0 - y = 0$
$-y = 0$
$y = 0$
The solution is (0, 0).

23. (1) $2x - y = 2$
(2) $6x - 3y = 6$
Solve Equation (1) for y.
$2x - y = 2$
$-y = -2x + 2$
$y = 2x - 2$
Substitute in Equation (2).
$6x - 3y = 6$
$6x - 3(2x - 2) = 6$
$6x - 6x + 6 = 6$
$6 = 6$
The system is dependent. The solutions are the ordered pairs that satisfy the equation $2x - y = 2$.

25. (1) $x = 3y + 2$
(2) $y = 2x + 6$
Substitute in Equation (2).
$y = 2x + 6$
$y = 2(3y + 2) + 6$
$y = 6y + 4 + 6$
$-5y = 10$
$y = -2$
Substitute in Equation (1).
$x = 3y + 2$
$x = 3(-2) + 2$
$x = -6 + 2$
$x = -4$
The solution is (−4, −2).

27. (1) $y = 2x + 11$
(2) $y = 5x - 19$
Substitute in Equation (2).
$y = 5x - 19$
$2x + 11 = 5x - 19$
$-3x + 11 = -19$
$-3x = -30$
$x = 10$
Substitute in Equation (1).
$y = 2x + 11$
$y = 2(10) + 11$
$y = 20 + 11$
$y = 31$
The solution is (10, 31).

29. (1) $y = -4x + 2$
(2) $y = -3x - 1$
Substitute in Equation (2).
$y = -3x - 1$
$-4x + 2 = -3x - 1$
$-x + 2 = -1$
$-x = -3$
$x = 3$
Substitute in Equation (1).
$y = -4x + 2$
$y = -4(3) + 2$
$y = -12 + 2$
$y = -10$
The solution is (3, −10).

31. (1) $x = 4y - 2$
(2) $x = 6y + 8$
Substitute in Equation (2).
$x = 6y + 8$
$4y - 2 = 6y + 8$
$-2y - 2 = 8$
$-2y = 10$
$y = -5$
Substitute in Equation (1).
$x = 4y - 2$
$x = 4(-5) - 2$
$x = -20 - 2$
$x = -22$
The solution is (−22, −5).

Objective B Exercises

33. Strategy • Amount invested at 5%: x
Amount invested at 7.5%: y

	Principal	Rate	Interest
Amount at 5%	x	0.05	$0.05x$
Amount at 7.5%	y	0.075	$0.075x$

• The total amount invested is $3500. The total annual interest earned is $215.

Solution $x + y = 3500 \quad y = 3500 - x$
$0.05x + 0.075y = 215$
$0.05x + 0.075(3500 - x) = 215$
$0.05x + 262.50 - 0.075x = 215$
$-0.025x = -47.50$
$x = 1900$
$y = 3500 - x = 3500 - 1900 = 1600$
The amounts invested should be $1900 at 5% and $1600 at 7.5%.

35. Strategy • Amount invested at 9%: x
Amount invested at 6%: y

	Principal	Rate	Interest
Amount at 9%	x	0.09	$0.09x$
Amount at 6%	y	0.06	$0.06y$

• The total amount invested is $6000. The interest earned in each account is the same.

Solution $x + y = 6000 \quad y = 6000 - x$
$0.09x = 0.06y$
$0.09x = 0.06(6000 - x)$
$0.09x = 360 - 0.06x$
$0.15x = 360$
$x = 2400$
$y = 6000 - x = 6000 - 2400 = 3600$
The amounts invested were $2400 at 9% and $3600 at 6%.

37. Strategy • Amount invested at 8%: x
Amount invested at 11%: y

	Principal	Rate	Interest
Amount at 8%	x	0.08	$0.08x$
Amount at 11%	y	0.11	$0.11y$

• The total amount invested is $6000. The interest earned on the 8% account is twice the amount of interest earned on the 11% account.

Solution $x + y = 6000 \quad y = 6000 - x$
$0.08x = 2(0.11y)$
$0.08x = 0.22y$
$0.08x = 0.22(6000 - x)$
$0.08x = 1320 - 0.22x$
$0.30x = 1320$
$x = 4400$
$y = 6000 - x = 6000 - 4400 = 1600$
The amounts invested should be $4400 at 8% and $1600 at 11%.

39. Strategy • Amount invested at 6.5%: x
Amount invested at 8.5%: y

	Principal	Rate	Interest
Amount at 6.5%	x	0.065	$0.065x$
Amount at 8.5%	y	0.085	$0.085y$

• The amount invested at 8.5% is twice the amount invested at 6.5%. The total annual interest earned is $4935.

Solution $y = 2x$
$0.065x + 0.085y = 4935$
$0.065x + 0.085(2x) = 4935$
$0.065x + 0.17x = 4935$
$0.235x = 4935$
$x = 21{,}000$
The amount invested was 6.5% is $21,000.

41. Strategy • Amount invested at 8%: x
Amount invested at 7%: y

	Principal	Rate	Interest
Amount at 8%	x	0.08	$0.08x$
Amount at 7%	y	0.07	$0.07y$

• The total amount invested is 10($2000) = $20,000. The total annual interest earned is $1520.

Solution $x + y = 20{,}000 \quad y = 20{,}000 - x$
$0.08x + 0.07y = 1520$
$0.08x + 0.07(20{,}000 - x) = 1520$
$0.08x + 1400 - 0.07x = 1520$
$0.01x = 120$
$x = 12{,}000$
$y = 20{,}000 - x = 20{,}000 - 12{,}000$
$= 8000$
The amounts invested were $12,000 at 8% and $8000 at 7%.

43. **Strategy** • Amount invested at 7.5%: x
Amount invested at 9%: y

	Principal	Rate	Interest
Amount at 7.5%	x	0.075	$0.075x$
Amount at 9%	y	0.09	$0.09y$

• The amount invested at 9% (trust deed) is one-half the amount invested at 7.5% (real estate trust). The total annual interest earned is $900.

Solution $y = \frac{1}{2}x$

$$0.075x + 0.09y = 900$$
$$0.075x + 0.09\left(\frac{1}{2}x\right) = 900$$
$$0.075x + 0.045x = 900$$
$$0.12x = 900$$
$$x = 7500$$

$y = \frac{1}{2}x = \frac{1}{2}(7500) = 3750$

The amount invested in the trust deed was $3750.

Applying the Concepts

45. **Strategy** • Number of loads equals number of hours drying clothes: x
Cost of drying clothes using gas dryer: $0.45x + 240$
Cost of drying clothes at laundromat: $1.75x$
• Purchasing the dryer becomes more economical when the cost of drying clothes using the dryer equals the cost of drying clothes at the laundromat.

Solution $1.75x = 0.45x + 240$
$1.30x = 240$
$x \approx 184.61$
The gas dryer becomes more economical after 185 loads of clothes.

47. (1) $8x - 4y = 1 \Rightarrow -4y = -8x + 1 \Rightarrow y = 2x - \frac{1}{4}$

(2) $2x - ky = 3 \Rightarrow -ky = -2x + 3 \Rightarrow y = \frac{2}{k}x - \frac{3}{k}$

For there to be no solution, the slopes must be the same and the intercepts different.

$2 = \frac{2}{k}$
$2k = 2$
$k = 1$
k must be 1.

49. The assertion is not correct. The system of equations is independent. The correct solution is

$x = 0$ and $y = \frac{1}{2}(0) + 2$
$y = 2$

The solution is (0, 2).

51. **Strategy** • Amount invested at 9%: x
Amount invested at 8%: $x + 5000$

	Principal	Rate	Interest
Amount at 9%	x	0.09	$0.09x$
Amount at 8%	$x + 5000$	0.08	$0.08(x + 5000)$

• Interest earned is the same.

Solution
$$0.09x = 0.08(x + 5000)$$
$$0.09x = 0.08x + 400$$
$$0.09x - 0.08x = 0.08x - 0.08x + 400$$
$$0.01x = 400$$
$$x = \frac{400}{0.01}$$
$$x = 40{,}000$$
$$x + 5000 = 45{,}000$$

The research consultant's investment is $45,000.

53. $I = P\left[\left(1 + \frac{r}{n}\right)^n - 1\right]$

For $P = 5000$, $r = 0.08$, $n = 1$

$I = 5000\left[\left(1 + \frac{0.08}{1}\right)^1 - 1\right]$

$I = 5000(0.08) = 400$

For $P = 5000$, $r = 0.08$, $n = 12$

$I = 5000\left\{\left(1 + \frac{0.08}{12}\right)^{12} - 1\right\}$

$I = 5000(0.082999) \approx 415.00$

For $P = 5000$, $r = 0.08$, $n = 365$

$I = 5000\left[\left(1 + \frac{0.08}{365}\right)^{365} - 1\right]$

$I = 5000(0.083277) \approx 416.39$

Simple interest: $400
Compounded monthly: $415
Compounded daily: $416.39

Section 8.3

Objective A Exercises

1. (1) $x + y = 4$
(2) $x - y = 6$
Add the equations.
$2x = 10$
$x = 5$
Replace x in Equation (1).
$x + y = 4$
$5 + y = 4$
$y = -1$
The solution is (5, −1).

3. (1) $x + y = 4$
(2) $2x + y = 5$
Eliminate y.
$-1(x + y) = -1 \cdot 4$
$2x + y = 5$
$-x - y = -4$
$2x + y = 5$
Add the equations.
$x = 1$
Replace x in Equation (2).
$2x + y = 5$
$2(1) + y = 5$
$2 + y = 5$
$y = 3$
The solution is (1, 3).

5. (1) $2x - y = 1$
(2) $x + 3y = 4$
Eliminate y.
$3(2x - y) = 3 \cdot 1$
$x + 3y = 4$
$6x - 3y = 3$
$x + 3y = 4$
Add the equations.
$7x = 7$
$x = 1$
Replace x in Equation (2).
$x + 3y = 4$
$1 + 3y = 4$
$3y = 3$
$y = 1$
The solution is (1, 1).

7. (1) $4x - 5y = 22$
(2) $x + 2y = -1$
Eliminate x.
$4x - 5y = 22$
$-4(x + 2y) = -4(-1)$
$4x - 5y = 22$
$-4x - 8y = 4$
Add the equations.
$-13y = 26$
$y = -2$
Replace y in Equation (1).
$4x - 5y = 22$
$4x - 5(-2) = 22$
$4x + 10 = 22$
$4x = 12$
$x = 3$
The solution is (3, −2).

9. (1) $2x - y = 1$
(2) $4x - 2y = 2$
Eliminate y.
$-2(2x - y) = -2(1)$
$4x - 2y = 2$
$-4x + 2y = -2$
$4x - 2y = 2$
Add the equations.
$0 = 0$
The system is dependent. The solutions are the ordered pairs that satisfy the equation $2x - y = 1$.

11. (1) $4x + 3y = 15$
(2) $2x - 5y = 1$
Eliminate y.
$5(4x + 3y) = 5 \cdot 15$
$3(2x - 5y) = 3 \cdot 1$
$20x + 15y = 75$
$6x - 15y = 3$
Add the equations.
$26x = 78$
$x = 3$
Replace x in Equation (1).
$4x + 3y = 15$
$4(3) + 3y = 15$
$12 + 3y = 15$
$3y = 3$
$y = 1$
The solution is (3, 1).

13. (1) $2x - 3y = 1$
(2) $4x - 6y = 2$
Eliminate x.
$-2(2x - 3y) = -2 \cdot 1$
$4x - 6y = 2$
$-4x + 6y = -2$
$4x - 6y = 2$
Add the equations.
$0 = 0$
The system is dependent. The solutions are the ordered pairs that satisfy the equation $2x - 3y = 1$.

15. (1) $3x - 6y = -1$
(2) $6x - 4y = 2$
Eliminate x.
$-2(3x - 6y) = -2(-1)$
$6x - 4y = 2$
$-6x + 12y = 2$
$6x - 4y = 2$
Add the equations.
$8y = 4$
$y = \frac{1}{2}$
Replace y in Equation (1).
$3x - 6y = -1$
$3x - 6\left(\frac{1}{2}\right) = -1$
$3x - 3 = -1$
$3x = 2$
$x = \frac{2}{3}$
The solution is $\left(\frac{2}{3}, \frac{1}{2}\right)$.

17. (1) $5x + 7y = 10$
(2) $3x - 14y = 6$
Eliminate y.
$2(5x + 7y) = 2 \cdot 10$
$3x - 14y = 6$
$10x + 14y = 20$
$3x - 14y = 6$
Add the equations.
$13x = 26$
$x = 2$
Replace x in Equation (2).
$3x - 14y = 6$
$3(2) - 14y = 6$
$6 - 14y = 6$
$-14y = 0$
$y = 0$
The solution is (2, 0).

19. (1) $3x - 2y = 0$
(2) $6x + 5y = 0$
Eliminate x.
$-2(3x - 2y) = -2(0)$
$6x + 5y = 0$
$-6x + 4y = 0$
$6x + 5y = 0$
Add the equations.
$9y = 0$
$y = 0$
Replace y in Equation (2).
$6x + 5y = 0$
$6x + 5(0) = 0$
$6x = 0$
$x = 0$
The solution is (0, 0).

21. (1) $2x - 3y = 16$
(2) $3x + 4y = 7$
Eliminate y.
$4(2x - 3y) = 4 \cdot 16$
$3(3x + 4y) = 3 \cdot 7$
$8x - 12y = 64$
$9x + 12y = 21$
Add the equations.
$17x = 85$
$x = 5$
Replace x in Equation (1).
$2x - 3y = 16$
$2(5) - 3y = 16$
$10 - 3y = 16$
$-3y = 6$
$y = -2$
The solution is (5, −2).

23. (1) $5x + 3y = 7$
(2) $2x + 5y = 1$
Eliminate x.
$-2(5x + 3y) = -2 \cdot 7$
$5(2x + 5y) = 5 \cdot 1$
$-10x - 6y = -14$
$10x + 25y = 5$
Add the equations.
$19y = -9$
$y = -\frac{9}{19}$
Replace y in Equation (2).
$2x + 5y = 1$
$2x + 5\left(-\frac{9}{19}\right) = 1$
$2x - \frac{45}{19} = 1$
$2x = 1 + \frac{45}{19}$
$2x = \frac{19}{19} + \frac{45}{19}$
$2x = \frac{64}{19}$
$x = \frac{32}{19}$
The solution is $\left(\frac{32}{19}, -\frac{9}{19}\right)$.

25. (1) $3x + 4y = 4$
(2) $5x + 12y = 5$
Eliminate y.
$-3(3x + 4y) = -3(4)$
$5x + 12y = 5$
$-9x - 12y = -12$
$5x + 12y = 5$
Add the equations.
$-4x = -7$
$x = \frac{7}{4}$
Replace x in Equation (1).
$3x + 4y = 4$
$3\left(\frac{7}{4}\right) + 4y = 4$
$\frac{21}{4} + 4y = 4$
$4y = -\frac{5}{4}$
$y = -\frac{5}{16}$
The solution is $\left(\frac{7}{4}, -\frac{5}{16}\right)$.

27. (1) $8x - 3y = 11$
(2) $6x - 5y = 11$
Eliminate y.
$5(8x - 3y) = 5 \cdot 11$
$-3(6x - 5y) = -3 \cdot 11$
$40x - 15y = 55$
$-18x + 15y = -33$
Add the equations.
$22x = 22$
$x = 1$
Replace x in Equation (1).
$8x - 3y = 11$
$8(1) - 3y = 11$
$8 - 3y = 11$
$-3y = 3$
$y = -1$
The solution is $(1, -1)$.

29. (1) $5x + 15y = 20$
(2) $2x + 6y = 12$
Eliminate x.
$2(5x + 15y) = 2 \cdot 20$
$-5(2x + 6y) = -5 \cdot 12$
$10x + 30y = 40$
$-10x - 30y = -60$
Add the equations.
$0 = -20$
This is not a true equation. The system is inconsistent. There is no solution.

31. (1) $3x = 2y + 7$
(2) $5x - 2y = 13$
Write Equation (1) in the form $Ax + By = C$.
$3x = 2y + 7$
$3x - 2y = 7$
Eliminate y.
$-1(3x - 2y) = -1 \cdot 7$
$5x - 2y = 13$
$-3x + 2y = -7$
$5x - 2y = 13$
Add the equations.
$2x = 6$
$x = 3$
Replace x in Equation (2).
$5x - 2y = 13$
$5(3) - 2y = 13$
$15 - 2y = 13$
$-2y = -2$
$y = 1$
The solution is $(3, 1)$.

33. (1) $2x + 9y = 16$
(2) $5x = 1 - 3y$
Write Equation (2) in the form $Ax + By = C$.
$5x = 1 - 3y$
$5x + 3y = 1$
Eliminate y.
$2x + 9y = 16$
$-3(5x + 3y) = -3 \cdot 1$
$2x + 9y = 16$
$-15x - 9y = -3$
Add the equations.
$-13x = 13$
$x = -1$
Replace x in Equation (1).
$2x + 9y = 16$
$2(-1) + 9y = 16$
$-2 + 9y = 16$
$9y = 18$
$y = 2$
The solution is $(-1, 2)$.

35. (1) $2x + 3y = 7 - 2x$
(2) $7x + 2y = 9$
Write Equation (1) in the form $Ax + By = C$.
$2x + 3y = 7 - 2x$
$4x + 3y = 7$
Eliminate y.
$2(4x + 3y) = 2 \cdot 7$
$-3(7x + 2y) = -3 \cdot 9$
$8x + 6y = 14$
$-21x - 6y = -27$
Add the equations.
$-13x = -13$
$x = 1$
Replace x in Equation (2).
$7x + 2y = 9$
$7(1) + 2y = 9$
$7 + 2y = 9$
$2y = 2$
$y = 1$
The solution is $(1, 1)$.

Applying the Concepts

37. Student descriptions should include the following steps:
(1) If necessary, multiply one or both of the equations by a constant so that the coefficients of one variable will be opposites.
(2) Add the two equations and solve for the variable.
(3) Substitute the value of the variable into either equation in the system and solve for the second variable.
(4) Write the ordered-pair solution.
(5) Check the solution.

39. (1) $Ax - 4y = 9$
$A(-1) - 4(-3) = 9$
$-A + 12 = 9$
$-A = -3$
$A = 3$

(2) $4x + By = -1$
$4(-1) + B(-3) = -1$
$-4 - 3B = -1$
$-3B = 3$
$B = -1$

41. For the system to be inconsistent, the slopes must be the same and the intercepts different. Write each equation in slope-intercept form and solve for k.

a. (1) $x + y = 7 \Rightarrow y = -x + 7$
(2) $kx + y = 3 \Rightarrow y = -kx + 3$
$-1 = -k$
$1 = k$

b. (1) $x + 2y = 4 \Rightarrow 2y = -x + 4 \Rightarrow y = -\frac{1}{2}x + 2$
(2) $kx + 3y = 2 \Rightarrow 3y = -kx + 2 \Rightarrow y = -\frac{k}{3}x + \frac{2}{3}$
$-\frac{1}{2} = -\frac{k}{3}$
$2k = 3$
$k = \frac{3}{2}$

c. (1) $2x + ky = 1 \Rightarrow ky = -2x + 1 \Rightarrow y = -\frac{2}{k}x + \frac{1}{k}$
(2) $x + 2y = 2 \Rightarrow 2y = -x + 2 \Rightarrow y = -\frac{1}{2}x + 1$
$-\frac{2}{k} = -\frac{1}{2}$
$k = 4$

Section 8.4

Objective A Exercises

1. **Strategy** • Rate of the whale in calm water: r
Rate of the ocean current: c

	Rate	Time	Distance
With current	$r + c$	1.5	$1.5(r + c)$
Against current	$r - c$	2	$2(r - c)$

• The distance traveled with the current is 60 mi.
The distance traveled against the current is 60 mi.

Solution
$$1.5(r+c) = 60 \qquad \frac{1}{1.5}\cdot 1.5(r+c) = \frac{1}{1.5}\cdot 60$$
$$2(r-c) = 60 \qquad \frac{1}{2}\cdot 2(r-c) = \frac{1}{2}\cdot 60$$
$$r+c = 40$$
$$r-c = 30$$
$$2r = 70$$
$$r = 35$$
$$r+c = 40$$
$$35+c = 40$$
$$c = 5$$
The rate of the whale in calm water is 35 mph.
The rate of the current is 5 mph.

3. **Strategy** • Rate of rowing in calm water: r
Rate of the current: c

	Rate	Time	Distance
With current	$r + c$	2	$2(r + c)$
Against current	$r - c$	2	$2(r - c)$

• The distance traveled with the current is 40 km.
The distance traveled against the current is 16 km.

Solution
$$2(r+c) = 40 \qquad \frac{1}{2}\cdot 2(r+c) = 40\cdot\frac{1}{2}$$
$$2(r-c) = 16 \qquad \frac{1}{2}\cdot 2(r-c) = 16\cdot\frac{1}{2}$$
$$r+c = 20$$
$$r-c = 8$$
$$2r = 28$$
$$r = 14$$
$$r+c = 20$$
$$14+c = 20$$
$$c = 6$$
The rate rowing in calm water is 14 km/h.
The rate of the current is 6 km/h.

5. **Strategy** • The rate of the Learjet in calm air: r
The rate of the wind: w

	Rate	Time	Distance
With wind	$r + w$	2	$2(r + w)$
Against wind	$r - w$	2	$2(r - w)$

• The distance traveled with the wind is 1120 mi.
The distance traveled against the wind is 980 mi.

Solution
$$2(r+w) = 1120 \qquad 2(r+w) = 1120$$
$$2(r-w) = 980 \qquad 2(r-w) = 980$$
$$2r+2w = 1120$$
$$2r-2w = 980$$
$$4r = 2100$$
$$r = 525$$
$$r+w = 560$$
$$525+w = 560$$
$$w = 35$$
The rate of the Learjet is 525 mph.
The rate of the wind is 35 mph.

7. Strategy • Rate of the helicopter in calm air: r
Rate of the wind: w

	Rate	Time	Distance
With wind	$r + w$	$\frac{5}{3}$	$\frac{5}{3}(r + w)$
Against wind	$r - w$	2.5	$2.5(r - w)$

• The distance traveled with the wind is 450 mi.
The distance traveled against the wind is 450 mi.

Solution

$$\frac{5}{3}(r+w) = 450 \qquad \frac{3}{5}\cdot\frac{5}{3}(r+w) = \frac{3}{5}\cdot 450$$

$$2.5(r-w) = 450 \qquad \frac{1}{2.5}\cdot 2.5(r-w) = \frac{1}{2.5}\cdot 450$$

$$r + w = 270$$
$$r - w = 180$$

$$2r = 450$$
$$r = 225$$

$$r + w = 270$$
$$225 + w = 270$$
$$w = 45$$

The rate of the helicopter in calm air is 225 mph. The rate of the wind is 45 mph.

9. Strategy • Rate of the canoeist in calm water: r
Rate of the current: c

	Rate	Time	Distance
With current	$r + c$	2	$2(r + c)$
Against current	$r - c$	2	$2(r - c)$

• The distance with the current is 14 mi. The distance against the current is 10 mi.

Solution

$$2(r+c) = 14 \qquad \frac{1}{2}\cdot 2(r+c) = 14\cdot\frac{1}{2}$$

$$2(r-c) = 10 \qquad \frac{1}{2}\cdot 2(r-c) = 10\cdot\frac{1}{2}$$

$$r + c = 7$$
$$r - c = 5$$

$$2r = 12$$
$$r = 6$$

$$r + c = 7$$
$$6 + c = 7$$
$$c = 1$$

The rate of the canoeist in calm water is 6 mph. The rate of the current is 1 mph.

Objective B Exercises

11. Strategy

- Cost per pound of the wheat flour: x
  Cost per pound of the rye flour: y

First Purchase

	Number	Cost	Total Cost
Wheat	12	x	$12x$
Rye	15	y	$15y$

Second Purchase

	Number	Cost	Total Cost
Wheat	15	x	$15x$
Rye	10	y	$10y$

- The first purchase cost \$18.30. The second purchase cost \$16.75.

Solution

$$12x + 15y = 18.30 \qquad 5(12x + 15y) = 18.30(5)$$
$$15x + 10y = 16.75 \qquad -4(15x + 10y) = 16.75(-4)$$

$$60x + 75y = 91.50$$
$$-60x - 40y = -67.00$$

$$35y = 24.50$$
$$y = 0.70$$

$$12x + 15y = 18.30$$
$$12x + 10.50 = 18.30$$
$$12x = 7.80$$
$$x = 0.65$$

The cost per pound of the wheat flour is \$.65. The cost per pound of the rye flour is \$.70.

13. Strategy

- % hydrochloric acid in reagent I: x
  % hydrochloric acid in reagent II: y

First solution

	Amt of Reagent	% Hydrochloric acid	Amt of Hydrochloric acid
Reagent I	40	x	$40x$
Reagent II	60	y	$60y$

Second solution

	Amt of Reagent	% Hydrochloric acid	Amt of Hydrochloric acid
Reagent I	35	x	$35x$
Reagent II	65	y	$65y$

- The first solution is 31% hydrochloric acid. The second solution is 31.5% hydrochloric acid.

Solution

$$40x + 60y = .31(40 + 60) \qquad -7(40x + 60y) = -7(31)$$
$$35x + 65y = .315(35 + 65) \qquad 8(35x + 65y) = 8(31.5)$$
$$-280x - 420y = -217$$
$$280x + 520y = 252$$
$$100y = 35$$
$$y = 0.35$$

$$40x + 60y = 31$$
$$40x + 60(0.35) = 31$$
$$40x + 21 = 31$$
$$40x = 10$$
$$x = 0.25$$

Reagent I is 25% hydrochloric acid. Reagent II is 35% hydrochloric acid.

15. Strategy • Amount of 87-octane gasoline: x
Amount of 93-octane gasoline: y

	Amount	Percent	Quantity
87-octane	x	0.87	$0.87x$
93-octane	y	0.93	$0.93y$
89-octane	18	0.89	0.89(18)

• The sum of the quantities before mixing is equal to the sum of the quantities after mixing. The total amount of the two gasolines is 18 gallons.

Solution

$$x + y = 18 \qquad -0.87(x+y) = -0.87(18)$$
$$0.87x + 0.93y = 16.02 \qquad 0.87x + 0.93y = 16.02$$

$$-0.87x - 0.87y = -15.66$$
$$0.87x + 0.93y = 16.02$$
$$0.06y = 0.36$$
$$y = 6$$

$$x + y = 18$$
$$x + 6 = 18$$
$$x = 12$$

12 gal of 87-octane gasoline and 6 gal of 93-octane fuel must be used.

Applying the Concepts

17. Strategy • Number of nickels: x
Number of dimes: y

Original

	Value	Number	Total Value
Nickels	5	x	$5x$
Dimes	10	y	$10y$

Double

	Value	Number	Total Value
Nickels	5	$2x$	$10x$
Dimes	10	$2y$	$20y$

• The original value of the coins is \$0.25.
The doubled value is \$0.50.

Solution

$$5x + 10y = 25$$
$$10x + 20y = 50$$
$$-10x - 20y = -50$$
$$10x + 20y = 50$$
$$0 = 0$$

The system is dependent. There is more than one solution. Based on the restrictions of the problem, there could be 1 nickel and 2 dimes or 3 nickels and 1 dime.

19. Strategy • Acres of good land: x
Acres of bad land: y

	Acres	Price	Total Price
Good land	x	300	$300x$
Bad land	y	$\frac{500}{7}$	$\frac{500y}{7}$

• Total number of acres bought is 100.
Total price is \$10,000.

Solution

$$x + y = 100$$
$$300x + \frac{500y}{7} = 10{,}000$$
$$x + y = 100$$
$$2100x + 500y = 70{,}000$$
$$-2100x - 2100y = -210{,}000$$
$$2100x + 500y = 70{,}000$$
$$-1600y = -140{,}000$$
$$y = 87.5$$
$$x + 87.5 = 100$$
$$x = 12.5$$

12.5 acres of good land and 87.5 acres of bad land were bought.

Chapter 8 Review Exercises

1.

$$5x + 4y = -17 \qquad 2x - y = 1$$
$$5(-1) + 4(-3) \;|\; -17 \qquad 2(-1) - (-3) \;|\; 1$$
$$-5 - 12 \;|\; -17 \qquad -2 + 3 \;|\; 1$$
$$-17 = -17 \qquad 1 = 1$$

Yes, $(-1, -3)$ is a solution of the system of equations.

2.

$$-x + 9y = 2 \qquad 6x - 4y = 12$$
$$-(-2) + 0 \;|\; 2 \qquad 6(-2) - 0 \;|\; 12$$
$$2 = 2 \qquad -12 \neq 12$$

No, $(-2, 0)$ is not a solution of the system of equations.

3.

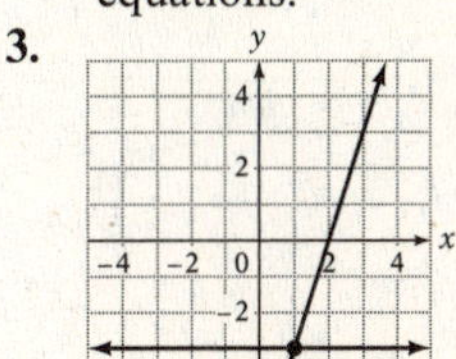

4. 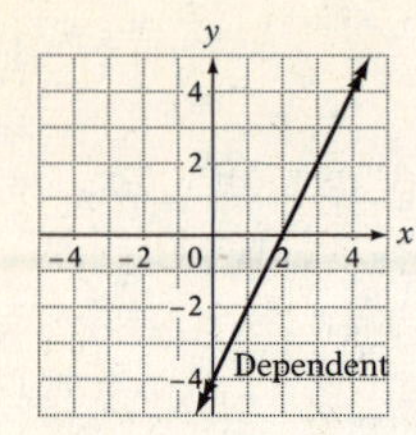

The solutions are the ordered-pair solutions of $y = 2x - 4$.

5. 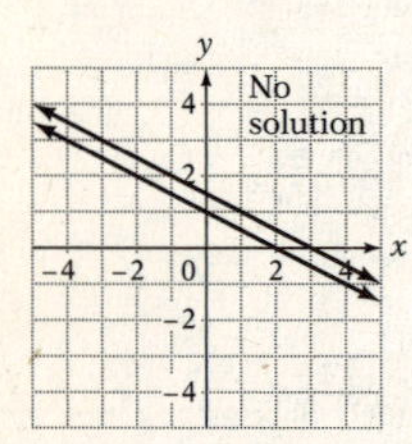

6. (1) $4x + 7y = 3$
(2) $x = y - 2$
Substitute in Equation (1).
$4(y - 2) + 7y = 3$
$4y - 8 + 7y = 3$
$11y = 11$
$y = 1$
Substitute in Equation (2).
$x = y - 2$
$x = 1 - 2$
$x = -1$
The solution is $(-1, 1)$.

7. (1) $6x - y = 0$
(2) $7x - y = 1$
Substitute in Equation (2).
$7x - 6x = 1$
$x = 1$
Substitute in Equation (1).
$6(1) - y = 0$
$-y = -6$
$y = 6$
The solution is $(1, 6)$.

8. (1) $3x + 8y = -1$
(2) $x - 2y = -5$
Eliminate y and add the equation.

$3x + 8y = -1 \qquad 3x + 8y = -1$
$4(x - 2y) = -5(4) \qquad 4x - 8y = -20$
$7x = -21$
$x = -3$

Replace x in Equation (2).
$x - 2y = -5$
$-3 - 2y = -5$
$-2y = -2$
$y = 1$
The solution is $(-3, 1)$.

9. (1) $6x + 4y = -3$
(2) $12x - 10y = -15$
Eliminate x and add the equations.
$-2(6x + 4y) = -3(-2) \qquad -12x - 8y = 6$
$12x - 10y = -15 \qquad 12x - 10y = -15$
$-18y = -9$
$y = \frac{1}{2}$
Replace y in Equation (1).
$6x + 4y = -3$
$6x + 2 = -3$
$6x = -5$
$x = -\frac{5}{6}$
The solution is $\left(-\frac{5}{6}, \frac{1}{2}\right)$.

10. (1) $12x - 9y = 18$
(2) $y = \frac{4}{3}x - 3$
Substitute in Equation (1).
$12x - 9\left(\frac{4}{3}x - 3\right) = 18$
$12x - 12x + 27 = 18$
$27 \neq 18$
The system is inconsistent and has no solution.

11. (1) $8x - y = 2$
(2) $y = 5x + 1$
Substitute in Equation (1).
$8x - (5x + 1) = 2$
$8x - 5x - 1 = 2$
$3x = 3$
$x = 1$
Substitute in Equation (2).
$y = 5x + 1$
$y = 5 + 1$
$y = 6$
The solution is $(1, 6)$.

12. (1) $4x - y = 9$
(2) $2x + 3y = -13$
Eliminate y and add the equations.
$3(4x - y) = 9(3) \qquad 12x - 3y = 27$
$2x + 3y = -13 \qquad 2x + 3y = -13$
$14x = 14$
$x = 1$
Replace x in Equation (1).
$4x - y = 9$
$4 - y = 9$
$-y = 5$
$y = -5$
The solution is $(1, -5)$.

13. (1) $5x + 7y = 21$
(2) $20x + 28y = 63$
Eliminate x and add the equations.

$$-4(5x + 7y) = 21(-4) \quad -20x - 28y = -84$$
$$20x + 28y = 63 \quad 20x + 28y = 63$$
$$0 \neq -21$$

The system is inconsistent and has no solution.

14. (1) $4x + 3y = 12$
(2) $y = -\frac{4}{3}x + 4$
Substitute in Equation (1).

$$4x + 3\left(-\frac{4}{3}x + 4\right) = 12$$
$$4x - 4x + 12 = 12$$
$$12 = 12$$

The system of equations is dependent. The solutions are the ordered pairs that satisfy the equation $y = -\frac{4}{3}x + 4$.

15. (1) $7x + 3y = -16$
(2) $x - 2y = 5$
Substitute in Equation (1).

$$7(2y + 5) + 3y = -16$$
$$14y + 35 + 3y = -16$$
$$17y = -51$$
$$y = -3$$

Substitute in Equation (2).

$$x - 2y = 5$$
$$x - 2(-3) = 5$$
$$x + 6 = 5$$
$$x = -1$$

The solution is $(-1, -3)$.

16. (1) $3x + y = -2$
(2) $-9x - 3y = 6$
Eliminate x and add the equations.

$$3(3x + y) = -2(3) \quad 9x + 3y = -6$$
$$-9x - 3y = 6 \quad -9x - 3y = 6$$
$$0 = 0$$

The system is dependent. The solutions are the ordered pairs that satisfy the equation $3x + y = -2$.

17. (1) $6x - 18y = 7$
(2) $9x + 24y = 2$
Eliminate x and add the equations.

$$9(6x - 18y) = 7(9) \quad 54x - 162y = 63$$
$$-6(9x + 24y) = 2(-6) \quad -54x - 144y = -12$$
$$-306y = 51$$
$$y = -\frac{1}{6}$$

Replace y in Equation (1).

$$6x - 18y = 7$$
$$6x - 18\left(-\frac{1}{6}\right) = 7$$
$$6x + 3 = 7$$
$$6x = 4$$
$$x = \frac{2}{3}$$

The solution is $\left(\frac{2}{3}, -\frac{1}{6}\right)$.

18. **Strategy** • Rate of the sculling team in calm water: r
Rate of the current: c

	Rate	Time	Distance
With current	$r + c$	2	$2(r + c)$
Against current	$r - c$	3	$3(r - c)$

• The distance with the current is 24 mi.
The distance against the current is 18 mi.

Solution

$$2(r + c) = 24 \quad \frac{1}{2} \cdot 2(r + c) = 24 \cdot \frac{1}{2}$$
$$3(r - c) = 18 \quad \frac{1}{3} \cdot 3(r - c) = 18 \cdot \frac{1}{3}$$
$$r + c = 12$$
$$r - c = 6$$
$$2r = 18$$
$$r = 9$$
$$r + c = 12$$
$$9 + c = 12$$
$$c = 3$$

The rate of the sculling team in calm water is 9 mph.
The rate of the current is 3 mph.

19. Strategy • Number of shares at \$6 per share: x
Number of shares at \$25 per share: y

$$x + y = 1500$$
$$6x + 25y = 12{,}800$$

• The total number of shares is 1500.
The total purchase price is \$12,800.

Solution

$$x + y = 1500 \qquad -6(x + y) = 1500(-6)$$
$$6x + 25y = 12{,}800 \qquad 6x + 25y = 12{,}800$$

$$-6x - 6y = -9000$$
$$6x + 25y = 12{,}800$$

$$19y = 3800$$
$$y = 200$$

$$x + y = 1500$$
$$x + 200 = 1500$$
$$x = 1300$$

Number of \$6 shares purchased is 1300.
Number of \$25 shares purchased is 200.

20. Strategy • Rate of the flight crew in calm air: p
Rate of the wind: w

	Rate	Time	Distance
With wind	$p + w$	3	$3(p + w)$
Against wind	$p - w$	4	$4(p - w)$

• The distance traveled with the wind is 420 km. The distance traveled against the wind is 440 km.

Solution

$$3(p + w) = 420 \qquad \frac{1}{3} \cdot 3(p + w) = 420 \cdot \frac{1}{3}$$
$$4(p - w) = 440 \qquad \frac{1}{4} \cdot 4(p - w) = 400 \cdot \frac{1}{4}$$

$$p + w = 140$$
$$p - w = 110$$

$$2p = 250$$
$$p = 125$$

$$p + w = 140$$
$$125 + w = 140$$
$$w = 15$$

The rate of the flight crew in calm air is 125 km/h.
The rate of the wind is 15 km/h.

21. Strategy • Rate of the plane in calm air: p
Rate of the wind: w

	Rate	Time	Distance
With wind	$p + w$	3	$3(p + w)$
Against wind	$p - w$	4	$4(p - w)$

• The distance traveled with the wind and against the wind is 360.

Solution

$$3(p + w) = 360 \qquad \frac{1}{3} \cdot 3(p + w) = 360 \cdot \frac{1}{3}$$
$$4(p - w) = 360 \qquad \frac{1}{4} \cdot 4(p - w) = 360 \cdot \frac{1}{4}$$

$$p + w = 120$$
$$p - w = 90$$

$$2p = 210$$
$$p = 105$$

$$p + w = 120$$
$$105 + w = 120$$
$$w = 15$$

The rate of the plane in calm air is 105 mph.
The rate of the wind is 15 mph.

22. Strategy • Number of advertisements requiring \$.25 postage: x
Number of advertisements requiring \$.45 postage: y

$$x + y = 190$$
$$25x + 45y = 5950$$

• The total number of advertisements is 190. The total cost of the mailing is 5950 cents.

Solution

$$x + y = 190 \qquad -25(x + y) = 190(-25)$$
$$25x + 45y = 5950 \qquad 25x + 45y = 5950$$

$$-25x - 25y = -4750$$
$$25x + 45y = 5950$$

$$20y = 1200$$
$$y = 60$$

$$x + y = 190$$
$$x + 60 = 190$$
$$x = 130$$

The number of ads requiring \$.25 postage is 130.
The number of ads requiring \$.45 postage is 60.

23. Strategy • Amount invested at 8.5%: x
Amount invested at 7%: y

	Principal	Rate	Interest
Amount at 8.5%	x	0.085	$0.085x$
Amount at 7%	y	0.07	$0.07y$

• The total amount invested is \$12,000.
The total annual interest earned is \$915.

Solution

$$x + y = 12{,}000 \qquad y = 12{,}000 - x$$
$$0.085x + 0.07y = 915$$
$$0.085x + 0.07(12{,}000 - x) = 915$$
$$0.085x + 840 - 0.07x = 915$$
$$0.015x = 75$$
$$x = 5000$$

$$y = 12{,}000 - x = 12{,}000 - 5000 = 7000$$

The amounts invested are \$7000 at 7% and \$5000 at 8.5%.

24. Strategy • Number of bushels of lentils: x
Number of bushels of corn: y

Solution

$$x + 50 = 2y \qquad x - 2y = -50$$
$$y + 150 = x \qquad -x + y = -150$$

$$-y = -200$$
$$y = 200$$

$$y + 150 = x$$
$$200 + 150 = x$$
$$350 = x$$

There were originally 350 bushels of lentils and 200 bushels of corn in the silo.

25. Strategy • The amount invested at 5.4%: x
The amount invested at 6.6%: y

	Principal	Rate	Interest
Amount at 5.4%	x	0.054	$0.054x$
Amount at 6.6%	y	0.066	$0.066y$

• The total amount invested is \$300,000.
The interest earned in each account is the same.

Solution

$$x + y = 300{,}000 \qquad y = 300{,}000 - x$$
$$0.054x = 0.066y$$
$$0.054x = 0.066(300{,}000 - x)$$
$$0.054x = 19{,}800 - 0.066x$$
$$0.12x = 19{,}800$$
$$x = 165{,}000$$
$$y = 300{,}000 - x = 300{,}000 - 165{,}000 = 135{,}000$$

The amounts invested were \$165,000 at 5.4% and \$135,000 at 6.6%.

Chapter 8 Test

1.

$$\begin{array}{r|l} 2x + 5y = 11 & \\ \hline 2(-2) + 5(3) & 11 \\ -4 + 15 & 11 \\ 11 = 11 & \end{array} \qquad \begin{array}{r|l} x + 3y = 7 & \\ \hline -2 + (3)(3) & 7 \\ -2 + 9 & 7 \\ 7 = 7 & \end{array}$$

Yes, $(-2, 3)$ is a solution.

2.

$$\begin{array}{r|l} 3x - 2y = 9 & \\ \hline 3(1) - 2(-3) & 9 \\ 3 + 6 & 9 \\ 9 = 9 & \end{array} \qquad \begin{array}{r|l} 4x + y = 1 & \\ \hline 4(1) - 3 & 1 \\ 4 - 3 & 1 \\ 1 = 1 & \end{array}$$

Yes, $(1, -3)$ is a solution.

3.

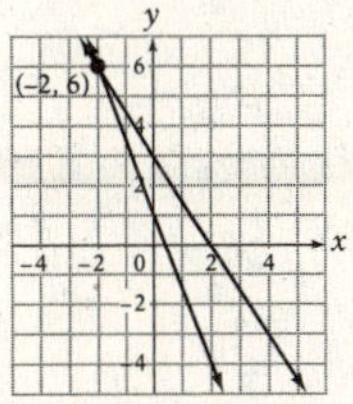

(-2, 6)

4. (1) $4x - y = 11$
(2) $y = 2x - 5$

Substitute in Equation (1).

$$4x - (2x - 5) = 11$$
$$4x - 2x + 5 = 11$$
$$2x = 6$$
$$x = 3$$

Substitute in Equation (2).

$$y = 2x - 5$$
$$y = 6 - 5$$
$$y = 1$$

The solution is $(3, 1)$.

5. (1) $x = 2y + 3$
(2) $3x - 2y = 5$

Substitute in Equation (2).

$$3(2y + 3) - 2y = 5$$
$$6y + 9 - 2y = 5$$
$$4y = -4$$
$$y = -1$$

Substitute in Equation (1).

$$x = 2y + 3$$
$$x = -2 + 3$$
$$x = 1$$

The solution is $(1, -1)$.

6. (1) $3x + 5y = 1$
(2) $2x - y = 5 \quad 2x - 5 = y$
Substitute in Equation (1).
$3x + 5(2x - 5) = 1$
$3x + 10x - 25 = 1$
$13x = 26$
$x = 2$
Substitute in Equation (2).
$y = 2x - 5$
$y = 4 - 5$
$y = -1$
The solution is $(2, -1)$.

7. (1) $3x - 5y = 13$
(2) $x + 3y = 1 \quad x = -3y + 1$
Substitute in Equation (1).
$3(-3y + 1) - 5y = 13$
$-9y + 3 - 5y = 13$
$-14y = 10$
$y = -\frac{5}{7}$
Substitute in Equation (2).
$x = -3y + 1$
$x = -3\left(-\frac{5}{7}\right) + 1$
$x = \frac{15}{7} + 1$
$x = \frac{15}{7} + \frac{7}{7}$
$x = \frac{22}{7}$
The solution is $\left(\frac{22}{7}, -\frac{5}{7}\right)$.

8. (1) $2x - 4y = 1$
(2) $y = \frac{1}{2}x + 3$
Substitute in Equation (1).
$2x - 4\left(\frac{1}{2}x + 3\right) = 1$
$2x - 2x - 12 = 1$
$-12 \neq 1$
The system is inconsistent and has no solution.

9. Add the equation: (1) $4x + 3y = 11$
(2) $5x - 3y = 7$
$9x = 18$
$x = 2$

Replace x in Equation (1).
$4x + 3y = 11$
$8 + 3y = 11$
$3y = 3$
$y = 1$
The solution is $(2, 1)$.

10. (1) $2x - 5y = 6$
(2) $4x + 3y = -1$
Eliminate x and add the equation.
$-2(2x - 5y) = 6(-2) \quad -4x + 10y = -12$
$4x + 3y = -1 \quad 4x + 3y = -1$
$13y = -13$
$y = -1$

Replace y in Equation (1).
$2x - 5y = 6$
$2x - 5(-1) = 6$
$2x + 5 = 6$
$2x = 1$
$x = \frac{1}{2}$

The solution is $\left(\frac{1}{2}, -1\right)$.

11. (1) $x + 2y = 8$
(2) $3x + 6y = 24$
Eliminate x and add the equation.
$-3(x + 2y) = 8(-3) \quad -3x - 6y = -24$
$3x + 6y = 24 \quad 3x + 6y = 24$
$0 = 0$
The system is dependent. The solutions are the ordered pairs that satisfy the equation $x + 2y = 8$.

12. (1) $7x + 3y = 11$
(2) $2x - 5y = 9$
Eliminate y and add the equations.
$5(7x + 3y) = 11(5) \quad 35x + 15y = 55$
$3(2x - 5y) = 9(3) \quad 6x - 15y = 27$
$41x = 82$
$x = 2$

Replace x in Equation (1).
$7x + 3y = 11$
$14 + 3y = 11$
$3y = -3$
$y = -1$
The solution is $(2, -1)$.

13. (1) $5x + 6y = -7$
(2) $3x + 4y = -5$
Eliminate x and add the equations.
$3(5x + 6y) = -7(3) \quad 15x + 18y = -21$
$-5(3x + 4y) = -5(-5) \quad -15x - 20y = 25$
$-2y = 4$
$y = -2$

Replace y in Equation (1).
$5x + 6y = -7$
$5x - 12 = -7$
$5x = 5$
$x = 1$
The solution is $(1, -2)$.

14. Strategy

- Rate of the plane in calm air: p
  Rate of the wind: w

	Rate	Time	Distance
With wind	$p + w$	2	$2(p + w)$
Against wind	$p - w$	3	$3(p - w)$

- The distance traveled with the wind is 240 mi.
  The distance traveled against the wind is 240 mi.

Solution

$2(p + w) = 240 \quad \frac{1}{2}(2(p + w) = 240 \cdot \frac{1}{2} \quad p + w = 120$

$3(p - w) = 240 \quad \frac{1}{3} \cdot 3(p - w) = 240 \cdot \frac{1}{3} \quad p - w = 80$

$2p = 200$
$p = 100$

$p + w = 120$
$100 + w = 120$
$w = 20$

The rate of the plane in calm air is 100 mph.
The rate of the wind is 20 mph.

15. Strategy

- Price of a reserved-seat ticket: x
  Price of a general-admission ticket: y

First Performance

	Number	Cost	Total Cost
Reserved seat	50	x	$50x$
General admission	80	y	$80y$

Second Performance

	Number	Cost	Total Cost
Reserved seat	60	x	$60x$
General admission	90	y	$90y$

- The total receipts for the first performance were $980. The total receipts for the second performance were $1140.

Solution

$50x + 80y = 980 \quad 5x + 8y = 98 \quad 6(5x + 8y) = 98(6) \quad 30x + 48y = 588$

$60x + 90y = 1140 \quad 6x + 9y = 114 \quad -5(6x + 9y) = 114(-5) \quad -30x - 45y = -570$

$3y = 18$
$y = 6$

$5x + 8y = 98$
$5x + 48 = 98$
$5x = 50$
$x = 10$

The price of a reserved-seat ticket is $10.
The price of a general-admission ticket is $6.

16. Strategy
- Amount invested at 7.6%: x
  Amount invested at 6.4%: y

	Principal	Rate	Interest
Amount at 7.6%	x	0.076	$0.076x$
Amount at 6.4%	y	0.064	$0.064y$

- The total amount invested is \$28,000. The interest earned in each account is the same.

Solution $x + y = 28{,}000 \quad y = 28{,}000 - x$

$0.076x = 0.064y$
$0.076x = 0.064(28{,}000 - x)$
$0.076x = 1792 - 0.064x$
$0.14x = 1792$
$x = 12{,}800$
$y = 28{,}000 - x = 28{,}000 - 12{,}800 = 15{,}200$

The amounts invested were \$15,200 at 6.4% and \$12,800 at 7.6%.

Cumulative Review Exercises

1. $\dfrac{a^2 - b^2}{2a} = \dfrac{4^2 - (-2)^2}{2(4)} = \dfrac{16 - 4}{8} = \dfrac{12}{8} = \dfrac{3}{2}$

2. $-\dfrac{3}{4}x = \dfrac{9}{8}$

$-\dfrac{4}{3} \cdot -\dfrac{3}{4}x = \dfrac{9}{8} \cdot -\dfrac{4}{3}$

$x = -\dfrac{3}{2}$

3. $f(x) = x^2 + 2x - 1$
$f(2) = 2^2 + 2(2) - 1$
$= 4 + 4 - 1$
$= 7$

4. $(2a^2 - 3a + 1)(2 - 3a)$
$(2 - 3a)(2a^2 - 3a + 1)$
$(-3a + 2)(2a^2 - 3a + 1)$
$-3a(2a^2 - 3a + 1) + 2(2a^2 - 3a + 1)$

$$\begin{array}{r} -6a^3 + 9a^2 - 3a \\ 4a^2 - 6a + 2 \\ \hline -6a^3 + 13a^2 - 9a + 2 \end{array}$$

5. $\dfrac{(-2x^2y)^4}{-8x^3y^2} = \dfrac{16x^8y^4}{-8x^3y^2} = -2x^5y^2$

6.
$$\begin{array}{r} 2b - 1 \\ 2b - 3\overline{)4b^2 - 8b + 4} \\ \underline{4b^2 - 6b} \\ -2b + 4 \\ \underline{-2b + 3} \\ 1 \end{array}$$

$(4b^2 - 8b + 4) \div (2b - 3)$
$= 2b - 1 + \dfrac{1}{2b - 3}$

7. $\dfrac{8x^{-2}y^5}{-2xy^4} = -\dfrac{4y}{x^3}$

8. $4x^2y^4 - 64y^2$
The GCF is $4y^2 : 4y^2(x^2y^2 - 16)$
$4y^2((xy)^2 - (4)^2) = 4y^2(xy + 4)(xy - 4)$

9. $(x - 5)(x + 2) = -6$
$x^2 - 3x - 10 = -6$
$x^2 - 3x - 4 = 0$
$(x - 4)(x + 1) = 0$
$x - 4 = 0 \quad x + 1 = 0$
$x = 4 \quad x = -1$
The solutions are 4 and -1.

10. $\dfrac{x^2 - 6x + 8}{2x^3 + 6x^2} \div \dfrac{2x - 8}{4x^3 + 12x^2}$
$= \dfrac{x^2 - 6x + 8}{2x^3 + 6x^2} \cdot \dfrac{4x^3 + 12x^2}{2x - 8}$
$= \dfrac{(x - 4)(x - 2)}{2x^2(x + 3)} \cdot \dfrac{4x^2(x + 3)}{2(x - 4)} = x - 2$

11. $\dfrac{x - 1}{x + 2} + \dfrac{2x + 1}{x^2 + x - 2}$
$\dfrac{x - 1}{x + 2} + \dfrac{2x + 1}{(x + 2)(x - 1)}$
$\text{LCM} = (x + 2)(x - 1)$
$\dfrac{(x - 1)(x - 1)}{(x + 2)(x - 1)} + \dfrac{2x + 1}{(x + 2)(x - 1)}$
$= \dfrac{x^2 - 2x + 1 + 2x + 1}{(x + 2)(x - 1)}$
$= \dfrac{x^2 + 2}{(x + 2)(x - 1)}$

12. $\dfrac{x + 4 - \dfrac{7}{x - 2}}{x + 8 + \dfrac{21}{x - 2}}$

$\dfrac{(x - 2)\left(x + 4 - \dfrac{7}{x - 2}\right)}{(x - 2)\left(x + 8 + \dfrac{21}{x - 2}\right)} = \dfrac{(x - 2)(x + 4) - 7}{(x - 2)(x + 8) + 21}$
$= \dfrac{x^2 + 2x - 8 - 7}{x^2 + 6x - 16 + 21}$
$= \dfrac{x^2 + 2x - 15}{x^2 + 6x + 5}$
$= \dfrac{(x + 5)(x - 3)}{(x + 5)(x + 1)} = \dfrac{x - 3}{x + 1}$

13. $\frac{x}{2x-3}+2=\frac{-7}{2x-3}$

$(2x-3)\left(\frac{x}{2x-3}+2\right)=(2x-3)\left(\frac{-7}{2x-3}\right)$

$x+2(2x-3)=-7$

$x+4x-6=-7$

$5x=-1$

$x=-\frac{1}{5}$

14. $A=P+Prt$

$A-P=Prt$

$\frac{A-P}{Pt}=r$

15. $2x-3y=12$

x-intercept:	y-intercept:
$2x=12$	$-3y=12$
$x=6$	$y=-4$
(6, 0)	(0, −4)

16. $m=\frac{y_2-y_1}{x_2-x_1}=\frac{4-(-3)}{-3-2}=\frac{7}{-5}=-\frac{7}{5}$

The slope is $-\frac{7}{5}$.

17. $m=-\frac{3}{2}$, $(x_1, y_1)=(-2, 3)$

$y-y_1=m(x-x_1)$

$y-3=-\frac{3}{2}(x+2)$

$y-3=-\frac{3}{2}x-3$

$y=-\frac{3}{2}x$

The equation of the line is $y=-\frac{3}{2}x$.

18.

$5x-3y=10$		$4x+7y=8$	
$5(2)-0$	10	$4(2)+0$	8
$10=10$		$8=8$	

Yes, (2, 0) is a solution of the system.

19. (1) $3x-5y=-23$

(2) $x+2y=-4$ $\quad x=-2y-4$

Substitute in Equation (1).

$3(-2y-4)-5y=-23$

$-6y-12-5y=-23$

$-11y=-11$

$y=1$

Substitute in Equation (2).

$x=-2y-4$

$x=-2-4$

$x=-6$

The solution is (−6, 1).

20. (1) $5x-3y=29$

(2) $4x+7y=-5$

Eliminate y and add the equations.

$7(5x-3y)=29(7) \qquad 35x-21y=203$

$3(4x+7y)=-5(3) \qquad 12x+21y=-15$

$47x=188$

$x=4$

Replace x in Equation (1).

$5x-3y=29$

$20-3y=29$

$-3y=9$

$y=-3$

The solution is (4, −3).

21. Strategy

- Amount invested at 9.6%: x
  Amount invested at 7.2%: y

	Principal	Rate	Interest
Amount at 9.6%	x	0.096	$0.096x$
Amount at 7.2%	y	0.072	$0.072y$

- The total amount invested at $8750. The interest earned in each account is the same.

Solution $x+y=8750 \qquad y=8750-x$

$0.096x=0.072y$

$0.096x=0.072(8750-x)$

$0.096x=630-0.072x$

$0.168x=630$

$x=3750$

$y=8750-x=8750-3750=5000$

The amounts invested should be $3750 at 9.6% and $5000 at 7.2%.

22. Strategy

- Rate of passenger train: x
  Rate of freight train: $x-8$

	Rate	Time	Distance
Passenger train	x	3	$3x$
Freight train	$x-8$	3.5	$3.5(x-8)$

- The distances the trains travel are the same.

Solution $3x=3.5(x-8)$

$3x=3.5x-28$

$-0.5x=-28$

$x=56$

$x-8=56-8=48$

The rate of the passenger train is 56 mph.

The rate of the freight train is 48 mph.

23. Strategy • Side of the original square: x
Side of the new square: $x+4$
• Use the equation for the area of a square.

Solution

$$\begin{aligned} s^2 &= A \\ (x+4)^2 &= 144 \\ x^2+8x+16 &= 144 \\ x^2+8x-128 &= 0 \\ (x+16)(x-8) &= 0 \\ x-8 &= 0 \\ x &= 8 \end{aligned}$$

The side of the original square is 8 in.

24. Strategy • Rate of the wind: x

	Rate	Time	Distance
With wind	$160+x$	$\frac{570}{160+x}$	570
Against wind	$160-x$	$\frac{390}{160-x}$	390

• The time with the wind and the time against the wind are equal.

Solution

$$\begin{aligned} \frac{570}{160+x} &= \frac{390}{160-x} \\ 570(160-x) &= 390(160+x) \\ 91{,}200-570x &= 62{,}400+390x \\ 28{,}800 &= 960x \\ 30 &= x \end{aligned}$$

The rate of the wind is 30 mph.

25.

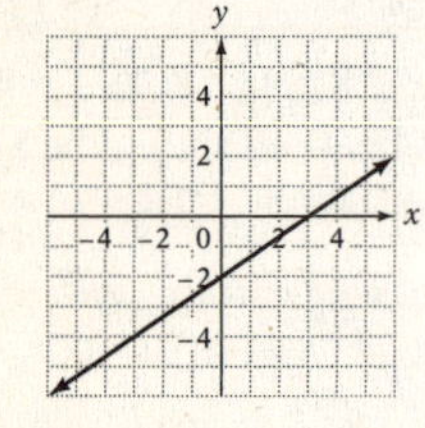

26.

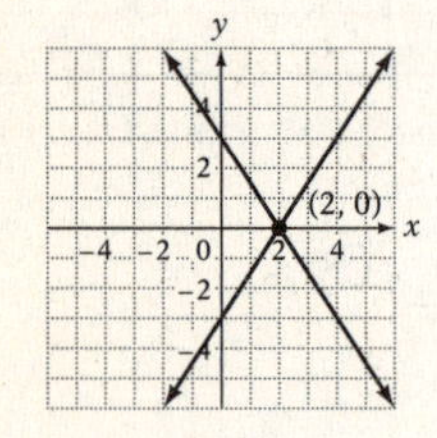

27. Strategy • Rate of the motorboat in calm water: r
Rate of the current: c

	Rate	Time	Distance
With current	$r+c$	3	$3(r+c)$
Against current	$r-c$	4	$4(r-c)$

• The distance with the current is the same as the distance against the current.

Solution

$$3(r+c) = 48 \qquad \frac{1}{3}\cdot 3(r+c) = 48\cdot\frac{1}{3}$$

$$4(r-c) = 48 \qquad \frac{1}{4}\cdot 4(r-c) = 48\cdot\frac{1}{4}$$

$$\begin{aligned} r+c &= 16 \\ r-c &= 12 \end{aligned}$$

$$\begin{aligned} 2r &= 28 \\ r &= 14 \end{aligned}$$

The rate of the motorboat in calm water is 14 mph.

28. Strategy • The percent concentration of sugar in the mixture: x

	Amount	Percent	Quantity
Cereal	50	0.25	0.25(50)
Sugar	8	1.00	1.00(8)
Mixture	58	x	$58x$

• The sum of the quantities before mixing is equal to the quantity after mixing.

Solution

$$\begin{aligned} 0.25(50)+1.00(8) &= 58x \\ 12.5+8 &= 58x \\ 20.5 &= 58x \\ 0.353 &\approx x \end{aligned}$$

The percent concentration of sugar in the mixture is 35.3%.

Chapter 9: Inequalities

Prep Test

1. $-45 < -27$

2. $3x - 5(2x - 3) = 3x - 10x + 15$
$= -7x + 15$

3. The same number can be added to each side of an equation without changing the solution of the equation.

4. Each side of an equation can be multiplied by the same nonzero number without changing the solution of the equation.

5. **Strategy** To find the amount of fat in the hamburger, multiply the number of pounds by 0.15.

Solution $(0.15)3 = 0.45$ lbs
There are 0.45 lbs of fat in 3 lb of this grade of hamburger.

6. $$\begin{aligned} 4x - 5 &= -7 \\ 4x &= -2 \\ x &= -\frac{1}{2} \end{aligned}$$

7. $$\begin{aligned} 4 &= 2 - \frac{3}{4}x \\ 2 &= -\frac{3}{4}x \\ x &= -\frac{8}{3} \end{aligned}$$

8. $$\begin{aligned} 7 - 2(2x - 3) &= 3x - 1 \\ 7 - 4x + 6 &= 3x - 1 \\ 13 - 4x &= 3x - 1 \\ -7x &= -14 \\ x &= 2 \end{aligned}$$

9.

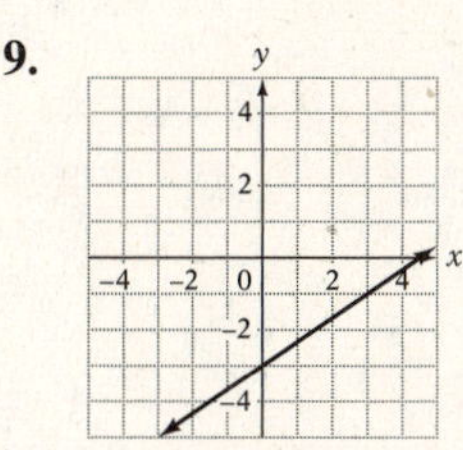

10.

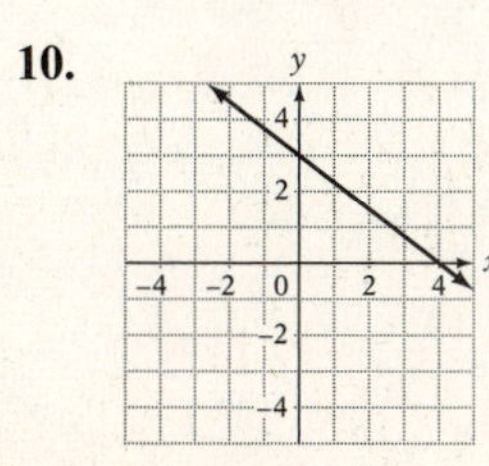

Go Figure

Since

$$\begin{aligned} 2^{150} &= (2^3)^{50} = 8^{50}, \\ 3^{100} &= (3^2)^{50} = 9^{50}, \\ \text{and } 5^{50} &= 5^{50}, \end{aligned}$$

3^{100} is the largest number.

Section 9.1

Objective A Exercises

1. Student explanations should include the idea that to find the union of two sets, we list all the elements of the first set and then list all the elements of the second set that are not elements of the first set.

3. $A = \{16, 17, 18, 19, 20, 21\}$

5. $A = \{9, 11, 13, 15, 17\}$

7. $A = \{\text{b}, \text{c}\}$

9. $A \cup B = \{3, 4, 5, 6\}$

11. $A \cup B = \{-10, -9, -8, 8, 9, 10\}$

13. $A \cup B = \{\text{a}, \text{b}, \text{c}, \text{d}, \text{e}, \text{f}\}$

15. $A \cup B = \{1, 3, 7, 9, 11, 13\}$

17. $A \cap B = \{4, 5\}$

19. $A \cap B = \varnothing$

21. $A \cap B = \{\text{c}, \text{d}, \text{e}\}$

Objective B Exercises

23. $\{x | x > -5, x \in \text{negative integers}\}$

25. $\{x | x > 30, x \in \text{integers}\}$

27. $\{x | x > 5, x \in \text{even integers}\}$

29. $\{x | x > 8, x \in \text{real integers}\}$

Objective C Exercises

31. $\{x | x > 2\}$

−5 −4 −3 −2 −1 0 1 2 3 4 5

33. $\{x | x \le 0\}$

−5 −4 −3 −2 −1 0 1 2 3 4 5

35. $\{x | x > -2\} \cup \{x | x < -4\}$

−5 −4 −3 −2 −1 0 1 2 3 4 5

37. $\{x|x > -2\} \cap \{x|x < 4\}$

−5 −4 −3 −2 −1 0 1 2 3 4 5

39. $\{x|x \geq -2\} \cup \{x|x < 4\}$

−5 −4 −3 −2 −1 0 1 2 3 4 5

Applying the Concepts

41. a. Never true. A positive number times a negative number is always negative.

b. Always true. The square of a nonzero number is always positive.

c. Always true. The square of a nonzero number is always positive and is therefore greater than a negative number.

43. Let $A = \{1, 2, 3, 4\}$ $B = \{1, 3, 5\}$ $C = \{2, 4, 6\}$

a. $A \cap B = \{1, 3\}$ $B \cap C = \{\ \}$
$B \cap A = \{1, 3\}$ $C \cap B = \{\ \}$
Yes, the intersection of two sets is commutative.

b. $(A \cap B) \cap C = \{1, 3\} \cap \{2, 4, 6\} = \{\ \}$
$A \cap (B \cap C) = \{1, 2, 3, 4\} \cap \{\ \} = \{\ \}$
Yes, the intersection of two sets is associative.

Section 9.2

Objective A Exercises

1.
$$\begin{aligned} x + 1 &< 3 \\ x + 1 + (-1) &< 3 + (-1) \\ x &< 2 \end{aligned}$$

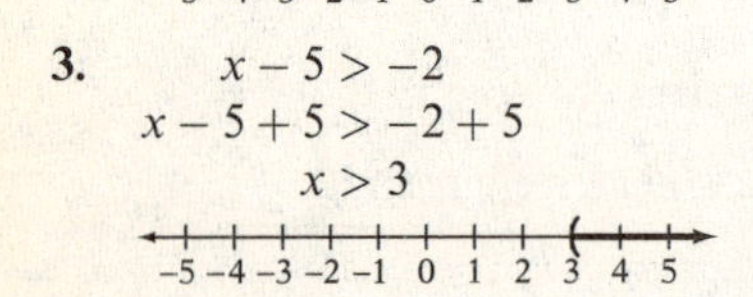

3.
$$\begin{aligned} x - 5 &> -2 \\ x - 5 + 5 &> -2 + 5 \\ x &> 3 \end{aligned}$$

5.
$$\begin{aligned} 7 &\leq n + 4 \\ 7 + (-4) &\leq n + 4 + (-4) \\ n &\geq 3 \end{aligned}$$

−5 −4 −3 −2 −1 0 1 2 3 4 5

7.
$$\begin{aligned} x - 6 &\leq -10 \\ x - 6 + 6 &\leq -10 + 6 \\ x &\leq -4 \end{aligned}$$

−5 −4 −3 −2 −1 0 1 2 3 4 5

9. $x < 1$

11. $x \leq -3$

13.
$$\begin{aligned} y - 3 &\geq -12 \\ y - 3 + 3 &\geq -12 + 3 \\ y &\geq -9 \end{aligned}$$

15.
$$\begin{aligned} 3x - 5 &< 2x + 7 \\ 3x + (-2x) - 5 &< 2x + (-2x) + 7 \\ x - 5 &< 7 \\ x - 5 + 5 &< 7 + 5 \\ x &< 12 \end{aligned}$$

17.
$$\begin{aligned} 8x - 7 &\geq 7x - 2 \\ 8x + (-7x) - 7 &\geq 7x + (-7x) - 2 \\ x - 7 &\geq -2 \\ x - 7 + 7 &\geq -2 + 7 \\ x &\geq 5 \end{aligned}$$

19.
$$\begin{aligned} 2x + 4 &< x - 7 \\ 2x + (-x) + 4 &< x + (-x) - 7 \\ x + 4 &< -7 \\ x + 4 + (-4) &< -7 + (-4) \\ x &< -11 \end{aligned}$$

21.
$$\begin{aligned} 4x - 8 &\leq 2 + 3x \\ 4x + (-3x) - 8 &\leq 2 + 3x + (-3x) \\ x - 8 &\leq 2 \\ x - 8 + 8 &\leq 2 + 8 \\ x &\leq 10 \end{aligned}$$

23.
$$\begin{aligned} 6x + 4 &\geq 5x - 2 \\ 6x + (-5x) + 4 &\geq 5x + (-5x) - 2 \\ x + 4 &\geq -2 \\ x + 4 + (-4) &\geq -2 + (-4) \\ x &\geq -6 \end{aligned}$$

25.
$$\begin{aligned} 2x - 12 &> x - 10 \\ 2x + (-x) - 12 &> x + (-x) - 10 \\ x - 12 &> -10 \\ x - 12 + 12 &> -10 + 12 \\ x &> 2 \end{aligned}$$

27.
$$\begin{aligned} d + \frac{1}{2} &< \frac{1}{3} \\ d + \frac{1}{2} + \left(-\frac{1}{2}\right) &< \frac{1}{3} + \left(-\frac{1}{2}\right) \\ d &< \frac{2}{6} - \frac{3}{6} \\ d &< -\frac{1}{6} \end{aligned}$$

29.
$$\begin{aligned} x + \frac{5}{8} &\geq -\frac{2}{3} \\ x + \frac{5}{8} + \left(-\frac{5}{8}\right) &\geq -\frac{2}{3} + \left(-\frac{5}{8}\right) \\ x &\geq -\frac{16}{24} - \frac{15}{24} \\ x &\geq -\frac{31}{24} \end{aligned}$$

31.
$$x - \frac{3}{8} < \frac{1}{4}$$
$$x - \frac{3}{8} + \frac{3}{8} < \frac{1}{4} + \frac{3}{8}$$
$$x < \frac{2}{8} + \frac{3}{8}$$
$$x < \frac{5}{8}$$

33.
$$2x - \frac{1}{2} < x + \frac{3}{4}$$
$$2x + (-x) - \frac{1}{2} < x + (-x) + \frac{3}{4}$$
$$x - \frac{1}{2} < \frac{3}{4}$$
$$x - \frac{1}{2} + \frac{1}{2} < \frac{3}{4} + \frac{1}{2}$$
$$x < \frac{3}{4} + \frac{2}{4}$$
$$x < \frac{5}{4}$$

35.
$$3x + \frac{5}{8} > 2x + \frac{5}{6}$$
$$3x + (-2x) + \frac{5}{8} > 2x + (-2x) + \frac{5}{6}$$
$$x + \frac{5}{8} > \frac{5}{6}$$
$$x + \frac{5}{8} + \left(-\frac{5}{8}\right) > \frac{5}{6} + \left(-\frac{5}{8}\right)$$
$$x > \frac{20}{24} - \frac{15}{24}$$
$$x > \frac{5}{24}$$

37.
$$3.8x < 2.8x - 3.8$$
$$3.8x + (-2.8x) < 2.8x + (-2.8x) - 3.8$$
$$x < -3.8$$

39.
$$x + 5.8 \le 4.6$$
$$x + 5.8 + (-5.8) \le 4.6 + (-5.8)$$
$$x \le -1.2$$

41.
$$x - 3.5 < 2.1$$
$$x - 3.5 + 3.5 < 2.1 + 3.5$$
$$x < 5.6$$

Objective B Exercises

43.
$$3x < 12$$
$$\frac{1}{3}(3x) < \frac{1}{3}(12)$$
$$x < 4$$

–5 –4 –3 –2 –1 0 1 2 3 4 5

45.
$$15 \le 5y$$
$$\frac{1}{5}(15) \le \frac{1}{5}(5y)$$
$$y \ge 3$$

–5 –4 –3 –2 –1 0 1 2 3 4 5

47.
$$16x \le 16$$
$$\frac{1}{16}(16x) \le \frac{1}{16}(16)$$
$$x \le 1$$

–5 –4 –3 –2 –1 0 1 2 3 4 5

49.
$$-8x > 8$$
$$-\frac{1}{8}(-8x) < -\frac{1}{8}(8)$$
$$x < -1$$

–5 –4 –3 –2 –1 0 1 2 3 4 5

51.
$$-6b > 24$$
$$-\frac{1}{6}(-6b) < -\frac{1}{6}(24)$$
$$b < -4$$

–5 –4 –3 –2 –1 0 1 2 3 4 5

53.
$$-5y \ge 0$$
$$-\frac{1}{5}(-5y) \le -\frac{1}{5}(0)$$
$$y \le 0$$

55.
$$7x > 2$$
$$\frac{1}{7}(7x) > \frac{1}{7}(2)$$
$$x > \frac{2}{7}$$

57.
$$2x \le -5$$
$$\frac{1}{2}(2x) \le \frac{1}{2}(-5)$$
$$x \le -\frac{5}{2}$$

59.
$$\frac{3}{4}x < 12$$
$$\frac{4}{3}\left(\frac{3}{4}x\right) < \frac{4}{3}(12)$$
$$x < 16$$

61.
$$10 \le \frac{5}{8}x$$
$$\frac{8}{5}(10) \le \frac{8}{5}\left(\frac{5}{8}x\right)$$
$$x \ge 16$$

63.
$$-\frac{3}{7}x \le 6$$
$$-\frac{7}{3}\left(-\frac{3}{7}x\right) \ge -\frac{7}{3}(6)$$
$$x \ge -14$$

65.
$$-\frac{4}{7}x \ge -12$$
$$-\frac{7}{4}\left(-\frac{4}{7}x\right) \le -\frac{7}{4}(-12)$$
$$x \le 21$$

67.
$$-\frac{3}{5}x < 0$$
$$-\frac{5}{3}\left(-\frac{3}{5}x\right) > -\frac{5}{3}(0)$$
$$x > 0$$

69.
$$-\frac{3}{8}x \ge \frac{9}{14}$$
$$-\frac{8}{3}\left(-\frac{3}{8}x\right) \le -\frac{8}{3}\left(\frac{9}{14}\right)$$
$$x \le -\frac{12}{7}$$

71.
$$-\frac{4}{5}x < -\frac{8}{15}$$
$$-\frac{5}{4}\left(-\frac{4}{5}x\right) > -\frac{5}{4}\left(-\frac{8}{15}\right)$$
$$x > \frac{2}{3}$$

73.
$$-\frac{8}{9}x \ge -\frac{16}{27}$$
$$-\frac{9}{8}\left(-\frac{8}{9}x\right) \le -\frac{9}{8}\left(-\frac{16}{27}\right)$$
$$x \le \frac{2}{3}$$

75.
$$2.3x \le 5.29$$
$$\frac{1}{2.3}(2.3x) \le \frac{1}{2.3}(5.29)$$
$$x \le 2.3$$

77.
$$-0.24 > 0.768$$
$$-\frac{1}{0.24}(-0.24x) < -\frac{1}{0.24}(0.768)$$
$$x < -3.2$$

79.
$$-3.9x \ge -19.5$$
$$-\frac{1}{3.9}(-3.9x) \le -\frac{1}{3.9}(-19.5)$$
$$x \le 5$$

Objective C Exercises

81. Strategy
- Number of games to qualify for the tournament: x

The team must win 60% of the remaining games to be eligible for the tournament.

Solution
$$x \ge 60\%(17)$$
$$x \ge 0.60(17)$$
$$x \ge 10.2$$

The team must win at least 11 games to be eligible for the tournament.

83. Strategy
- Number of pounds of aluminum to collect on the fourth collection drive: x

The service organization must collect over 1850 lb.

Solution
$$505 + 493 + 412 + x > 1850$$
$$1410 + x > 1850$$
$$x > 440$$

The service organization must collect more than 440 lb on the fourth drive to collect the bonus.

85. Strategy
- Number of additional milligrams of vitamin C: x

Minimum daily allowance of vitamin C is 60 mg.

Solution
$$x + 10 \ge 60$$
$$x \ge 50$$

The person needs at least 50 mg of additional vitamin C to satisfy the recommended daily allowance.

87. Strategy
- Score on fourth test: x

A minimum grade of 92 points is required to earn an A grade.

Solution
$$\frac{89 + 86 + 90 + x}{4} \ge 92$$
$$\frac{265 + x}{4} \ge 92$$
$$265 + x \ge 368$$
$$x \ge 103$$

No, the student cannot earn an A grade because it is impossible to score more than the maximum of 100 points.

Applying the Concepts

89. $ac > bc$ when c is a positive number.
$\{c|c > 0\}$

91. $a + c > b + c$ for all real numbers c.
$\{c|c > \in$ real numbers$\}$

93. $\frac{a}{c} > \frac{b}{c}$ when c is a positive number.
$\{c|c > 0\}$

95. The Addition Property of Inequalities states that the same term can be added to each side of an inequality without changing the solution set of the inequality.

Section 9.3

Objective A Exercises

1.
$$\begin{aligned} 4x - 8 &< 2x \\ 4x + (-2x) - 8 &< 2x + (-2x) \\ 2x - 8 &< 0 \\ 2x - 8 + 8 &< 0 + 8 \\ 2x &< 8 \\ \frac{1}{2}(2x) &< \frac{1}{2}(8) \\ x &< 4 \end{aligned}$$

3.
$$\begin{aligned} 2x - 8 &> 4x \\ 2x + (-4x) - 8 &> 4x + (-4x) \\ -2x - 8 &> 0 \\ -2x - 8 + 8 &> 0 + 8 \\ -2x &> 8 \\ -\frac{1}{2}(-2x) &< -\frac{1}{2}(8) \\ x &< -4 \end{aligned}$$

5.
$$\begin{aligned} 8 - 3x &\le 5x \\ 8 - 3x + (-5x) &\le 5x + (-5x) \\ 8 - 8x &\le 0 \\ 8 + (-8) - 8x &\le 0 + (-8) \\ -8x &\le -8 \\ -\frac{1}{8}(-8x) &\ge -\frac{1}{8}(-8) \\ x &\ge 1 \end{aligned}$$

7.
$$\begin{aligned} 3x + 2 &> 5x - 8 \\ 3x + (-5x) + 2 &> 5x + (-5x) - 8 \\ -2x + 2 &> -8 \\ -2x + 2 + (-2) &> -8 + (-2) \\ -2x &> -10 \\ -\frac{1}{2}(-2x) &< -\frac{1}{2}(-10) \\ x &< 5 \end{aligned}$$

9.
$$\begin{aligned} 5x - 2 &< 3x - 2 \\ 5x + (-3x) - 2 &< 3x + (-3x) - 2 \\ 2x - 2 &< -2 \\ 2x - 2 + 2 &< -2 + 2 \\ 2x &< 0 \\ \frac{1}{2}(2x) &< \frac{1}{2}(0) \\ x &< 0 \end{aligned}$$

11.
$$\begin{aligned} 0.1(180 + x) &> x \\ 18 + 0.1x &> x \\ 18 + 0.1x + (-x) &> x + (-x) \\ 18 - 0.9x &> 0 \\ 18 + (-18) - 0.9x &> 0 + (-18) \\ -0.9x &> -(18) \\ -\frac{1}{0.9}(-0.9x) &< -\frac{1}{0.9}(-18) \\ x &< 20 \end{aligned}$$

13.
$$\begin{aligned} 2(2y - 5) &\le 3(5 - 2y) \\ 4y - 10 &\le 15 - 6y \\ 4y + 6y - 10 &\le 15 - 6y + 6y \\ 10y - 10 &\le 15 \\ 10y - 10 + 10 &\le 15 + 10 \\ 10y &\le 25 \\ \frac{1}{10}(10y) &\le \frac{1}{10}(25) \\ y &\le \frac{5}{2} \end{aligned}$$

15.
$$\begin{aligned} 5(2 - x) &> 3(2x - 5) \\ 10 - 5x &> 6x - 15 \\ 10 - 5x + (-6x) &> 6x + (-6x) - 15 \\ 10 - 11x &> -15 \\ 10 + (-10) - 11x &> -15 + (-10) \\ -11x &> -25 \\ -\frac{1}{11}(-11x) &< -\frac{1}{11}(-25) \\ x &< \frac{25}{11} \end{aligned}$$

17.
$$\begin{aligned} 4 - 3(3 - n) &\le 3(2 - 5n) \\ 4 - 9 + 3n &\le 6 - 15n \\ -5 + 3n &\le 6 - 15n \\ -5 + 3n + 15n &\le 6 - 15n + 15n \\ -5 + 18n &\le 6 \\ -5 + 5 + 18n &\le 6 + 5 \\ 18n &\le 11 \\ \frac{1}{18}(18n) &\le \frac{1}{18}(11) \\ n &\le \frac{11}{18} \end{aligned}$$

19.
$$\begin{aligned} 2x - 3(x - 4) &\ge 4 - 2(x - 7) \\ 2x - 3x + 12 &\ge 4 - 2x + 14 \\ -x + 12 &\ge 18 - 2x \\ -x + 2x + 12 &\ge 18 - 2x + 2x \\ x + 12 &\ge 18 \\ x + 12 + (-12) &\ge 18 + (-12) \\ x &\ge 6 \end{aligned}$$

Objective B Exercises

21. Strategy • To find the dollar amount, write and solve an inequality using x to represent the total sales the agent made.
The agent's salary must be the greater of \$3200 or \$1000 plus an 11% commission on the selling price of each item.

Solution
$$1000 + 11\%x \le 3200$$
$$1000 + 0.11x \le 3200$$
$$0.11x \le 2200$$
$$x \le 20{,}000$$
In one month the agent expects to make sales totaling at most \$20,000.

23. Strategy • Required number of minutes: x
The flat fee must exceed \$10 per month or \$4 per month plus \$0.10 for each minute of service.

Solution
$$4 + 0.10x > 10$$
$$0.10x > 6$$
$$x > 60$$
A person must use more than 60 min to exceed \$10.

25. Strategy • Number of ounces of artificial flavors: x
The number of ounces of artificial flavors that can be added to 32 oz of real orange juice so that the mixture is 80% real orange juice must be less than or equal to 32 oz.

Solution
$$0.80(x + 32) \le 32$$
$$x + 32 \le 40$$
$$x \le 8$$
The amount of artificial flavors that can be added is at most 8 oz.

27. Strategy • Number of miles: x
The car rental cost must exceed \$64.

Solution
$$45 + 0.25x > 64$$
$$0.25x > 19$$
$$x > 76$$
The round-trip distance is more than 76 mi. Thus the distance to the ski resort must be more than 38 mi.

Applying the Concepts

29.
$$7 - 2b \le 15 - 5b$$
$$7 - 2b + 5b \le 15 - 5b + 5b$$
$$7 + 3b \le 15$$
$$7 + (-7) + 3b \le 15 + (-7)$$
$$3b \le 8$$
$$\frac{3b}{3} \le \frac{8}{3}$$
$$b \le \frac{8}{3}$$
The solution is $\{1, 2\}$.

31.
$$5x - 12 \le x + 8 \qquad 3x - 4 \ge 2 + x$$
$$5x + (-x) - 12 \le x + (-x) + 8 \qquad 3x + (-x) - 4 \ge 2 + x + (-x)$$
$$4x - 12 \le 8 \qquad 2x - 4 \ge 2$$
$$4x - 12 + 12 \le 8 + 12 \qquad 2x - 4 + 4 \ge 2 + 4$$
$$4x \le 20 \qquad 2x \ge 6$$
$$\frac{4x}{4} \le \frac{20}{4} \qquad \frac{2x}{2} \ge \frac{6}{2}$$
$$x \le 5 \qquad x \ge 3$$
The solution is $\{\ldots, 1, 2, 3, 4, 5\} \cap \{3, 4, 5, \ldots\} = \{3, 4, 5\}$.

33.
$$2 - 3(x + 4) < 5 - 3x$$
$$2 - 3x - 12 < 5 - 3x$$
$$-3x - 10 < 5 - 3x$$
$$-3x + 3x - 10 < 5 - 3x + 3x$$
$$-10 < 5$$
This is always true.
The solution is $\{x | x \in \text{real numbers}\}$.

Section 9.4

Objective A Exercises

1. $$x + y > 4$$
$$x + (-x) + y > -x + 4$$
$$y > -x + 4$$
Graph $y = -x + 4$ as a dotted line.
Shade the upper half-plane.

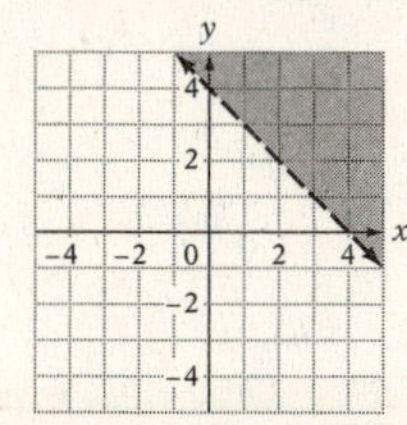

3. $$2x - y < -3$$
$$2x - 2x - y < -2x - 3$$
$$-y < -2x - 3$$
$$(-1)(-y) > (-1)(-2x - 3)$$
$$y > 2x + 3$$
Graph $y = 2x + 3$ as a dotted line.
Shade the upper half-plane.

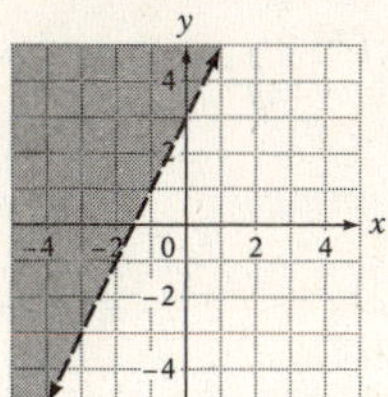

5. $$2x + y \geq 4$$
$$2x - 2x + y \geq -2x + 4$$
$$y \geq -2x + 4$$
Graph $y = -2x + 4$ as a solid line.
Shade the upper half-plane.

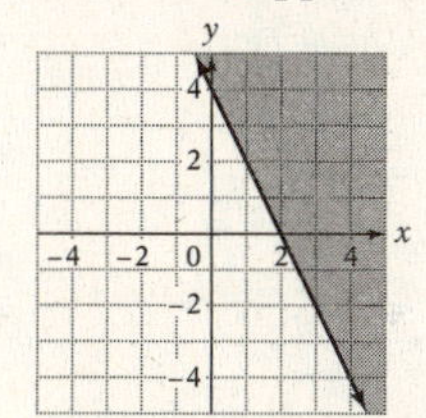

7. $y \leq -2$
Graph $y = -2$ as a solid line.
Shade the lower half-plane.

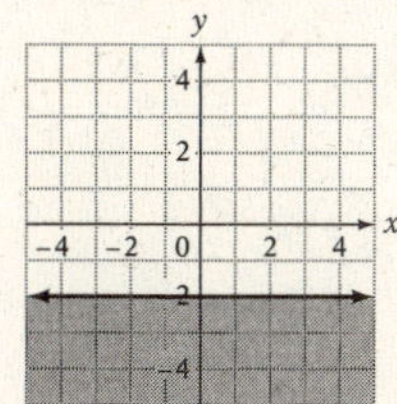

9. $$3x - 2y < 8$$
$$3x - 3x - 2y < -3x + 8$$
$$-2y < -3x + 8$$
$$-\frac{1}{2}(-2y) > -\frac{1}{2}(-3x + 8)$$
$$y > \frac{3}{2}x - 4$$
Graph $y = \frac{3}{2}x - 4$ as a dotted line.
Shade the upper half-plane.

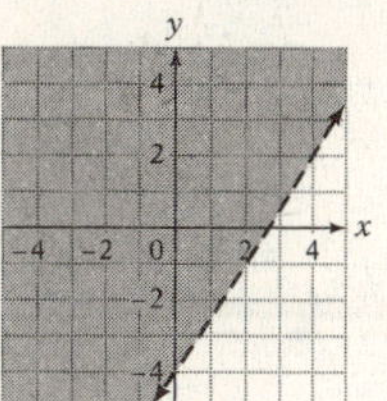

11. $$-3x - 4y \geq 4$$
$$-3x + 3x - 4y \geq 3x + 4$$
$$-4y \geq 3x + 4$$
$$-\frac{1}{4}(-4y) \leq -\frac{1}{4}(3x + 4)$$
$$y \leq -\frac{3}{4}x - 1$$
Graph $y = -\frac{3}{4}x - 1$ as a solid line.
Shade the lower half-plane.

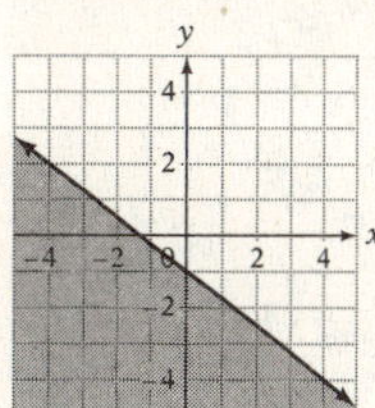

13. $$6x + 5y \leq -10$$
$$6x - 6x + 5y \leq -6x - 10$$
$$5y \leq -6x - 10$$
$$\frac{1}{5}(5y) \leq \frac{1}{5}(-6x - 10)$$
$$y \leq -\frac{6}{5}x - 2$$
Graph $y = -\frac{6}{5}x - 2$ as a solid line.
Shade the lower half-plane.

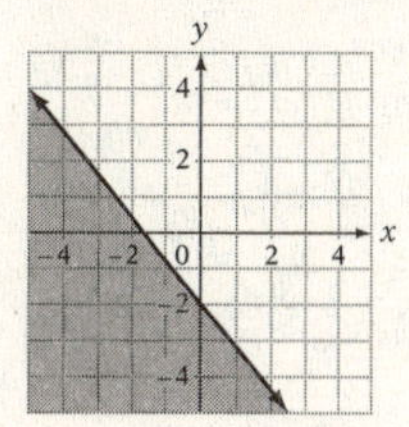

15.
$$\begin{aligned} -4x + 3y &< -12 \\ -4x + 4x + 3y &< 4x - 12 \\ 3y &< 4x - 12 \\ \frac{1}{3}(3y) &< \frac{1}{3}(4x - 12) \\ y &< \frac{4}{3}x - 4 \end{aligned}$$

Graph $y = \frac{4}{3}x - 4$ as a dotted line.
Shade the lower half-plane.

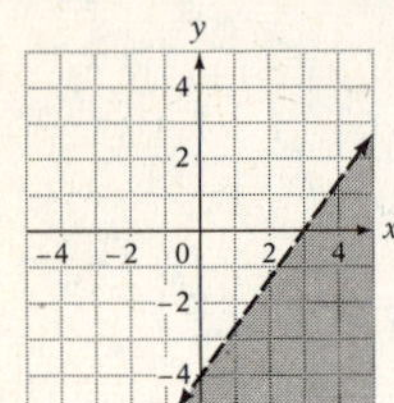

17.
$$\begin{aligned} -2x + 3y &\le 6 \\ -2x + 2x + 3y &\le 2x + 6 \\ 3y &\le 2x + 6 \\ \frac{1}{3}(3y) &\le \frac{1}{3}(2x + 6) \\ y &\le \frac{2}{3}x + 2 \end{aligned}$$

Graph $y = \frac{2}{3}x + 2$ as a solid line.
Shade the lower half-plane.

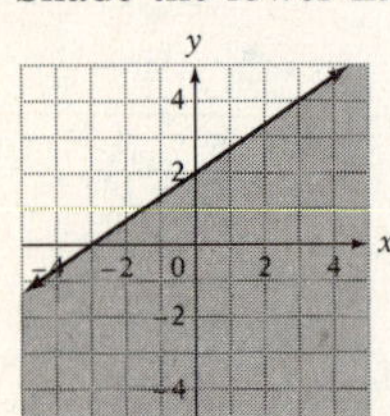

Applying the Concepts

19. $\frac{x}{4} + \frac{y}{2} > 1$

Graph $\frac{x}{4} + \frac{y}{2} = 1$ as a dotted line.
x-intercept: (4, 0)
y-intercept: (0, 2)
Shade the upper half-plane.

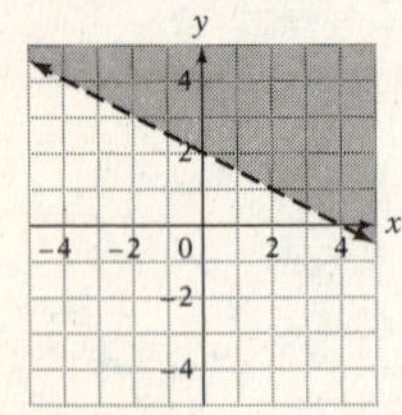

21.
$$\begin{aligned} 4y - 2(x + 1) &\ge 3(y - 1) + 3 \\ 4y - 2x - 2 &\ge 3y - 3 + 3 \\ 4y - 2x - 2 &\ge 3y \\ 4y - 3y &\ge 2x + 2 \\ y &\ge 2x + 2 \end{aligned}$$

Graph $y = 2x + 2$ as a solid line.
Shade the upper half-plane.

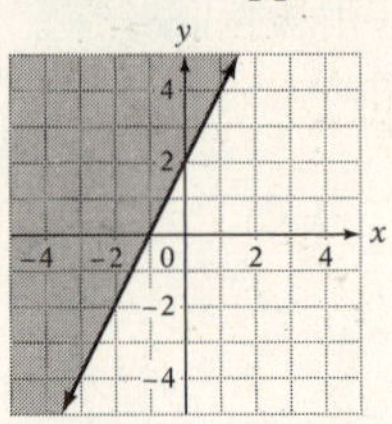

23. The line is $x = 3$, solid.
The inequality is $x \le 3$.

Chapter 9 Review Exercises

1.
$$\begin{aligned} 2x - 3 &> x + 15 \\ x &> 18 \end{aligned}$$

2. $A \cap B = \varnothing$

3. $\{x | x > -8, x \in \text{odd integers}\}$

4. $A \cup B = \{2, 4, 6, 8, 10\}$

5. $A = \{1, 3, 5, 7\}$

6.
$$\begin{aligned} 12 - 4(x - 1) &\le 5(x - 4) \\ 12 - 4x + 4 &\le 5x - 20 \\ -4x + 16 &\le 5x - 20 \\ -9x &\le -36 \\ x &\ge 4 \end{aligned}$$

7. −5 −4 −3 −2 −1 0 1 2 3 4 5

8.
$$\begin{aligned} 3x + 4 &\ge -8 \\ 3x &\ge -12 \\ x &\ge -4 \end{aligned}$$

9.

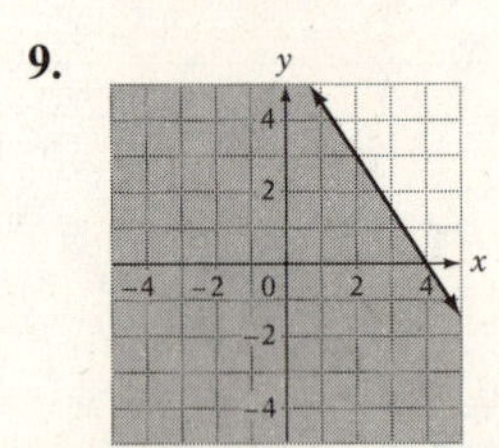

10.

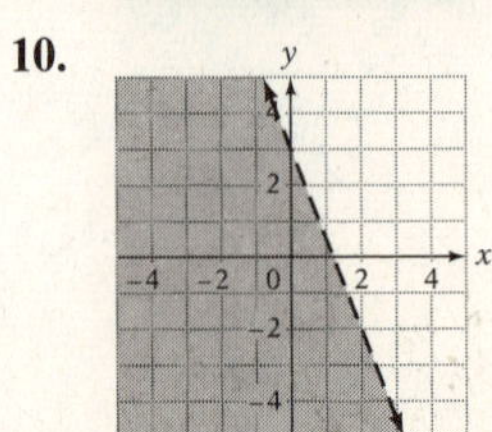

11. $\{x | x > 3, x \in \text{real numbers}\}$

12. $x > 2$

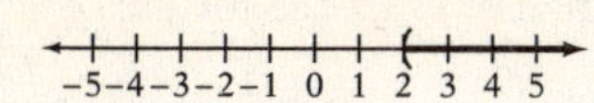

13. $A \cap B = \{1, 5, 9\}$

14. –5 –4 –3 –2 –1 0 1 2 3 4 5

15. –5 –4 –3 –2 –1 0 1 2 3 4 5

16.
$$-15x \leq 45$$
$$-\frac{1}{15} \cdot -15x \geq 45 \cdot -\frac{1}{15}$$
$$x \geq -3$$

17.
$$6x - 9 < 4x + 3(x + 3)$$
$$6x - 9 < 4x + 3x + 9$$
$$6x - 9 < 7x + 9$$
$$-x < 18$$
$$x > -18$$

18.
$$5 - 4(x + 9) > 11(12x - 9)$$
$$5 - 4x - 36 > 132x - 99$$
$$-4x - 31 > 132x - 99$$
$$-136x > -68$$
$$x < \frac{1}{2}$$

19.
$$-\frac{3}{4}x > \frac{2}{3}$$
$$-\frac{4}{3} \cdot -\frac{3}{4}x < \frac{2}{3} \cdot -\frac{4}{3}$$
$$x < -\frac{8}{9}$$

20.

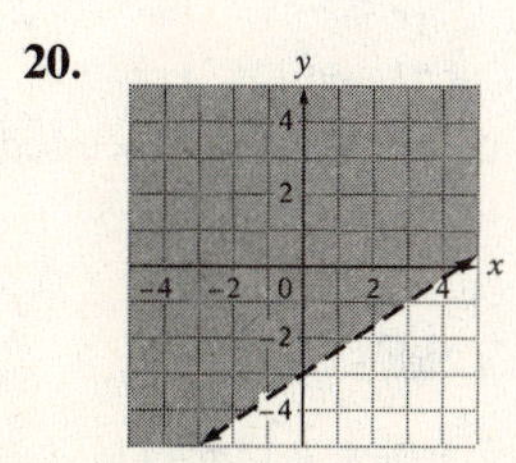

21.
$$7x - 2(x + 3) \geq x + 10$$
$$7x - 2x - 6 \geq x + 10$$
$$5x - 6 \geq x + 10$$
$$4x \geq 16$$
$$x \geq 4$$

22. Strategy • Number of residents in nursing home: x
Cost of florist B must be less than cost of florist A.

Solution
$$15 + 18x < 3 + 21x$$
$$-3x < -12$$
$$x > 4$$
For florist B to be more economical, there must be at least 5 residents in the nursing home.

23. Strategy To find the minimum length:
• Replace the variables in the area formula by the given values and solve for x.
• Replace the variable in the expression $3x + 5$ with the value found for x.

Solution
$$LW > A$$
$$(3x + 5)12 > 276$$
$$36x + 60 > 276$$
$$36x > 216$$
$$x > 6$$
$$3x + 5 = \text{length}$$
$$18 + 5 = \text{length}$$
$$23 = \text{length}$$
$$3x + 5 > 23$$
The minimum length is 24 ft.

24. Strategy • The smallest integer that satisfies the inequality: x

Solution
$$x - 6 > 25$$
$$x > 31$$
32 is the smallest integer that satisfies the inequality.

25. Strategy • Required score: x

Solution
$$68 + 82 + 90 + 73 + 95 + x \geq 480$$
$$408 + x \geq 480$$
$$x \geq 72$$
72 is the lowest score that the student can make and still achieve a minimum of 480 points.

Chapter 9 Test

1. –5 –4 –3 –2 –1 0 1 2 3 4 5

2. $\{x | x < 50, x \in \text{positive integers}\}$

3. $A = \{4, 6, 8\}$

4.
$$3(2x - 5) \geq 8x - 9$$
$$6x - 15 \geq 8x - 9$$
$$-2x \geq 6$$
$$x \leq -3$$

5.
$$x + \frac{1}{2} > \frac{5}{8}$$
$$x > \frac{5}{8} - \frac{1}{2}$$
$$x > \frac{5}{8} - \frac{4}{8}$$
$$x > \frac{1}{8}$$

6. –5 –4 –3 –2 –1 0 1 2 3 4 5

7. $5 - 3x > 8$
$-3x > 3$
$x < -1$

8. $\{x | x > -23, x \in \text{real numbers}\}$

9.

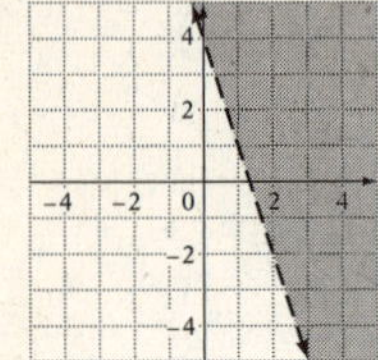

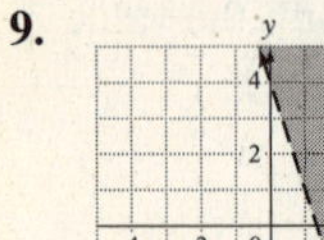

10.

11. $A \cap B = \{12\}$

12. $4 + x < 1$
$x < -3$

−5 −4 −3 −2 −1 0 1 2 3 4 5

13. $-\frac{3}{8}x \leq 5$

$-\frac{8}{3} \cdot -\frac{3}{8}x \geq 5 \cdot -\frac{8}{3}$

$x \geq -\frac{40}{3}$

14. $6x - 3(2 - 3x) < 4(2x - 7)$
$6x - 6 + 9x < 8x - 28$
$15x - 6 < 8x - 28$
$7x < -22$
$x < -\frac{22}{7}$

15. $\frac{2}{3}x \geq 2$

$\frac{3}{2} \cdot \frac{2}{3}x \geq \frac{3}{2} \cdot 2$

$x \geq 3$

−5 −4 −3 −2 −1 0 1 2 3 4 5

16. $2x - 7 \leq 6x + 9$
$-4x \leq 16$
$x \geq -4$

17. Strategy • Additional inches to grow: x
The child must be at least 48 in. tall.

Solution $x + 43 \geq 48$
$x \geq 5$
The child must grow at least 5 in.

18. Strategy • To find the width, solve an inequality using the formula $LW < A$.

Solution $15(2x - 4) < 180$
$30x - 60 < 180$
$30x < 240$
$x < 8$
$2x - 4 = 16 - 4 = 12$
$2x - 4 < 12$
The width must be less than or equal to 11 ft.

19. Strategy • Required diameter: d
Circumference: πd
The circumference must be between 0.1220 in. and 0.1240 in.

Solution $0.1220 < \pi d < 0.1240$
$0.1220 < 3.14d < 0.1240$
$\frac{0.1220}{3.14} < \frac{3.14d}{3.14} < \frac{0.1240}{3.14}$
$0.0389 < d < 0.0395$
The diameter must be between 0.0389 in. and 0.0395 in.

20. Strategy • Total value of stock: x
The salary of \$2500 must be greater than or equal to \$1000 plus 2% of the total value of the stock.

Solution $1000 + 2\%x \leq 2500$
$1000 + 0.02x \leq 2500$
$0.02x \leq 1500$
$x \leq 75{,}000$
The total value of the stock processed by the broker was at most \$75,000.

Cumulative Review Exercises

1. $2[5a - 3(2 - 5a) - 8] = 2[5a - 6 + 15a - 8]$
$= 2[20a - 14]$
$= 40a - 28$

2. $\frac{5}{8} - 4x = \frac{1}{8}$

$-4x = -\frac{4}{8}$

$-\frac{1}{4} \cdot -4x = -\frac{4}{8} \cdot -\frac{1}{4}$

$x = \frac{1}{8}$

3. $2x - 3[x - 2(x - 3)] = 2$
$2x - 3[x - 2x + 6] = 2$
$2x - 3[-x + 6] = 2$
$2x + 3x - 18 = 2$
$5x = 20$
$x = 4$

4. $(-3a)(-2a^3b^2)^2 = (-3a)(4a^6b^4)$
$= -12a^7b^4$

5. $\dfrac{27a^3b^2}{(-3ab^2)^3} = \dfrac{27a^3b^2}{-27a^3b^6} = -\dfrac{1}{b^4}$

6.
$$\begin{array}{r} 4x - 2 \\ 4x-1\overline{)16x^2 - 12x - 2} \\ \underline{16x^2 - 4x} \\ -8x - 2 \\ \underline{-8x + 2} \\ -4 \end{array}$$

$(16x^2 - 12x - 2) \div (4x - 1) = 4x - 2 - \dfrac{4}{4x - 1}$

7. $f(x) = x^2 - 4x - 5$
$f(-1) = (-1)^2 - 4(-1) - 5$
$= 1 + 4 - 5$
$= 0$

8. $27a^2x^2 - 3a^2$
The GCF is $3a^2$:
$3a^2(9x^2 - 1) = 3a^2[(3x)^2 - 1^2]$
$= 3a^2(3x + 1)(3x - 1)$

9. $\dfrac{x^2 - 2x}{x^2 - 2x - 8} \div \dfrac{x^3 - 5x^2 + 6x}{x^2 - 7x + 12}$
$= \dfrac{x(x-2)}{(x-4)(x+2)} \cdot \dfrac{(x-4)(x-3)}{x(x^2 - 5x + 6)}$
$= \dfrac{\not{x}(\not{x}-\not{2})}{(\not{x}-\not{4})(x+2)} \cdot \dfrac{(\not{x}-\not{4})(\not{x}-\not{3})}{\not{x}(\not{x}-\not{3})(\not{x}-\not{2})}$
$= \dfrac{1}{x+2}$

10. $\dfrac{4a}{2a-3} - \dfrac{2a}{a+3}$
LCM is $(2a - 3)(a + 3)$
$\dfrac{4a(a+3) - 2a(2a-3)}{(2a-3)(a+3)}$
$= \dfrac{4a^2 + 12a - 4a^2 + 6a}{(2a-3)(a+3)}$
$= \dfrac{18a}{(2a-3)(a+3)}$

11. $\dfrac{5y}{6} - \dfrac{5}{9} = \dfrac{y}{3} - \dfrac{5}{6}$
$18\left(\dfrac{5y}{6} - \dfrac{5}{9}\right) = 18\left(\dfrac{y}{3} - \dfrac{5}{6}\right)$
$15y - 10 = 6y - 15$
$9y = -5$
$y = -\dfrac{5}{9}$

12. $R = \dfrac{C - S}{t}$
$t \cdot R = \dfrac{C - S}{t} \cdot t$
$tR = C - S$
$tR + S = C$
$C = S + Rt$

13. $m = \dfrac{y_2 - y_1}{x_2 - x_1} = \dfrac{4 - (-3)}{-1 - 2}$
$= \dfrac{7}{-3} = -\dfrac{7}{3}$
The slope is $-\dfrac{7}{3}$.

14. $m = -\dfrac{3}{2}, (x_1, y_1) = (1, -3)$
$y - y_1 = m(x - x_1)$
$y + 3 = -\dfrac{3}{2}(x - 1)$
$y + 3 = -\dfrac{3}{2}x + \dfrac{3}{2}$
$y = -\dfrac{3}{2}x + \dfrac{3}{2} - 3$
$y = -\dfrac{3}{2}x + \dfrac{3}{2} - \dfrac{6}{2}$
$y = -\dfrac{3}{2}x - \dfrac{3}{2}$

15. (1) $x = 3y + 1$
(2) $2x + 5y = 13$
Substitute in Equation (2).
$2(3y + 1) + 5y = 13$
$6y + 2 + 5y = 13$
$11y = 11$
$y = 1$
Substitute in Equation (1).
$x = 3y + 1$
$x = 3 + 1$
$x = 4$
The solution is (4, 1).

16. (1) $9x - 2y = 17$
(2) $5x + 3y = -7$
Eliminate y and add the equation.

$$\begin{array}{r} 3(9x - 2y = 17) = 27x - 6y = 51 \\ \underline{2(5x + 3y = -7) = 10x + 6y = -14} \\ 37x = 37 \\ x = 1 \end{array}$$

Replace x in Equation (1).

$9(1) - 2y = 17$
$9 - 2y = 17$
$-2y = 8$
$y = -4$

The solution is (1, −4).

17. $A \cup B = \{-10, -2, 0, 1, 2\}$

18. $\{x \mid x < 48, x \in \text{real numbers}\}$

19.

−5 −4 −3 −2 −1 0 1 2 3 4 5

20.

$$\frac{3}{8}x > -\frac{3}{4}$$

$$\frac{8}{3}\left(\frac{3}{8}x\right) > \frac{8}{3}\left(-\frac{3}{4}\right)$$

$$x > -2$$

−5 −4 −3 −2 −1 0 1 2 3 4 5

21.

$$-\frac{4}{5}x > 12$$

$$-\frac{5}{4} \cdot -\frac{4}{5}x < 12 \cdot -\frac{5}{4}$$

$$x < -15$$

22.

$$15 - 3(5x - 7) < 2(7 - 2x)$$

$$15 - 15x + 21 < 14 - 4x$$

$$-15x + 36 < 14 - 4x$$

$$-11x < -22$$

$$x > 2$$

23. Strategy Unknown number: x

Solution

$$\frac{3}{5}x < -15$$

$$\frac{5}{3} \cdot \frac{3}{5}x < -15 \cdot \frac{5}{3}$$

$$x < -25$$

$$\{x \mid x \leq -26, x \in \text{integers}\}$$

24. Strategy • Number of miles: x

Company A's cost must be less than company B's cost.

Solution

$$6 + 0.25x < 15 + 0.10x$$

$$0.15x < 9$$

$$x < 60$$

For 6 days' rental, the maximum number of miles is 359 mi.

25. Strategy • Write and solve a proportion using x to represent the number of fish in the lake.

Solution

$$\frac{3}{150} = \frac{100}{x}$$

$$3x = 15000$$

$$x = 5000$$

There are an estimated 5000 fish in the lake.

26. Strategy • First angle: $x + 30$

Second angle: x

Third angle: $2x + 10$

Sum of the angle measures must equal 180.

Solution

$$(x + 30) + x + (2x + 10) = 180$$

$$4x + 40 = 180$$

$$4x = 140$$

$$x = 35$$

$$x + 30 = 65$$

$$2x + 10 = 80$$

The angle measures are 35°, 65°, and 80°.

27.

y, x; −4, −2, 0, 2, 4; 4, 2, −2, −4

28.

y, x; −4, −2, 0, 2, 4; 4, 2, −2, −4

Chapter 10: Radical Expressions

Prep Test

1. $-|-14| = -(14) = -14$
2. $3x^2y - 4xy^2 - 5x^2y = -2x^2y - 4xy^2$
3. $1.5h = 21$
   $h = \frac{21}{1.5} = 14$
4. $3x - 2 = 5 - 2x$
   $5x = 7$
   $x = \frac{7}{5}$
5. $x^3 \cdot x^3 = x^{3+3} = x^6$
6. $(x+y)^2 = x^2 + 2xy + y^2$
7. $(2x-3)^2 = (2x)^2 - 2(2x)(3) + (3)^2$
   $= 4x^2 - 12x + 9$
8. $(2-3v)(2+3v) = 4 - 9v^2$
9. $(a-5)(a+5) = a^2 - 25$
10. $\frac{2x^4y^3}{18x^2y} = \frac{x^{4-2}y^{3-1}}{9} = \frac{x^2y^2}{9}$

Go Figure

Strategy
- Prove ΔABC is similar to ΔEBD using the fact that the two walls are parallel.
- Use proportions to find the heights of ΔABC and ΔEBD.
- Prove ΔBFC is similar to ΔDEC.
- Use proportions to solve for x.

Solution Since the walls are parallel, $\angle$CDE and $\angle$ACD are alternate interior angles. Likewise, $\angle$CAE and $\angle$DEA are alternate interior angles. So, $\angle CDE \cong \angle ACD$ and $\angle CAE \cong \angle DEA$. Also $\angle$ABC and $\angle$DBE are vertical angles. So $\angle ABC \cong \angle DBE$. Therefore, ΔABC is similar to ΔEBD.

Let y denote the length of CF. Since the height of ΔABC is the same length as y and the height of ΔEBD is the same as $8 - y$, y and $8 - y$ will be used to denote the heights of ΔABC and ΔEBD, respectively.

$\frac{y}{10} = \frac{8-y}{12}$
$12y = 80 - 10y$
$22y = 80$
$y = \frac{80}{22} = \frac{40}{11}$

Now ΔBFC and ΔDEC are right triangles. So, $\angle BFC \cong \angle DEC$. Furthermore $\angle$DCE and $\angle$BCF are the same angle, so $\angle DCE \cong \angle BCF$.

$\frac{x}{12} = \frac{\frac{40}{11}}{8}$
$8x = \frac{480}{11}$
$x = \frac{60}{11} \approx 5.45$ ft

The pieces of lumber cross approximately 5.45 ft above the ground.

Section 10.1

Objective A Exercises

1. Students should explain that to simplify a radical expression, we first write the prime factorization of the radicand in exponential form. Write the radicand as the product of the perfect-square factors and the factors that do not contain a perfect square. Take the square root of the perfect-square factors; write the square root in front of the radical sign. The other factors remain under the radical sign.
3. $\sqrt{16} = \sqrt{2^4} = 2^2 = 4$
5. $\sqrt{49} = \sqrt{7^2} = 7$
7. $\sqrt{32} = \sqrt{2^5} = \sqrt{2^4 \cdot 2} = \sqrt{2^4}\sqrt{2} = 2^2\sqrt{2} = 4\sqrt{2}$
9. $\sqrt{8} = \sqrt{2^3} = \sqrt{2^2 \cdot 2} = \sqrt{2^2}\sqrt{2} = 2\sqrt{2}$
11. $6\sqrt{18} = 6\sqrt{2 \cdot 3^2} = 6\sqrt{3^2}\sqrt{2} = 6 \cdot 3\sqrt{2} = 18\sqrt{2}$
13. $5\sqrt{40} = 5\sqrt{2^3 \cdot 5}$
    $= 5\sqrt{2^2(2 \cdot 5)}$
    $= 5\sqrt{2^2}\sqrt{2 \cdot 5}$
    $= 5 \cdot 2\sqrt{10}$
    $= 10\sqrt{10}$
15. $\sqrt{15} = \sqrt{3 \cdot 5} = \sqrt{15}$
17. $\sqrt{29}$
19. $-9\sqrt{72} = -9\sqrt{2^3 \cdot 3^2}$
    $= -9\sqrt{2^2 \cdot 3^2 \cdot 2}$
    $= -9\sqrt{2^2 \cdot 3^2}\sqrt{2}$
    $= -9 \cdot 2 \cdot 3\sqrt{2}$
    $= -54\sqrt{2}$
21. $\sqrt{45} = \sqrt{3^2 \cdot 5} = \sqrt{3^2}\sqrt{5} = 3\sqrt{5}$
23. $\sqrt{0} = 0$

25. $6\sqrt{128} = 6\sqrt{2^7}$
$= 6\sqrt{2^6 \cdot 2}$
$= 6\sqrt{2^6}\sqrt{2}$
$= 6 \cdot 2^3\sqrt{2}$
$= 48\sqrt{2}$

27. $\sqrt{240} \approx 15.492$

29. $\sqrt{288} \approx 16.971$

31. $\sqrt{256} = \sqrt{2^8} = 2^4 = 16$

Objective B Exercises

33. $\sqrt{x^6} = x^3$

35. $\sqrt{y^{15}} = \sqrt{y^{14} \cdot y} = \sqrt{y^{14}}\sqrt{y} = y^7\sqrt{y}$

37. $\sqrt{a^{20}} = a^{10}$

39. $\sqrt{x^4y^4} = x^2y^2$

41. $\sqrt{4x^4} = \sqrt{2^2x^4} = 2x^2$

43. $\sqrt{24x^2} = \sqrt{2^3 \cdot 3 \cdot x^2}$
$= \sqrt{2^2x^2(2 \cdot 3)}$
$= \sqrt{2^2x^2}\sqrt{2 \cdot 3}$
$= 2x\sqrt{6}$

45. $\sqrt{60x^5} = \sqrt{2^2 \cdot 3 \cdot 5x^5}$
$= \sqrt{2^2x^4(3 \cdot 5x)}$
$= \sqrt{2^2x^4}\sqrt{3 \cdot 5x}$
$= 2x^2\sqrt{15x}$

47. $\sqrt{49a^4b^8} = \sqrt{7^2a^4b^8} = 7a^2b^4$

49. $\sqrt{18x^5y^7} = \sqrt{2 \cdot 3^2x^5y^7}$
$= \sqrt{3^2x^4y^6(2xy)}$
$= \sqrt{3^2x^4y^6}\sqrt{2xy}$
$= 3x^2y^3\sqrt{2xy}$

51. $\sqrt{40x^{11}y^7} = \sqrt{2^3 \cdot 5x^{11}y^7}$
$= \sqrt{2^2x^{10}y^6(2 \cdot 5xy)}$
$= \sqrt{2^2x^{10}y^6}\sqrt{2 \cdot 5xy}$
$= 2x^5y^3\sqrt{10xy}$

53. $\sqrt{80a^9b^{10}} = \sqrt{2^4 \cdot 5a^9b^{10}}$
$= \sqrt{2^4a^8b^{10}(5a)}$
$= \sqrt{2^4a^8b^{10}}\sqrt{5a}$
$= 2^2a^4b^5\sqrt{5a}$
$= 4a^4b^5\sqrt{5a}$

55. $2\sqrt{16a^2b^3} = 2\sqrt{2^4a^2b^3}$
$= 2\sqrt{2^4a^2b^2(b)}$
$= 2\sqrt{2^4a^2b^2}\sqrt{b}$
$= 2 \cdot 2^2ab\sqrt{b}$
$= 8ab\sqrt{b}$

57. $x\sqrt{x^4y^2} = x \cdot x^2y = x^3y$

59. $4\sqrt{20a^4b^7} = 4\sqrt{2^2 \cdot 5a^4b^7}$
$= 4\sqrt{2^2a^4b^6(5b)}$
$= 4\sqrt{2^2a^4b^6}\sqrt{5b}$
$= 4 \cdot 2a^2b^3\sqrt{5b} = 8a^2b^3\sqrt{5b}$

61. $3x\sqrt{12x^2y^7} = 3x\sqrt{2^2 \cdot 3x^2y^7}$
$= 3x\sqrt{2^2x^2y^6(3y)}$
$= 3x\sqrt{2^2x^2y^6}\sqrt{3y}$
$= 3x \cdot 2xy^3\sqrt{3y}$
$= 6x^2y^3\sqrt{3y}$

63. $2x^2\sqrt{8x^2y^3} = 2x^2\sqrt{2^3x^2y^3}$
$= 2x^2\sqrt{2^2x^2y^2(2y)}$
$= 2x^2\sqrt{2^2x^2y^2}\sqrt{2y}$
$= 2x^2 \cdot 2xy\sqrt{2y}$
$= 4x^3y\sqrt{2y}$

65. $\sqrt{25(a+4)^2} = \sqrt{5^2(a+4)^2} = 5(a+4) = 5a+20$

67. $\sqrt{4(x+2)^4} = \sqrt{2^2(x+2)^4}$
$= 2(x+2)^2$
$= 2(x^2+4x+4)$
$= 2x^2+8x+8$

69. $\sqrt{x^2+4x+4} = \sqrt{(x+2)^2} = x+2$

71. $\sqrt{y^2+2y+1} = \sqrt{(y+1)^2} = y+1$

Applying the Concepts

73. a. **Strategy** Evaluate the expression $S = \sqrt{30fl}$ when $f = 1.2$ and $l = 60$.

Solution $S = \sqrt{30fl}$
$= \sqrt{30(1.2)(60)}$
$= \sqrt{2160}$
$= \sqrt{9 \cdot 16 \cdot 15}$
$= 12\sqrt{15}$

The speed of the car was $12\sqrt{15}$ mph.

b. $12\sqrt{15} \approx 46$ mph rounded to the nearest integer.

75. No.
For example, let $a = 16$ and $b = 9$.
$\sqrt{a+b} = \sqrt{16+9} = \sqrt{25} = 5$
$\sqrt{a} + \sqrt{b} = \sqrt{16} + \sqrt{9} = 4+3 = 7$

77. No, the solution is incomplete. 18 contains a perfect-square factor.
The correct solution is
$\sqrt{72} = \sqrt{2^3 \cdot 3^2}$
$= \sqrt{2^2 \cdot 3^2 \cdot 2}$
$= \sqrt{2^2 \cdot 3^2}\sqrt{2} = 2 \cdot 3\sqrt{2} = 6\sqrt{2}$

79. a. $f(1) = \sqrt{2(1) - 1} = \sqrt{2 - 1} = \sqrt{1} = 1$

b. $f(5) = \sqrt{2(5) - 1} = \sqrt{10 - 1} = \sqrt{9} = 3$

c. $f(14) = \sqrt{2(14) - 1}$
$= \sqrt{28 - 1}$
$= \sqrt{27} = \sqrt{9 \cdot 3} = 3\sqrt{3}$

Section 10.2

Objective A Exercises

1. 2, 20, and 50 are not perfect squares.

3. A perfect square is the square of a number. A perfect-square factor of a number is a perfect square that divides the number evenly.

5. $3\sqrt{2}$

7. $-\sqrt{7}$

9. $-11\sqrt{11}$

11. $10\sqrt{x}$

13. $-2\sqrt{y}$

15. $-11\sqrt{3b}$

17. $2x\sqrt{2}$

19. $-3a\sqrt{3a}$

21. $-5\sqrt{xy}$

23. $\sqrt{45} + \sqrt{125} = \sqrt{3^2 \cdot 5} + \sqrt{5^3}$
$= \sqrt{3^2}\sqrt{5} + \sqrt{5^2}\sqrt{5}$
$= 3\sqrt{5} + 5\sqrt{5} = 8\sqrt{5}$

25. $2\sqrt{2} + 3\sqrt{8} = 2\sqrt{2} + 3\sqrt{2^3}$
$= 2\sqrt{2} + 3\sqrt{2^2}\sqrt{2}$
$= 2\sqrt{2} + 3 \cdot 2\sqrt{2}$
$= 2\sqrt{2} + 6\sqrt{2} = 8\sqrt{2}$

27. $5\sqrt{18} - 2\sqrt{75} = 5\sqrt{2 \cdot 3^2} - 2\sqrt{3 \cdot 5^2}$
$= 5\sqrt{3^2}\sqrt{2} - 2\sqrt{5^2}\sqrt{3}$
$= 5 \cdot 3\sqrt{2} - 2 \cdot 5\sqrt{3} = 15\sqrt{2} - 10\sqrt{3}$

29. $5\sqrt{4x} - 3\sqrt{9x} = 5\sqrt{2^2x} - 3\sqrt{3^2x}$
$= 5\sqrt{2^2}\sqrt{x} - 3\sqrt{3^2}\sqrt{x}$
$= 5 \cdot 2\sqrt{x} - 3 \cdot 3\sqrt{x}$
$= 10\sqrt{x} - 9\sqrt{x} = \sqrt{x}$

31. $3\sqrt{3x^2} - 5\sqrt{27x^2} = 3\sqrt{3x^2} - 5\sqrt{3^3x^2}$
$= 3\sqrt{x^2}\sqrt{3} - 5\sqrt{3^2x^2}\sqrt{3}$
$= 3x\sqrt{3} - 5 \cdot 3x\sqrt{3}$
$= 3x\sqrt{3} - 15x\sqrt{3} = -12x\sqrt{3}$

33. $2x\sqrt{xy^2} - 3y\sqrt{x^2y} = 2x\sqrt{y^2}\sqrt{x} - 3y\sqrt{x^2}\sqrt{y}$
$= 2xy\sqrt{x} - 3xy\sqrt{y}$

35. $3x\sqrt{12x} - 5\sqrt{27x^3} = 3x\sqrt{2^2 \cdot 3x} - 5\sqrt{3^3x^3}$
$= 3x\sqrt{2^2}\sqrt{3x} - 5\sqrt{3^2x^2}\sqrt{3x}$
$= 3x \cdot 2\sqrt{3x} - 5 \cdot 3x\sqrt{3x}$
$= 6x\sqrt{3x} - 15x\sqrt{3x} = -9x\sqrt{3x}$

37. $4y\sqrt{8y^3} - 7\sqrt{18y^5} = 4y\sqrt{2^3y^3} - 7\sqrt{2 \cdot 3^2y^5}$
$= 4y\sqrt{2^2y^2}\sqrt{2y} - 7\sqrt{3^2y^4}\sqrt{2y}$
$= 4y \cdot 2y\sqrt{2y} - 7 \cdot 3y^2\sqrt{2y}$
$= 8y^2\sqrt{2y} - 21y^2\sqrt{2y}$
$= -13y^2\sqrt{2y}$

39. $b^2\sqrt{a^5b} + 3a^2\sqrt{ab^5} = b^2\sqrt{a^4}\sqrt{ab} + 3a^2\sqrt{b^4}\sqrt{ab}$
$= a^2b^2\sqrt{ab} + 3a^2b^2\sqrt{ab}$
$= 4a^2b^2\sqrt{ab}$

41. $7\sqrt{2}$

43. $6\sqrt{x}$

45. $-3\sqrt{y}$

47. $8\sqrt{8} - 4\sqrt{32} - 9\sqrt{50} = 8\sqrt{2^3} - 4\sqrt{2^5} - 9\sqrt{2 \cdot 5^2}$
$= 8\sqrt{2^2}\sqrt{2} - 4\sqrt{2^4}\sqrt{2} - 9\sqrt{5^2}\sqrt{2}$
$= 8 \cdot 2\sqrt{2} - 4 \cdot 2^2\sqrt{2} - 9 \cdot 5\sqrt{2}$
$= 16\sqrt{2} - 16\sqrt{2} - 45\sqrt{2} = -45\sqrt{2}$

49. $-2\sqrt{3} + 5\sqrt{27} - 4\sqrt{45} = -2\sqrt{3} + 5\sqrt{3^3} - 4\sqrt{3^2 \cdot 5}$
$= -2\sqrt{3} + 5\sqrt{3^2}\sqrt{3} - 4\sqrt{3^2}\sqrt{5}$
$= -2\sqrt{3} + 5 \cdot 3\sqrt{3} - 4 \cdot 3\sqrt{5}$
$= -2\sqrt{3} + 15\sqrt{3} - 12\sqrt{5} = 13\sqrt{3} - 12\sqrt{5}$

51. $4\sqrt{75} + 3\sqrt{48} - \sqrt{99} = 4\sqrt{3 \cdot 5^2} + 3\sqrt{2^4 \cdot 3} - \sqrt{3^2 \cdot 11}$
$= 4\sqrt{5^2}\sqrt{3} + 3\sqrt{2^4}\sqrt{3} - \sqrt{3^2}\sqrt{11}$
$= 4 \cdot 5\sqrt{3} + 3 \cdot 2^2\sqrt{3} - 3\sqrt{11}$
$= 20\sqrt{3} + 12\sqrt{3} - 3\sqrt{11} = 32\sqrt{3} - 3\sqrt{11}$

53. $\sqrt{25x} - \sqrt{9x} + \sqrt{16x} = \sqrt{5^2x} - \sqrt{3^2x} + \sqrt{2^4x}$
$= \sqrt{5^2}\sqrt{x} - \sqrt{3^2}\sqrt{x} + \sqrt{2^4}\sqrt{x}$
$= 5\sqrt{x} - 3\sqrt{x} + 2^2\sqrt{x}$
$= 5\sqrt{x} - 3\sqrt{x} + 4\sqrt{x} = 6\sqrt{x}$

55. $3\sqrt{3x} + \sqrt{27x} - 8\sqrt{75x} = 3\sqrt{3x} + \sqrt{3^3x} - 8\sqrt{3 \cdot 5^2x}$
$= 3\sqrt{3x} + \sqrt{3^2}\sqrt{3x} - 8\sqrt{5^2}\sqrt{3x}$
$= 3\sqrt{3x} + 3\sqrt{3x} - 8 \cdot 5\sqrt{3x}$
$= 3\sqrt{3x} + 3\sqrt{3x} - 40\sqrt{3x} = -34\sqrt{3x}$

57. $2a\sqrt{75b} - a\sqrt{20b} + 4a\sqrt{45b} = 2a\sqrt{3 \cdot 5^2b} - a\sqrt{2^2 \cdot 5b} + 4a\sqrt{3^2 \cdot 5b}$
$= 2a\sqrt{5^2}\sqrt{3b} - a\sqrt{2^2}\sqrt{5b} + 4a\sqrt{3^2}\sqrt{5b}$
$= 2a \cdot 5\sqrt{3b} - 2a\sqrt{5b} + 4a \cdot 3\sqrt{5b}$
$= 10a\sqrt{3b} - 2a\sqrt{5b} + 12a\sqrt{5b} = 10a\sqrt{3b} + 10a\sqrt{5b}$

59. $x\sqrt{3y^2} - 2y\sqrt{12x^2} + xy\sqrt{3} = x\sqrt{3y^2} - 2y\sqrt{2^2 \cdot 3x^2} + xy\sqrt{3}$
$= x\sqrt{y^2}\sqrt{3} - 2y\sqrt{2^2x^2}\sqrt{3} + xy\sqrt{3}$
$= xy\sqrt{3} - 2y \cdot 2x\sqrt{3} + xy\sqrt{3}$
$= xy\sqrt{3} - 4xy\sqrt{3} + xy\sqrt{3} = -2xy\sqrt{3}$

Applying the Concepts

61. $G(3) = \sqrt{3+5} + \sqrt{5(3)+3}$
$= \sqrt{8} + \sqrt{15+3}$
$= \sqrt{8} + \sqrt{18}$
$= 2\sqrt{2} + 3\sqrt{2} = 5\sqrt{2}$

63. $4\sqrt{2a^3b} + 5\sqrt{8a^3b}$
Write each radicand as the product of a perfect square and factors that do not contain a perfect square.
$= 4\sqrt{a^2 \cdot 2ab} + 5\sqrt{4a^2 \cdot 2ab}$
Use the Product Property of Square Roots. Write the perfect square under the first radical sign and all the remaining factors under the second radical sign.
$= 4\sqrt{a^2}\sqrt{2ab} + 5 \cdot \sqrt{4a^2}\sqrt{2ab}$
Take the square roots of the perfect squares.
$= 4a\sqrt{2ab} + 5 \cdot 2a\sqrt{2ab}$
Simplify.
$= 4a\sqrt{2ab} + 10a\sqrt{2ab}$
Combine like terms.
$= 14a\sqrt{2ab}$

Section 10.3

Objective A Exercises

1. The Product Property of Square Roots states that the square root of the product of two positive numbers equals the product of the square roots of the two numbers.
$\sqrt{ab} = \sqrt{a}\sqrt{b},\ a > 0,\ b > 0$

3. $\sqrt{5} \cdot \sqrt{5} = \sqrt{5^2} = 5$

5. $\sqrt{3} \cdot \sqrt{12} = \sqrt{36} = \sqrt{2^2 \cdot 3^2} = 2 \cdot 3 = 6$

7. $\sqrt{x} \cdot \sqrt{x} = \sqrt{x^2} = x$

9. $\sqrt{xy^3} \cdot \sqrt{x^5y} = \sqrt{x^6y^4} = x^3y^2$

11. $\sqrt{3a^2b^5} \cdot \sqrt{6ab^7} = \sqrt{18a^3b^{12}}$
$= \sqrt{2 \cdot 3^2a^3b^{12}}$
$= \sqrt{3^2a^2b^{12}}\sqrt{2a}$
$= 3ab^6\sqrt{2a}$

13. $\sqrt{6a^3b^2} \cdot \sqrt{24a^5b} = \sqrt{144a^8b^3}$
$= \sqrt{2^4 \cdot 3^2a^8b^3}$
$= \sqrt{2^4 \cdot 3^2a^8b^2}\sqrt{b}$
$= 2^2 \cdot 3a^4b\sqrt{b} = 12a^4b\sqrt{b}$

15. $\sqrt{2}(\sqrt{2} - \sqrt{3}) = \sqrt{2^2} - \sqrt{6} = 2 - \sqrt{6}$

17. $\sqrt{x}(\sqrt{x} - \sqrt{y}) = \sqrt{x^2} - \sqrt{xy} = x - \sqrt{xy}$

19. $\sqrt{5}(\sqrt{10} - \sqrt{x}) = \sqrt{50} - \sqrt{5x}$
$= \sqrt{2 \cdot 5^2} - \sqrt{5x}$
$= \sqrt{5^2}\sqrt{2} - \sqrt{5x}$
$= 5\sqrt{2} - \sqrt{5x}$

21. $\sqrt{8}(\sqrt{2} - \sqrt{5}) = \sqrt{16} - \sqrt{40}$
$= \sqrt{2^4} - \sqrt{2^3 \cdot 5}$
$= \sqrt{2^4} - \sqrt{2^2}\sqrt{2 \cdot 5}$
$= 2^2 - 2\sqrt{10} = 4 - 2\sqrt{10}$

23. $(\sqrt{x} - 3)^2 = \sqrt{x^2} - 2 \cdot 3\sqrt{x} + 3^2 = x - 6\sqrt{x} + 9$

25. $\sqrt{3a}(\sqrt{3a} - \sqrt{3b}) = \sqrt{3^2a^2} - \sqrt{3^2ab}$
$= \sqrt{3^2a^2} - \sqrt{3^2}\sqrt{ab}$
$= 3a - 3\sqrt{ab}$

27. $\sqrt{2ac} \cdot \sqrt{5ab} \cdot \sqrt{10cb} = \sqrt{100a^2b^2c^2}$
$= \sqrt{2^2 \cdot 5^2a^2b^2c^2}$
$= 2 \cdot 5abc = 10abc$

29. $(\sqrt{5} + 3)(2\sqrt{5} - 4) = 2\sqrt{25} - 4\sqrt{5} + 6\sqrt{5} - 12$
$= 2 \cdot 5 + 2\sqrt{5} - 12$
$= 10 + 2\sqrt{5} - 12 = -2 + 2\sqrt{5}$

31. $(4 + \sqrt{8})(3 + \sqrt{2}) = 12 + 4\sqrt{2} + 3\sqrt{8} + \sqrt{16}$
$= 12 + 4\sqrt{2} + 3 \cdot 2\sqrt{2} + 4$
$= 12 + 4\sqrt{2} + 6\sqrt{2} + 4$
$= 16 + 10\sqrt{2}$

33. $(2\sqrt{x} + 4)(3\sqrt{x} - 1) = 6\sqrt{x^2} - 2\sqrt{x} + 12\sqrt{x} - 4$
$= 6x + 10\sqrt{x} - 4$

35. $(3\sqrt{x} - 2y)(5\sqrt{x} - 4y)$
$= 15\sqrt{x^2} - 12y\sqrt{x} - 10y\sqrt{x} + 8y^2$
$= 15x - 22y\sqrt{x} + 8y^2$

37. $(\sqrt{x} - \sqrt{y})(\sqrt{x} + \sqrt{y}) = \sqrt{x^2} - \sqrt{y^2} = x - y$

Objective B Exercises

39. Because $\dfrac{\sqrt{5}}{\sqrt{5}} = 1$, multiplying $\dfrac{2}{\sqrt{5}}$ by $\dfrac{\sqrt{5}}{\sqrt{5}}$ is the same as multiplying by 1, and any number multiplied by 1 is the number itself.

41. $\dfrac{\sqrt{45}}{\sqrt{5}} = \sqrt{\dfrac{45}{5}} = \sqrt{9} = \sqrt{3^2} = 3$

43. $\dfrac{\sqrt{48}}{\sqrt{3}} = \sqrt{\dfrac{48}{3}} = \sqrt{16} = \sqrt{2^4} = 2^2 = 4$

45. $\dfrac{\sqrt{72x^5}}{\sqrt{2x}} = \sqrt{\dfrac{72x^5}{2x}} = \sqrt{36x^4} = \sqrt{2^2 \cdot 3^2x^4}$
$= 2 \cdot 3x^2 = 6x^2$

47. $\dfrac{\sqrt{40x^5y^2}}{\sqrt{5xy}} = \sqrt{\dfrac{40x^5y^2}{5xy}}$
$= \sqrt{8x^4y}$
$= \sqrt{2^3x^4y}$
$= \sqrt{2^2x^4}\sqrt{2y} = 2x^2\sqrt{2y}$

49. $\dfrac{\sqrt{48x^5y^2}}{\sqrt{3x^3y}} = \sqrt{\dfrac{48x^5y^2}{3x^3y}}$
$= \sqrt{16x^2y}$
$= \sqrt{2^4x^2y}$
$= \sqrt{2^4x^2}\sqrt{y} = 2^2x\sqrt{y} = 4x\sqrt{y}$

51. $\dfrac{\sqrt{4x^2y}}{\sqrt{3xy^3}} = \sqrt{\dfrac{4x^2y}{3xy^3}}$
$= \sqrt{\dfrac{4x}{3y^2}}$
$= \dfrac{\sqrt{4x}}{\sqrt{3y^2}}$
$= \dfrac{\sqrt{2^2x}}{\sqrt{3y^2}}$
$= \dfrac{\sqrt{2^2}\sqrt{x}}{\sqrt{y^2}\sqrt{3}}$
$= \dfrac{2\sqrt{x}}{y\sqrt{3}}$
$= \dfrac{2\sqrt{x}}{y\sqrt{3}} \cdot \dfrac{\sqrt{3}}{\sqrt{3}} = \dfrac{2\sqrt{3x}}{y\sqrt{3^2}} = \dfrac{2\sqrt{3x}}{3y}$

53. $\dfrac{\sqrt{2}}{\sqrt{8}+4} = \dfrac{\sqrt{2}}{\sqrt{8}+4} \cdot \dfrac{\sqrt{8}-4}{\sqrt{8}-4}$
$= \dfrac{\sqrt{16}-4\sqrt{2}}{8-16}$
$= \dfrac{4-4\sqrt{2}}{-8}$
$= \dfrac{1-\sqrt{2}}{-2}$ or $\dfrac{\sqrt{2}-1}{2}$

55. $\dfrac{5}{\sqrt{7}-3} = \dfrac{5}{\sqrt{7}-3} \cdot \dfrac{\sqrt{7}+3}{\sqrt{7}+3}$
$= \dfrac{5(\sqrt{7}+3)}{7-9}$
$= \dfrac{5\sqrt{7}+15}{-2} = -\dfrac{5\sqrt{7}+15}{2}$

57. $\dfrac{\sqrt{3}}{5-\sqrt{27}} = \dfrac{\sqrt{3}}{5-\sqrt{27}} \cdot \dfrac{5+\sqrt{27}}{5+\sqrt{27}}$
$= \dfrac{5\sqrt{3}+\sqrt{81}}{25-27}$
$= \dfrac{5\sqrt{3}+9}{-2} = -\dfrac{5\sqrt{3}+9}{2}$

59. $\dfrac{3-\sqrt{6}}{5-2\sqrt{6}} = \dfrac{3-\sqrt{6}}{5-2\sqrt{6}} \cdot \dfrac{5+2\sqrt{6}}{5+2\sqrt{6}}$
$= \dfrac{15+\sqrt{6}-12}{25-24}$
$= 3+\sqrt{6}$

61. $\dfrac{-6}{4+\sqrt{2}} = \dfrac{-6}{4+\sqrt{2}} \cdot \dfrac{4-\sqrt{2}}{4-\sqrt{2}}$
$= \dfrac{-24+6\sqrt{2}}{16-2}$
$= \dfrac{-24+6\sqrt{2}}{14} = \dfrac{-12+3\sqrt{2}}{7}$

63. $\dfrac{2\sqrt{3}-\sqrt{6}}{5\sqrt{3}+2\sqrt{6}} = \dfrac{2\sqrt{3}-\sqrt{6}}{5\sqrt{3}+2\sqrt{6}} \cdot \dfrac{5\sqrt{3}-2\sqrt{6}}{5\sqrt{3}-2\sqrt{6}}$
$= \dfrac{30-9\sqrt{18}+12}{75-24}$
$= \dfrac{42-9\sqrt{18}}{51}$
$= \dfrac{42-27\sqrt{2}}{51}$
$= \dfrac{3(14-9\sqrt{2})}{51}$
$= \dfrac{14-9\sqrt{2}}{17}$

65. $\dfrac{-\sqrt{15}}{3-\sqrt{12}} = \dfrac{-\sqrt{15}}{3-\sqrt{12}} \cdot \dfrac{3+\sqrt{12}}{3+\sqrt{12}}$
$= \dfrac{-3\sqrt{15}-\sqrt{180}}{9-12}$
$= \dfrac{-3\sqrt{15}-6\sqrt{5}}{-3}$
$= \sqrt{15}+2\sqrt{5}$

67. $\dfrac{\sqrt{xy}}{\sqrt{x}-\sqrt{y}} = \dfrac{\sqrt{xy}}{\sqrt{x}-\sqrt{y}} \cdot \dfrac{\sqrt{x}+\sqrt{y}}{\sqrt{x}+\sqrt{y}}$
$= \dfrac{\sqrt{x^2y}+\sqrt{xy^2}}{x-y}$
$= \dfrac{\sqrt{x^2}\sqrt{y}+\sqrt{y^2}\sqrt{x}}{x-y} = \dfrac{x\sqrt{y}+y\sqrt{x}}{x-y}$

69. $\dfrac{-12}{\sqrt{6}-3} = \dfrac{-12}{\sqrt{6}-3} \cdot \dfrac{\sqrt{6}+3}{\sqrt{6}+3}$
$= \dfrac{-12\sqrt{6}-36}{6-9}$
$= \dfrac{-12\sqrt{6}-36}{-3} = 4\sqrt{6}+12$

Applying the Concepts

71.
$$x^2-2x-5=0$$
$$(1+\sqrt{6})^2-2(1+\sqrt{6})-5=0$$
$$1+2\sqrt{6}+6-2-2\sqrt{6}-5=0$$
$$0=0$$
$$x^2-2x-5=0$$
$$(1-\sqrt{6})^2-2(1-\sqrt{6})-5=0$$
$$1-2\sqrt{6}+6-2+2\sqrt{6}-5=0$$
$$0=0$$

Section 10.4

Objective A Exercises

1. $\sqrt{x} = 5$ Check: $\sqrt{x} = 5$

$(\sqrt{x})^2 = 5^2$ $\sqrt{25} = 5$

$x = 25$ $\sqrt{5^2} = 5$

$5 = 5$

The solution is 25.

3. $\sqrt{a} = 12$ Check: $\sqrt{a} = 12$

$(\sqrt{a})^2 = 12^2$ $\sqrt{144} = 12$

$a = 144$ $\sqrt{2^4 \cdot 3^2} = 12$

$2^2 \cdot 3 = 12$

$12 = 12$

The solution is 144.

5. $\sqrt{5x} = 5$ Check: $\sqrt{5x} = 5$

$(\sqrt{5x})^2 = 5^2$ $\sqrt{5 \cdot 5} = 5$

$5x = 25$ $\sqrt{5^2} = 5$

$x = 5$ $5 = 5$

The solution is 5.

7. $\sqrt{3x} + 9 = 4$ Check: $\sqrt{3x} + 9 = 4$

$\sqrt{3x} = -5$ $\sqrt{3 \cdot \frac{25}{3}} + 9 = 4$

$(\sqrt{3x})^2 = -5^2$ $\sqrt{25} + 9 = 4$

$3x = 25$ $\sqrt{5^2} + 9 = 4$

$x = \frac{25}{3}$ $5 + 9 = 4$

$14 \neq 4$

The equation has no solution.

9. $\sqrt{5x+6} = 1$ Check: $\sqrt{5x+6} = 1$

$(\sqrt{5x+6})^2 = 1^2$ $\sqrt{5(-1)+6} = 1$

$5x + 6 = 1$ $\sqrt{-5+6} = 1$

$5x = -5$ $\sqrt{1} = 1$

$x = -1$ $1 = 1$

The solution is -1.

11. $\sqrt{5x+4} = 3$ Check: $\sqrt{5x+4} = 3$

$(\sqrt{5x+4})^2 = 3^2$ $\sqrt{5 \cdot 1 + 4} = 3$

$5x + 4 = 9$ $\sqrt{5+4} = 3$

$5x = 5$ $\sqrt{9} = 3$

$x = 1$ $\sqrt{3^2} = 3$

$3 = 3$

The solution is 1.

13. $0 = 5 - \sqrt{10+x}$ Check: $0 = 5 - \sqrt{10+x}$

$\sqrt{10+x} = 5$ $0 = 5 - \sqrt{10+15}$

$(\sqrt{10+x})^2 = 5^2$ $0 = 5 - \sqrt{25}$

$10 + x = 25$ $0 = 5 - \sqrt{5^2}$

$x = 15$ $0 = 5 - 5$

$0 = 0$

The solution is 15.

15. $\sqrt{3x-7}=0$ Check: $\sqrt{3x-7}=0$

$(\sqrt{3x-7})^2=0^2$ $\qquad$ $\sqrt{3\left(\frac{7}{3}\right)-7}=0$

$3x-7=0$ $\qquad$ $\sqrt{7-7}=0$

$3x=7$ $\qquad$ $\sqrt{0}=0$

$x=\frac{7}{3}$ $\qquad$ $0=0$

The solution is $\frac{7}{3}$.

17. $\sqrt{x^2+5}=x+1$ Check: $\sqrt{x^2+5}=x+1$

$(\sqrt{x^2+5})^2=(x+1)^2$ $\qquad$ $\sqrt{2^2+5}=2+1$

$x^2+5=x^2+2x+1$ $\qquad$ $\sqrt{9}=3$

$4=2x$ $\qquad$ $3=3$

$2=x$

The solution is 2.

19. $\sqrt{x+4}-\sqrt{x-1}=1$ Check: $\sqrt{x+4}-\sqrt{x-1}=1$

$\sqrt{x+4}=1+\sqrt{x-1}$ $\qquad$ $\sqrt{5+4}-\sqrt{5-1}=1$

$(\sqrt{x+4})^2=(1+\sqrt{x-1})^2$ $\qquad$ $3-2=1$

$x+4=1+2\sqrt{x-1}+x-1$ $\qquad$ $1=1$

$4=2\sqrt{x-1}$

$2=\sqrt{x-1}$

$2^2=(\sqrt{x-1})^2$

$4=x-1$

$5=x$

The solution is 5.

21. $\sqrt{2x+1}-\sqrt{2x-4}=1$ Check: $\sqrt{2x+1}-\sqrt{2x-4}=1$

$\sqrt{2x+1}=1+\sqrt{2x-4}$ $\qquad$ $\sqrt{2\cdot 4+1}-\sqrt{2\cdot 4-4}=1$

$(\sqrt{2x+1})^2=(1+\sqrt{2x-4})^2$ $\qquad$ $\sqrt{9}-\sqrt{4}=1$

$2x+1=1+2\sqrt{2x-4}+2x-4$ $\qquad$ $3-2=1$

$4=2\sqrt{2x-4}$ $\qquad$ $1=1$

$2=\sqrt{2x-4}$

$2^2=(\sqrt{2x-4})^2$

$4=2x-4$

$8=2x$

$4=x$

The solution is 4.

Objective B Exercises

23. **Strategy** • To find the length of the diagonal between home plate and second base, use the equation $c=\sqrt{a^2+b^2}$ and compare $\frac{1}{2}c$ with 60.5.

Solution

$c=\sqrt{a^2+b^2}$

$c=\sqrt{90^2+90^2}$

$c=\sqrt{8100+8100}$

$c=\sqrt{16200}$

$c\approx 127.28$

$\frac{1}{2}c\approx 63.64$

The pitcher's mound is less than halfway between home plate and second base.

25. Strategy • To find the height above the water, replace d in the equation with the given value and solve for h.

Solution

$$\sqrt{1.5h} = d$$
$$\sqrt{1.5h} = 5$$
$$(\sqrt{1.5h})^2 = (5)^2$$
$$1.5h = 25$$
$$h = \frac{25}{1.5}$$
$$h \approx 16.67$$

The periscope must be 16.67 ft above the water.

27. Strategy • To find the height of the screen, use the equation $a = \sqrt{c^2 - b^2}$, where $c = 36$ and $b = 28.8$.

Solution

$$a = \sqrt{c^2 - b^2}$$
$$a = \sqrt{36^2 - (28.8)^2}$$
$$a = \sqrt{1296 - 829.44}$$
$$a = \sqrt{466.56}$$
$$a = 21.6$$

The height of the screen is 21.6 in.

29. Strategy • To find the distance from the center of the merry-go-round, replace v in the equation with the given value and solve for r.

Solution

$$v = \sqrt{12r}$$
$$15 = \sqrt{12r}$$
$$(15)^2 = (\sqrt{12r})^2$$
$$225 = 12r$$
$$18.75 = r$$

The distance of the child from the center is 18.75 ft.

Applying the Concepts

31. $a = 5, b = 12$

$$c = \sqrt{a^2 + b^2}$$
$$= \sqrt{5^2 + 12^2} = \sqrt{25 + 144} = \sqrt{169} = 13$$

Perimeter $= a + b + c = 5 + 12 + 13 = 30$

The perimeter of the triangle is 30 units.

33. If a and b are real numbers and $a^2 = b^2$, then it does not necessarily follow that $a = b$. For example, if $a = 4$ and $b = -4$, then $a^2 = b^2$, but $a \neq b$.

35. $s = \frac{1}{2}(a + b + c) = \frac{1}{2}(28 + 31 + 37) = \frac{1}{2}(96) = 48$

$$r = \sqrt{\frac{(s-a)(s-b)(s-c)}{s}}$$
$$= \sqrt{\frac{(48-28)(48-31)(48-37)}{48}}$$
$$= \sqrt{\frac{(20)(17)(11)}{48}} = \sqrt{\frac{3740}{48}} \approx \sqrt{77.917} \approx 8.827$$

Area of fountain $= \pi r^2 = \pi(8.827)^2 \approx 244.78$

The area of the fountain is approximately 244.78 ft^2.

Chapter 10 Review Exercises

1. $\sqrt{3}(\sqrt{12} - \sqrt{3}) = \sqrt{36} - \sqrt{9} = 6 - 3 = 3$

2.
$$3\sqrt{18a^5b} = 3\sqrt{9a^4}\sqrt{2ab}$$
$$= 3 \cdot 3a^2\sqrt{2ab}$$
$$= 9a^2\sqrt{2ab}$$

3. $2\sqrt{36} = 2\sqrt{6^2} = 2 \cdot 6 = 12$

4.
$$\sqrt{6a}(\sqrt{3a} + \sqrt{2a}) = \sqrt{18a^2} + \sqrt{12a^2}$$
$$= 3a\sqrt{2} + 2a\sqrt{3}$$

5.
$$\frac{12}{\sqrt{6}} = \frac{12}{\sqrt{6}} \cdot \frac{\sqrt{6}}{\sqrt{6}}$$
$$= \frac{12\sqrt{6}}{6}$$
$$= 2\sqrt{6}$$

6.
$$2\sqrt{8} - 3\sqrt{32} = 2\sqrt{4 \cdot 2} - 3\sqrt{16 \cdot 2}$$
$$= 2\sqrt{4}\sqrt{2} - 3\sqrt{16}\sqrt{2}$$
$$= 2 \cdot 2\sqrt{2} - 3 \cdot 4\sqrt{2}$$
$$= 4\sqrt{2} - 12\sqrt{2}$$
$$= (4 - 12)\sqrt{2}$$
$$= -8\sqrt{2}$$

7.
$$(3 - \sqrt{7})(3 + \sqrt{7}) = 3^2 - (\sqrt{7})^2$$
$$= 9 - 7$$
$$= 2$$

8. $$\begin{aligned} \sqrt{x+3}-\sqrt{x}&=1\\ \sqrt{x+3}&=1+\sqrt{x}\\ (\sqrt{x+3})^2&=(1+\sqrt{x})^2\\ x+3&=1+2\sqrt{x}+x\\ 2&=2\sqrt{x}\\ 1&=\sqrt{x}\\ (1)^2&=(\sqrt{x})^2\\ 1&=x \end{aligned}$$

Check: $$\begin{aligned} \sqrt{x+3}-\sqrt{x}&=1\\ \sqrt{1+3}-\sqrt{1}&=1\\ \sqrt{4}-\sqrt{1}&=1\\ 2-1&=1\\ 1&=1 \end{aligned}$$

The solution is 1.

9. $$\begin{aligned} \frac{2x}{\sqrt{3}-\sqrt{5}}&=\frac{2x}{\sqrt{3}-\sqrt{5}}\cdot\frac{\sqrt{3}+\sqrt{5}}{\sqrt{3}+\sqrt{5}}\\ &=\frac{2x(\sqrt{3}+\sqrt{5})}{3-5}\\ &=\frac{2x(\sqrt{3}+\sqrt{5})}{-2}\\ &=-x\sqrt{3}-x\sqrt{5} \end{aligned}$$

10. $-3\sqrt{120}=-3\sqrt{4}\sqrt{30}=-3\cdot 2\sqrt{30}=-6\sqrt{30}$

11. $$\begin{aligned} \sqrt{5x}&=10\\ (\sqrt{5x})^2&=(10)^2\\ 5x&=100\\ x&=20 \end{aligned}$$

12. $5\sqrt{48}=5\sqrt{16}\sqrt{3}=5\cdot 4\sqrt{3}=20\sqrt{3}$

13. $\dfrac{\sqrt{98x^7y^9}}{\sqrt{2x^3y}}=\sqrt{49x^4y^8}=7x^2y^4$

14. $$\begin{aligned} 3-\sqrt{7x}&=5\\ -\sqrt{7x}&=2\\ \sqrt{7x}&=-2 \end{aligned}$$

The equation has no solution.

15. $$\begin{aligned} &6a\sqrt{80b}-\sqrt{180a^2b}+5a\sqrt{b}\\ &=6a\sqrt{16}\sqrt{5b}-\sqrt{36a^2}\sqrt{5b}+5a\sqrt{b}\\ &=6a\cdot 4\sqrt{5b}-6a\sqrt{5b}+5a\sqrt{b}\\ &=24a\sqrt{5b}-6a\sqrt{5b}+5a\sqrt{b}\\ &=18a\sqrt{5b}+5a\sqrt{b} \end{aligned}$$

16. $4\sqrt{250}=4\sqrt{25}\sqrt{10}=4\cdot 5\sqrt{10}=20\sqrt{10}$

17. $$\begin{aligned} &2x\sqrt{60x^3y^3}+3x^2y\sqrt{15xy}\\ &=2x\sqrt{4x^2y^2}\sqrt{15xy}+3x^2y\sqrt{15xy}\\ &=2x\cdot 2xy\sqrt{15xy}+3x^2y\sqrt{15xy}\\ &=4x^2y\sqrt{15xy}+3x^2y\sqrt{15xy}\\ &=7x^2y\sqrt{15xy} \end{aligned}$$

18. $$\begin{aligned} (4\sqrt{y}-\sqrt{5})(2\sqrt{y}+3\sqrt{5})&=8y+12\sqrt{5y}-2\sqrt{5y}-15\\ &=8y+10\sqrt{5y}-15 \end{aligned}$$

19. $$\begin{aligned} 3\sqrt{12x}+5\sqrt{48x}&=3\sqrt{4}\sqrt{3x}+5\sqrt{16}\sqrt{3x}\\ &=3\cdot 2\sqrt{3x}+5\cdot 4\sqrt{3x}\\ &=6\sqrt{3x}+20\sqrt{3x}=26\sqrt{3x} \end{aligned}$$

20. $$\begin{aligned} \sqrt{2x-3}+4&=0\\ \sqrt{2x-3}&=-4 \end{aligned}$$

The equation has no solution.

21. $\dfrac{8}{\sqrt{x}-3} = \dfrac{8}{\sqrt{x}-3} \cdot \dfrac{\sqrt{x}+3}{\sqrt{x}+3}$
$= \dfrac{8(\sqrt{x}+3)}{x-9}$
$= \dfrac{8\sqrt{x}+24}{x-9}$

22. $4y\sqrt{243x^{17}y^9} = 4y\sqrt{81x^{16}y^8}\sqrt{3xy}$
$= 4y \cdot 9x^8y^4\sqrt{3xy}$
$= 36x^8y^5\sqrt{3xy}$

23. $y\sqrt{24y^6} = y\sqrt{4y^6}\sqrt{6} = y \cdot 2y^3\sqrt{6} = 2y^4\sqrt{6}$

24. $2x+4 = \sqrt{x^2+3}$

$(2x+4)^2 = (\sqrt{x^2+3})^2$

$4x^2+16x+16 = x^2+3$
$3x^2+16x+13 = 0$
$(x+1)(3x+13) = 0$

$x+1=0 \quad 3x+13=0$

$x=-1 \qquad x=-\dfrac{13}{3}$

The solution is -1.

Check: $2x+4 = \sqrt{x^2+3}$
$2(-1)+4 = \sqrt{(-1)^2+3}$
$2 = 2$

$2x+4 = \sqrt{x^2+3}$
$2\left(-\dfrac{13}{3}\right)+4 = \sqrt{\left(-\dfrac{13}{3}\right)^2+3}$
$\text{neg} \neq \text{pos}$

25. $2x^2\sqrt{18x^2y^5} + 6y\sqrt{2x^6y^3} - 9xy^2\sqrt{8x^4y} = 2x^2\sqrt{9x^2y^4}\sqrt{2y} + 6y\sqrt{x^6y^2}\sqrt{2y} - 9xy^2\sqrt{4x^4}\sqrt{2y}$
$= 2x^2 \cdot 3xy^2\sqrt{2y} + 6y \cdot x^3y\sqrt{2y} - 9xy^2 \cdot 2x^2\sqrt{2y}$
$= 6x^3y^2\sqrt{2y} + 6x^3y^2\sqrt{2y} - 18x^3y^2\sqrt{2y}$
$= -6x^3y^2\sqrt{2y}$

26. $\dfrac{16}{\sqrt{a}} = \dfrac{16}{\sqrt{a}} \cdot \dfrac{\sqrt{a}}{\sqrt{a}} = \dfrac{16\sqrt{a}}{a}$

27. **Strategy** • To find the distance, d, across the pond, use the equation $c = \sqrt{a^2+b^2}$, where $c = d$, $a = 25$, and $b = 35$.

Solution $d = \sqrt{a^2+b^2}$
$d = \sqrt{25^2+35^2}$
$d = \sqrt{625+1225}$
$d = \sqrt{1850}$
$d \approx 43$
The distance across the pond is approximately 43 ft.

28. $d = 4000\sqrt{\dfrac{W_0}{W_d}} - 4000$

$4000 = 4000\sqrt{\dfrac{W_0}{36}} - 4000$

$8000 = 4000\sqrt{\dfrac{W_0}{36}}$

$2 = \sqrt{\dfrac{W_0}{36}}$

$4 = \dfrac{W_0}{36}$

$36 \cdot 4 = \dfrac{W_0}{36} \cdot 36$

$144 = W_0$

The explorer weighs 144 lb on the surface of Earth.

29.
$$\begin{aligned} V &= 3\sqrt{d} \\ 30 &= 3\sqrt{d} \\ 10 &= \sqrt{d} \\ (10)^2 &= (\sqrt{d})^2 \\ 100 &= d \end{aligned}$$
The depth of the water is 100 ft.

30.
$$\begin{aligned} V &= 4\sqrt{r} \\ 20 &= 4\sqrt{r} \\ 5 &= \sqrt{r} \\ (5)^2 &= (\sqrt{r})^2 \\ 25 &= r \end{aligned}$$
The radius of the corner is 25 ft.

Chapter 10 Test

1. $\sqrt{121x^8y^2} = 11x^4y$

2.
$$\begin{aligned} \sqrt{3x^2y}\sqrt{6xy^2}\sqrt{2x} &= \sqrt{36x^4y^3} \\ &= \sqrt{36x^4y^2}\sqrt{y} \\ &= 6x^2y\sqrt{y} \end{aligned}$$

3.
$$\begin{aligned} 5\sqrt{8} - 3\sqrt{50} &= 5\sqrt{4}\sqrt{2} - 3\sqrt{25}\sqrt{2} \\ &= 5\cdot 2\sqrt{2} - 3\cdot 5\sqrt{2} \\ &= 10\sqrt{2} - 15\sqrt{2} = -5\sqrt{2} \end{aligned}$$

4. $\sqrt{45} = \sqrt{9}\sqrt{5} = 3\sqrt{5}$

5. $\dfrac{\sqrt{162}}{\sqrt{2}} = \sqrt{81} = 9$

6.
$$\begin{aligned} \sqrt{9x} + 3 &= 18 \\ \sqrt{9x} &= 15 \\ 9x &= 225 \\ x &= 25 \end{aligned}$$

7.
$$\begin{aligned} \sqrt{32a^5b^{11}} &= \sqrt{16a^4b^{10}}\sqrt{2ab} \\ &= 4a^2b^5\sqrt{2ab} \end{aligned}$$

8. $\dfrac{\sqrt{98a^6b^4}}{\sqrt{2a^3b^2}} = \sqrt{49a^3b^2} = 7ab\sqrt{a}$

9.
$$\begin{aligned} \frac{2}{\sqrt{3}-1} &= \frac{2}{\sqrt{3}-1}\cdot\frac{\sqrt{3}+1}{\sqrt{3}+1} \\ &= \frac{2(\sqrt{3}+1)}{3-1} \\ &= \frac{2(\sqrt{3}+1)}{2} \\ &= \sqrt{3}+1 \end{aligned}$$

10.
$$\begin{aligned} \sqrt{8x^3y}\sqrt{10xy^4} &= \sqrt{80x^4y^5} \\ &= \sqrt{16x^4y^4}\sqrt{5y} \\ &= 4x^2y^2\sqrt{5y} \end{aligned}$$

11.
$$\begin{aligned} \sqrt{x-5} + \sqrt{x} &= 5 \\ \sqrt{x-5} &= 5 - \sqrt{x} \\ x - 5 &= 25 - 10\sqrt{x} + x \\ -30 &= -10\sqrt{x} \\ 3 &= \sqrt{x} \\ (3)^2 &= (\sqrt{x})^2 \\ 9 &= x \end{aligned}$$

Check:
$$\begin{aligned} \sqrt{x-5} + \sqrt{x} &= 5 \\ \sqrt{9-5} + \sqrt{9} &= 5 \\ \sqrt{4} + \sqrt{9} &= 5 \\ 2 + 3 &= 5 \\ 5 &= 5 \end{aligned}$$

The solution is 9.

12.
$$\begin{aligned} &3\sqrt{8y} - 2\sqrt{72x} + 5\sqrt{18y} \\ &= 3\sqrt{4}\sqrt{2y} - 2\sqrt{36}\sqrt{2x} + 5\sqrt{9}\sqrt{2y} \\ &= 3\cdot 2\sqrt{2y} - 2\cdot 6\sqrt{2x} + 5\cdot 3\sqrt{2y} \\ &= 6\sqrt{2y} - 12\sqrt{2x} + 15\sqrt{2y} \\ &= 21\sqrt{2y} - 12\sqrt{2x} \end{aligned}$$

13.
$$\begin{aligned} \sqrt{72x^7y^2} &= \sqrt{36x^6y^2}\sqrt{2x} \\ &= 6x^3y\sqrt{2x} \end{aligned}$$

14. $(\sqrt{y} - 3)(\sqrt{y} + 5) = y + 2\sqrt{y} - 15$

15.
$$\begin{aligned} &2x\sqrt{3xy^3} - 2y\sqrt{12x^3y} - 3xy\sqrt{xy} \\ &= 2x\sqrt{y^2}\sqrt{3xy} - 2y\sqrt{4x^2}\sqrt{3xy} - 3xy\sqrt{xy} \\ &= 2xy\sqrt{3xy} - 4xy\sqrt{3xy} - 3xy\sqrt{xy} \\ &= -2xy\sqrt{3xy} - 3xy\sqrt{xy} \end{aligned}$$

16.
$$\begin{aligned} \frac{2-\sqrt{5}}{6+\sqrt{5}} &= \frac{2-\sqrt{5}}{6+\sqrt{5}}\cdot\frac{6-\sqrt{5}}{6-\sqrt{5}} \\ &= \frac{12 - 8\sqrt{5} + 5}{36 - 5} \\ &= \frac{17 - 8\sqrt{5}}{31} \end{aligned}$$

17.
$$\begin{aligned} \sqrt{a}(\sqrt{a} - \sqrt{b}) &= \sqrt{a^2} - \sqrt{ab} \\ &= a - \sqrt{ab} \end{aligned}$$

18. $\sqrt{75} = \sqrt{25}\sqrt{3} = 5\sqrt{3}$

19.
$$\begin{aligned} T &= 2\pi\sqrt{\frac{L}{32}} \\ 3 &= 2(3.14)\sqrt{\frac{L}{32}} \\ 3 &= 6.28\sqrt{\frac{L}{32}} \\ \frac{3}{6.28} &= \frac{6.28\sqrt{\frac{L}{32}}}{6.28} \\ 0.4777 &\approx \sqrt{\frac{L}{32}} \\ (0.4777)^2 &= \left(\sqrt{\frac{L}{32}}\right)^2 \\ 0.2282 &\approx \frac{L}{32} \\ 32 \times 0.2282 &= \frac{L}{32} \times 32 \\ 7.30 &\approx L \end{aligned}$$
The length of the pendulum is 7.30 ft.

20. Strategy • To find the distance, x, from the base of the pole, use $a = \sqrt{c^2 - b^2}$, where $a = x$, $c = 8$, and $b = 4$.

Solution
$$a = \sqrt{c^2 - b^2}$$
$$x = \sqrt{8^2 - 4^2}$$
$$x = \sqrt{64 - 16}$$
$$x = \sqrt{48} \approx 6.93$$
The rope should be secured about 7 ft from the base of the pole.

Cumulative Review Exercises

1. $$\left(\frac{2}{3}\right)^2 \cdot \left(\frac{3}{4} - \frac{3}{2}\right) + \left(\frac{1}{2}\right)^2 = \frac{4}{9} \cdot \left(\frac{3}{4} - \frac{6}{4}\right) + \frac{1}{4}$$
$$= \frac{4}{9} \cdot \left(-\frac{3}{4}\right) + \frac{1}{4}$$
$$= -\frac{1}{3} + \frac{1}{4}$$
$$= -\frac{4}{12} + \frac{3}{12} = -\frac{1}{12}$$

2. $$-3[x - 2(3 - 2x) - 5x] + 2x$$
$$= -3[x - 6 + 4x - 5x] + 2x$$
$$= -3[-6] + 2x$$
$$= 2x + 18$$

3. $$2x - 4[3x - 2(1 - 3x)] = 2(3 - 4x)$$
$$2x - 4[3x - 2 + 6x] = 6 - 8x$$
$$2x - 4[9x - 2] = 6 - 8x$$
$$2x - 36x + 8 = 6 - 8x$$
$$-34x = -2 - 8x$$
$$-26x = -2$$
$$x = \frac{1}{13}$$

4. $(-3x^2y)(-2x^3y^4) = 6x^5y^5$

5. $$\frac{12b^4 - 6b^2 + 2}{-6b^2}$$
$$= \frac{12b^4}{-6b^2} - \frac{6b^2}{-6b^2} + \frac{2}{-6b^2}$$
$$= -2b^2 + 1 - \frac{1}{3b^2}$$

6. $$f(x) = \frac{2x}{x - 3}$$
$$f(-3) = \frac{2(-3)}{-3 - 3} = \frac{-6}{-6} = 1$$

7. $2a^3 - 16a^2 + 30a$
The GCF is $2a$: $2a(a^2 - 8a + 15)$.
$2a(a - 5)(a - 3)$

8. $$\frac{3x^3 - 6x^2}{4x^2 + 4x} \cdot \frac{3x - 9}{9x^3 - 45x^2 + 54x}$$
$$= \frac{3x^2(x - 2)}{4x(x + 1)} \cdot \frac{3(x - 3)}{9x(x^2 - 5x + 6)}$$
$$= \frac{3x^2(x - 2)}{4x(x + 1)} \cdot \frac{3(x - 3)}{9x(x - 3)(x - 2)}$$
$$= \frac{1}{4(x + 1)}$$

9. $$\frac{x + 2}{x - 4} - \frac{6}{(x - 4)(x - 3)}$$
$\text{LCM} = (x - 4)(x - 3)$
$$\frac{(x + 2)(x - 3)}{(x - 4)(x - 3)} - \frac{6}{(x - 4)(x - 3)} = \frac{x^2 - x - 6 - 6}{(x - 4)(x - 3)}$$
$$= \frac{x^2 - x - 12}{(x - 4)(x - 3)}$$
$$= \frac{(x - 4)(x + 3)}{(x - 4)(x - 3)}$$
$$= \frac{x + 3}{x - 3}$$

10. $$\frac{x}{2x - 5} - 2 = \frac{3x}{2x - 5}$$
$$(2x - 5)\left(\frac{x}{2x - 5} - 2\right) = (2x - 5)\left(\frac{3x}{2x - 5}\right)$$
$$x - 2(2x - 5) = 3x$$
$$x - 4x + 10 = 3x$$
$$-3x + 10 = 3x$$
$$10 = 6x$$
$$\frac{10}{6} = x$$
$$\frac{5}{3} = x$$

11. $m = \frac{1}{2}$, $(x_1, y_1) = (-2, -3)$
$$y - y_1 = m(x - x_1)$$
$$y + 3 = \frac{1}{2}(x + 2)$$
$$y + 3 = \frac{1}{2}x + 1$$
$$y = \frac{1}{2}x - 2$$

12. (1) $4x - 3y = 1$
(2) $2x + y = 3 \quad y = 3 - 2x$
Substitute y in Equation (1).
$$4x - 3(3 - 2x) = 1$$
$$4x - 9 + 6x = 1$$
$$10x = 10$$
$$x = 1$$
Substitute x in Equation (2).
$$y = 3 - 2x$$
$$y = 3 - 2(1)$$
$$y = 3 - 2$$
$$y = 1$$
The solution is (1, 1).

13. (1) $5x + 4y = 7$
(2) $3x - 2y = 13$
Eliminate y and add the equations.

$$\begin{array}{ll} 5x + 4y = 7 & 5x + 4y = 7 \\ 2(3x - 2y) = 13(2) & 6x - 4y = 26 \\ & \overline{11x = 33} \\ & x = 3 \end{array}$$

Replace x in Equation (1).

$$\begin{aligned} 5x + 4y &= 7 \\ 5(3) + 4y &= 7 \\ 15 + 4y &= 7 \\ 4y &= -8 \\ y &= -2 \end{aligned}$$

The solution is $(3, -2)$.

14.
$$\begin{aligned} 3(x - 7) &\geq 5x - 12 \\ 3x - 21 &\geq 5x - 12 \\ -2x &\geq 9 \\ x &\leq -\frac{9}{2} \end{aligned}$$

15. $\sqrt{108} = \sqrt{36}\sqrt{3} = 6\sqrt{3}$

16.
$$\begin{aligned} &3\sqrt{32} - 2\sqrt{128} \\ &= 3\sqrt{16}\sqrt{2} - 2\sqrt{64}\sqrt{2} \\ &= 3 \cdot 4\sqrt{2} - 2 \cdot 8\sqrt{2} \\ &= 12\sqrt{2} - 16\sqrt{2} \\ &= -4\sqrt{2} \end{aligned}$$

17.
$$\begin{aligned} &2a\sqrt{2ab^3} + b\sqrt{8a^3b} - 5ab\sqrt{ab} \\ &= 2a\sqrt{b^2}\sqrt{2ab} + b\sqrt{4a^2}\sqrt{2ab} - 5ab\sqrt{ab} \\ &= 2a \cdot b\sqrt{2ab} + 2a \cdot b\sqrt{2ab} - 5ab\sqrt{ab} \\ &= 2ab\sqrt{2ab} + 2ab\sqrt{2ab} - 5ab\sqrt{ab} \\ &= 4ab\sqrt{2ab} - 5ab\sqrt{ab} \end{aligned}$$

18.
$$\begin{aligned} \sqrt{2a^9b}\sqrt{98ab^3}\sqrt{2a} &= \sqrt{2^3 \cdot 49a^{11}b^4} \\ &= \sqrt{2^2 \cdot 49a^{10}b^4}\sqrt{2a} \\ &= 2 \cdot 7a^5b^2\sqrt{2a} \\ &= 14a^5b^2\sqrt{2a} \end{aligned}$$

19.
$$\begin{aligned} \sqrt{3}(\sqrt{6} - \sqrt{x^2}) &= \sqrt{18} - \sqrt{3x^2} \\ &= \sqrt{9}\sqrt{2} - \sqrt{x^2}\sqrt{3} = 3\sqrt{2} - x\sqrt{3} \end{aligned}$$

20. $\dfrac{\sqrt{320}}{\sqrt{5}} = \sqrt{64} = 8$

21.
$$\begin{aligned} \frac{3}{2 - \sqrt{5}} &= \frac{3}{2 - \sqrt{5}} \cdot \frac{2 + \sqrt{5}}{2 + \sqrt{5}} \\ &= \frac{3(2 + \sqrt{5})}{4 - 5} = -3(2 + \sqrt{5}) = -6 - 3\sqrt{5} \end{aligned}$$

22.
$$\begin{aligned} \sqrt{3x - 2} - 4 &= 0 \\ \sqrt{3x - 2} &= 4 \\ (\sqrt{3x - 2})^2 &= (4)^2 \\ 3x - 2 &= 16 \\ 3x &= 18 \\ x &= 6 \end{aligned}$$

23.
$$\begin{aligned} S &= C + rC \\ 29.40 &= C + 20\%C \\ 29.40 &= C + 0.20C \\ 29.40 &= 1.20C \\ 24.50 &= C \end{aligned}$$

The cost of the book is \$24.50.

24. Strategy • Amount of pure water: x

	Amount	Percent	Quantity
Water	x	0.00	$0.00x$
12% solution	40	0.12	0.12(40)
5% solution	$x + 40$	0.05	$0.05(x + 40)$

• The sum of the quantities before mixing is equal to the quantity after mixing.

Solution
$$\begin{aligned} 0.00x + 0.12(40) &= 0.05(x + 40) \\ 0.12(40) &= 0.05(x + 40) \\ 4.8 &= 0.05x + 2 \\ 2.8 &= 0.05x \\ 56 &= x \end{aligned}$$

56 oz of water must be added.

25. Strategy • First number: x
Second number: $21 - x$

The product of the two numbers	is	one hundred four

Solution
$$\begin{aligned} x(21 - x) &= 104 \\ 21x - x^2 &= 104 \\ x^2 - 21x + 104 &= 0 \\ (x - 8)(x - 13) &= 0 \end{aligned}$$

$$x - 8 = 0 \quad x - 13 = 0$$
$$x = 8 \quad\quad x = 13$$

The numbers are 8 and 13.

26. Strategy • Time for the large pipe: t
Time for the small pipe: $2t$

	Rate	Time	Part
Small pipe	$2t$	16	$\frac{16}{2t}$
Large pipe	t	16	$\frac{16}{t}$

• The sum of the parts of the task completed by each pipe must equal 1.

Solution
$$\begin{aligned} \frac{16}{2t} + \frac{16}{t} &= 1 \\ 2t\left(\frac{16}{2t} + \frac{16}{t}\right) &= 2t \cdot 1 \\ 16 + 32 &= 2t \\ 48 &= 2t \\ 24 &= t \end{aligned}$$

It would take the small pipe, working alone, 48 h.

27.

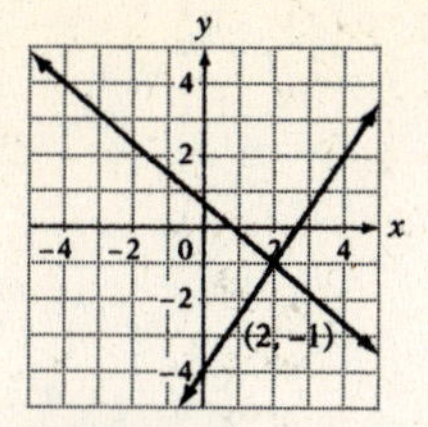

28.

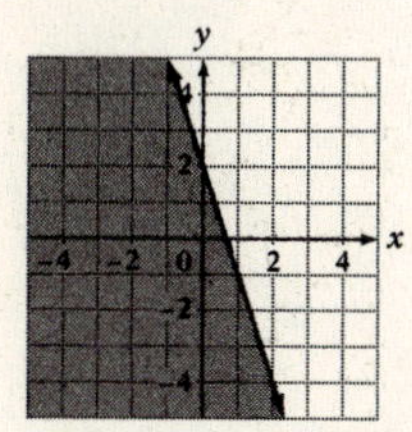

29. **Strategy** • First integer: n

Second integer: $n+1$

The square root of the sum of the two consecutive integers	is equal to	9

Solution

$$\sqrt{n+n+1} = 9$$
$$\sqrt{2n+1} = 9$$
$$(\sqrt{2n+1})^2 = (9)^2$$
$$2n+1 = 81$$
$$2n = 80$$
$$n = 40$$
$$n+1 = 41$$

The smaller integer is 40.

30.

$$T = \sqrt{\frac{d}{16}}$$
$$5 = \sqrt{\frac{d}{16}}$$
$$(5)^2 = \left(\sqrt{\frac{d}{16}}\right)^2$$
$$25 = \frac{d}{16}$$
$$16 \cdot 25 = \frac{d}{16} \cdot 16$$
$$400 = d$$

The height of the building is 400 ft.

Chapter 11: Quadratic Equations

Prep Test

1. $b^2 - 4ac$; $a = 2$, $b = -3$, $c = -4$

$$\begin{aligned} b^2 - 4ac &= (-3)^2 - 4(2)(-4) \\ &= 9 + 32 \\ &= 41 \end{aligned}$$

2.
$$\begin{aligned} 5x + 4 &= 3 \\ 5x &= -1 \\ x &= -\frac{1}{5} \end{aligned}$$

3. $x^2 + x - 12 = (x + 4)(x - 3)$

4. $4x^2 - 12x + 9 = (2x - 3)^2$

5. Yes; since $x^2 - 10x + 25 = (x - 5)^2$, $x^2 - 10x + 25$ is a perfect square trinomial.

6.
$$\begin{aligned} \frac{5}{x-2} &= \frac{15}{x} \\ 5x &= 15x - 30 \\ -10x &= -30 \\ x &= 3 \end{aligned}$$

7.

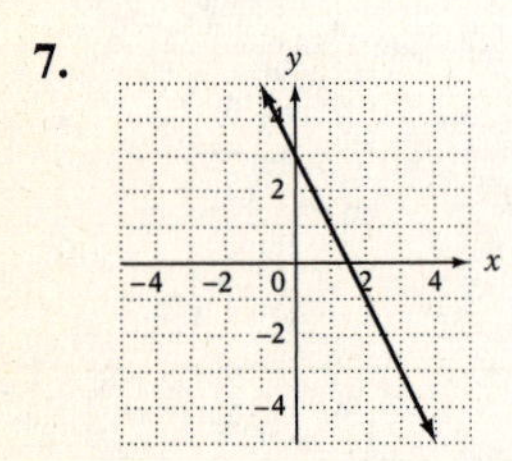

8.
$$\begin{aligned} \sqrt{28} &= \sqrt{4 \cdot 7} \\ &= \sqrt{4} \cdot \sqrt{7} \\ &= 2\sqrt{7} \end{aligned}$$

9. $\sqrt{a^2} = |a|$

10. **Strategy** • Time of first half of hike: t

	Rate	Time	Distance
1st half of hike	4.5	t	$4.5t$
2nd half of hike	3	$2 - t$	$3(2 - t)$

• The distance hiked at the different rates is the same.
• After finding the time, calculate the distance.

Solution
$$\begin{aligned} 4.5t &= 3(2 - t) \\ 4.5t &= 6 - 3t \\ 7.5t &= 6 \\ t &= 0.8 \text{ hr} \\ 4.5t &= 4.5(0.8) \\ &= 3.6 \text{ mi} \end{aligned}$$

The hiking trail is 3.6 mi long.

Go Figure

Strategy Rationalize the denominator of each addend, then add.

Solution

$$\frac{1}{\sqrt{1}+\sqrt{2}}+\frac{1}{\sqrt{2}+\sqrt{3}}+\cdots+\frac{1}{\sqrt{7}+\sqrt{8}}+\frac{1}{\sqrt{8}+\sqrt{9}}$$
$$=\frac{1}{\sqrt{1}+\sqrt{2}}\cdot\frac{\sqrt{1}-\sqrt{2}}{\sqrt{1}-\sqrt{2}}+\cdots+\frac{1}{\sqrt{8}+\sqrt{9}}\cdot\frac{\sqrt{8}-\sqrt{9}}{\sqrt{8}-\sqrt{9}}$$
$$=\frac{1-\sqrt{2}}{1-2}+\frac{\sqrt{2}-\sqrt{3}}{2-3}+\cdots+\frac{\sqrt{7}-\sqrt{8}}{7-8}+\frac{\sqrt{8}-3}{8-9}$$
$$=-(1-\sqrt{2})-(\sqrt{2}-\sqrt{3})-\cdots-(\sqrt{7}-\sqrt{8})-(\sqrt{8}-3)$$
$$=-1+\sqrt{2}-\sqrt{2}+\sqrt{3}-\sqrt{3}+\cdots-\sqrt{7}+\sqrt{8}-\sqrt{8}+3$$
$$=-1+3$$
$$=2$$

Section 11.1

Objective A Exercises

1. $(x+3)(x-5)=0$

$x+3=0 \quad x-5=0$

$x=-3 \quad x=5$

The solutions are -3 and 5.

3. $(2x+5)(3x-1)$

$2x+5=0 \quad 3x-1=0$

$2x=-5 \quad 3x=1$

$x=-\frac{5}{2} \quad x=\frac{1}{3}$

The solutions are $-\frac{5}{2}$ and $\frac{1}{3}$.

5. $x^2+2x-15=0$

$(x-3)(x+5)=0$

$x-3=0 \quad x+5=0$

$x=3 \quad x=-5$

The solutions are -5 and 3.

7. $z^2-4z+3=0$

$(z-1)(z-3)=0$

$z-1=0 \quad z-3=0$

$z=1 \quad z=3$

The solutions are 1 and 3.

9. $p^2+3p+2=0$

$(p+1)(p+2)=0$

$p+1=0 \quad p+2=0$

$p=-1 \quad p=-2$

The solutions are -2 and -1.

11. $x^2-6x+9=0$

$(x-3)(x-3)=0$

$x-3=0 \quad x-3=0$

$x=3 \quad x=3$

3 is a double root of the equation.
The solution is 3.

13. $12y^2+8y=0$

$4y(3y+2)=0$

$4y=0 \quad 3y+2=0$

$y=0 \quad 3y=-2$

$y=-\frac{2}{3}$

The solutions are $-\frac{2}{3}$ and 0.

15. $r^2-10=3r$

$r^2-3r-10=0$

$(r+2)(r-5)=0$

$r+2=0 \quad r-5=0$

$r=-2 \quad r=5$

The solutions are -2 and 5.

17. $3v^2-5v+2=0$

$(3v-2)(v-1)=0$

$3v-2=0 \quad v-1=0$

$3v=2 \quad v=1$

$v=\frac{2}{3}$

The solutions are $\frac{2}{3}$ and 1.

19. $3s^2+8s=3$

$3s^2+8s-3=0$

$(3s-1)(s+3)=0$

$3s-1=0 \quad s+3=0$

$3s=1 \quad s=-3$

$s=\frac{1}{3}$

The solutions are -3 and $\frac{1}{3}$.

21.
$$\frac{3}{4}z^2 - z = -\frac{1}{3}$$
$$\frac{3}{4}z^2 - z + \frac{1}{3} = 0$$
$$12\left(\frac{3}{4}z^2 - z + \frac{1}{3}\right) = 12 \cdot 0$$
$$9z^2 - 12z + 4 = 0$$
$$(3z - 2)(3z - 2) = 0$$
$$3z - 2 = 0 \quad 3z - 2 = 0$$
$$3z = 2 \quad 3z = 2$$
$$z = \frac{2}{3} \quad z = \frac{2}{3}$$
$\frac{2}{3}$ is a double root of the equation.
The solution is $\frac{2}{3}$.

23.
$$4t^2 = 4t + 3$$
$$4t^2 - 4t - 3 = 0$$
$$(2t + 1)(2t - 3) = 0$$
$$2t + 1 = 0 \quad 2t - 3 = 0$$
$$2t = -1 \quad 2t = 3$$
$$t = -\frac{1}{2} \quad t = \frac{3}{2}$$
The solutions are $-\frac{1}{2}$ and $\frac{3}{2}$.

25.
$$4v^2 - 4v + 1 = 0$$
$$(2v - 1)(2v - 1) = 0$$
$$2v - 1 = 0 \quad 2v - 1 = 0$$
$$2v = 1 \quad 2v = 1$$
$$v = \frac{1}{2} \quad v = \frac{1}{2}$$
$\frac{1}{2}$ is a double root of the equation.
The solution is $\frac{1}{2}$.

27.
$$x^2 - 9 = 0$$
$$(x - 3)(x + 3) = 0$$
$$x - 3 = 0 \quad x + 3 = 0$$
$$x = 3 \quad x = -3$$
The solutions are -3 and 3.

29.
$$4y^2 - 1 = 0$$
$$(2y - 1)(2y + 1) = 0$$
$$2y - 1 = 0 \quad 2y + 1 = 0$$
$$2y = 1 \quad 2y = -1$$
$$y = \frac{1}{2} \quad y = -\frac{1}{2}$$
The solutions are $-\frac{1}{2}$ and $\frac{1}{2}$.

31.
$$x + 15 = x(x - 1)$$
$$x + 15 = x^2 - x$$
$$15 = x^2 - 2x$$
$$0 = x^2 - 2x - 15$$
$$0 = (x + 3)(x - 5)$$
$$x + 3 = 0 \quad x - 5 = 0$$
$$x = -3 \quad x = 5$$
The solutions are -3 and 5.

33.
$$r^2 - r - 2 = (2r - 1)(r - 3)$$
$$r^2 - r - 2 = 2r^2 - 7r + 3$$
$$-r - 2 = r^2 - 7r + 3$$
$$-2 = r^2 - 6r + 3$$
$$0 = r^2 - 6r + 5$$
$$0 = (r - 1)(r - 5)$$
$$r - 1 = 0 \quad r - 5 = 0$$
$$r = 1 \quad r = 5$$
The solutions are 1 and 5.

Objective B Exercises

35.
$$x^2 = 36$$
$$\sqrt{x^2} = \sqrt{36}$$
$$x = \pm\sqrt{36} = \pm 6$$
The solutions are -6 and 6.

37.
$$v^2 - 1 = 0$$
$$v^2 = 1$$
$$\sqrt{v^2} = \sqrt{1}$$
$$v = \pm\sqrt{1} = \pm 1$$
The solutions are -1 and 1.

39.
$$4x^2 - 49 = 0$$
$$4x^2 = 49$$
$$x^2 = \frac{49}{4}$$
$$\sqrt{x^2} = \sqrt{\frac{49}{4}}$$
$$x = \pm\sqrt{\frac{49}{4}} = \pm\frac{7}{2}$$
The solutions are $-\frac{7}{2}$ and $\frac{7}{2}$.

41.
$$9y^2 = 4$$
$$y^2 = \frac{4}{9}$$
$$\sqrt{y^2} = \sqrt{\frac{4}{9}}$$
$$y = \pm\sqrt{\frac{4}{9}} = \pm\frac{2}{3}$$
The solutions are $-\frac{2}{3}$ and $\frac{2}{3}$.

43. $16v^2 - 9 = 0$

$16v^2 = 9$

$v^2 = \frac{9}{16}$

$\sqrt{v^2} = \sqrt{\frac{9}{16}}$

$v = \pm\sqrt{\frac{9}{16}} = \pm\frac{3}{4}$

The solutions are $-\frac{3}{4}$ and $\frac{3}{4}$.

45. $y^2 + 81 = 0$

$y^2 = -81$

$\sqrt{y^2} = \sqrt{-81}$

$\sqrt{-81}$ is not a real number.
The equation has no real number solution.

47. $w^2 - 24 = 0$

$w^2 = 24$

$\sqrt{w^2} = \sqrt{24}$

$\sqrt{w^2} = \pm\sqrt{24}$

$w = \pm\sqrt{2^3 \cdot 3}$

$w = \pm\sqrt{2^2}\sqrt{2 \cdot 3}$

$w = \pm 2\sqrt{6}$

The solutions are $-2\sqrt{6}$ and $2\sqrt{6}$.

49. $(x-1)^2 = 36$

$\sqrt{(x-1)^2} = \sqrt{36}$

$x - 1 = \pm\sqrt{36} = \pm 6$

$x - 1 = 6 \quad x - 1 = -6$

$x = 7 \quad x = -5$

The solutions are -5 and 7.

51. $2(x+5)^2 = 8$

$(x+5)^2 = 4$

$\sqrt{(x+5)^2} = \sqrt{4}$

$x + 5 = \pm\sqrt{4} = \pm 2$

$x + 5 = 2 \quad x + 5 = -2$

$x = -3 \quad x = -7$

The solutions are -7 and -3.

53. $9(x-1)^2 - 16 = 0$

$9(x-1)^2 = 16$

$(x-1)^2 = \frac{16}{9}$

$\sqrt{(x-1)^2} = \sqrt{\frac{16}{9}}$

$x - 1 = \pm\sqrt{\frac{16}{9}} = \pm\frac{4}{3}$

$x - 1 = \frac{4}{3} \quad x - 1 = -\frac{4}{3}$

$x = \frac{4}{3} + 1 \quad x = -\frac{4}{3} + 1$

$x = \frac{7}{3} \quad x = -\frac{1}{3}$

The solutions are $-\frac{1}{3}$ and $\frac{7}{3}$.

55. $49(v+1)^2 - 25 = 0$

$49(v+1)^2 = 25$

$(v+1)^2 = \frac{25}{49}$

$\sqrt{(v+1)^2} = \sqrt{\frac{25}{49}}$

$v + 1 = \pm\sqrt{\frac{25}{49}} = \pm\frac{5}{7}$

$v + 1 = \frac{5}{7} \quad v + 1 = -\frac{5}{7}$

$v = \frac{5}{7} - 1 \quad v = -\frac{5}{7} - 1$

$v = -\frac{2}{7} \quad v = -\frac{12}{7}$

The solutions are $-\frac{12}{7}$ and $-\frac{2}{7}$.

57. $(x-4)^2 - 20 = 0$

$(x-4)^2 = 20$

$\sqrt{(x-4)^2} = \sqrt{20}$

$x - 4 = \pm\sqrt{20}$

$x - 4 = \pm\sqrt{2^2 \cdot 5}$

$x - 4 = \pm\sqrt{2^2}\sqrt{5}$

$x - 4 = \pm 2\sqrt{5}$

$x - 4 = 2\sqrt{5} \quad x - 4 = -2\sqrt{5}$

$x = 4 + 2\sqrt{5} \quad x = 4 - 2\sqrt{5}$

The solutions are $4 - 2\sqrt{5}$ and $4 + 2\sqrt{5}$.

59. $(x+1)^2 + 36 = 0$

$(x+1)^2 = -36$

$\sqrt{(x+1)^2} = \sqrt{-36}$

$\sqrt{-36}$ is not a real number.
The equation has no real number solution.

61.
$$3\left(v+\frac{3}{4}\right)^2 = 36$$
$$\left(v+\frac{3}{4}\right)^2 = 12$$
$$\sqrt{\left(v+\frac{3}{4}\right)^2} = \pm\sqrt{12}$$
$$v+\frac{3}{4} = \pm\sqrt{12}$$
$$v+\frac{3}{4} = \pm\sqrt{2^2\cdot 3}$$
$$v+\frac{3}{4} = \pm\sqrt{2^2}\sqrt{3}$$
$$v+\frac{3}{4} = \pm 2\sqrt{3}$$
$$v+\frac{3}{4} = 2\sqrt{3} \qquad v+\frac{3}{4} = -2\sqrt{3}$$
$$v = -\frac{3}{4}+2\sqrt{3} \qquad v = -\frac{3}{4}-2\sqrt{3}$$
The solutions are $-\frac{3}{4}-2\sqrt{3}$ and $-\frac{3}{4}+2\sqrt{3}$.

63.
$$(6x^2-5)^2 = 1$$
$$\sqrt{(6x^2-5)^2} = \sqrt{1}$$
$$6x^2-5 = \pm 1$$
$$6x^2 = 5\pm 1$$
$$6x^2 = 5+1 \qquad 6x^2 = 5-1$$
$$6x^2 = 6 \qquad 6x^2 = 4$$
$$x^2 = 1 \qquad x^2 = \frac{4}{6}$$
$$\sqrt{x^2} = \sqrt{1} \qquad \sqrt{x^2} = \sqrt{\frac{4}{6}}$$
$$x = \pm 1 \qquad x = \pm\sqrt{\frac{4}{6}} = \pm\frac{2}{\sqrt{6}}$$
$$= \pm\frac{2\sqrt{6}}{6} = \pm\frac{\sqrt{6}}{3}$$
The solutions are -1, $-\frac{\sqrt{6}}{3}$, $\frac{\sqrt{6}}{3}$, and 1.

65.
$$P = A(1+r)^2$$
$$5832 = 5000(1+r)^2$$
$$1.1664 = (1+r)^2$$
$$\sqrt{1.1664} = \sqrt{(1+r)^2}$$
$$\pm 1.08 = 1+r$$
$$1+r = 1.08 \qquad 1+r = -1.08$$
$$r = 0.08 \qquad r = -2.08 \quad \text{not possible}$$
The annual percentage rate is 8%.

67.
$$d = 0.0074v^2$$
$$40 = 0.0074v^2$$
$$5405.41 = v^2$$
$$v \approx 73.52$$
Yes, because $v \approx 73.5$ mph is over the 65-mph speed limit.

Section 11.2

Objective A Exercises

1. $x^2 - 8x$
$$\left(\frac{8}{2}\right)^2 = (4)^2 = 16$$
$$x^2 - 8x + 16 = (x-4)^2$$

3. $x^2 + 5x$
$$\left(\frac{5}{2}\right)^2 = \frac{25}{4}$$
$$x^2 + 5x + \frac{25}{4} = \left(x+\frac{5}{2}\right)^2$$

5.
$$x^2 + 2x - 3 = 0$$
$$x^2 + 2x = 3$$
Complete the square.
$$x^2 + 2x + 1 = 3 + 1$$
$$(x+1)^2 = 4$$
$$\sqrt{(x+1)^2} = \sqrt{4}$$
$$x+1 = \pm 2$$
$$x+1 = 2 \qquad x+1 = -2$$
$$x = 1 \qquad x = -3$$
The solutions are -3 and 1.

7.
$$z^2 - 6z - 16 = 0$$
$$z^2 - 6z = 16$$
Complete the square.
$$z^2 - 6z + 9 = 16 + 9$$
$$(z-3)^2 = 25$$
$$\sqrt{(z-3)^2} = \sqrt{25}$$
$$z-3 = \pm 5$$
$$z-3 = 5 \qquad z-3 = -5$$
$$z = 8 \qquad z = -2$$
The solutions are -2 and 8.

9.
$$x^2 = 4x - 4$$
$$x^2 - 4x = -4$$
Complete the square.
$$x^2 - 4x + 4 = -4 + 4$$
$$(x-2)^2 = 0$$
$$\sqrt{(x-2)^2} = \sqrt{0}$$
$$x-2 = 0$$
$$x = 2$$
The solution is 2.

11.
$$v^2 - 6v + 13 = 0$$
$$v^2 - 6v = -13$$
Complete the square.
$$v^2 - 6v + 9 = -13 + 9$$
$$(v-3)^2 = -4$$
$$\sqrt{(v-3)^2} = \sqrt{-4}$$
$\sqrt{-4}$ is not a real number.
The quadratic equation has no real number solution.

13. $y^2 + 5y + 4 = 0$
$y^2 + 5y = -4$
Complete the square.
$y^2 + 5y + \frac{25}{4} = -4 + \frac{25}{4}$
$\left(y + \frac{5}{2}\right)^2 = \frac{9}{4}$
$\sqrt{\left(y + \frac{5}{2}\right)^2} = \sqrt{\frac{9}{4}}$
$y + \frac{5}{2} = \pm\frac{3}{2}$
$y + \frac{5}{2} = \frac{3}{2}$ $\quad y + \frac{5}{2} = -\frac{3}{2}$
$y = -1$ $\quad y = -4$
The solutions are -4 and -1.

15. $w^2 + 7w = 8$
Complete the square.
$w^2 + 7w + \frac{49}{4} = 8 + \frac{49}{4}$
$\left(w + \frac{7}{2}\right)^2 = \frac{81}{4}$
$\sqrt{\left(w + \frac{7}{2}\right)^2} = \sqrt{\frac{81}{4}}$
$w + \frac{7}{2} = \pm\frac{9}{2}$
$w + \frac{7}{2} = \frac{9}{2}$ $\quad w + \frac{7}{2} = -\frac{9}{2}$
$w = 1$ $\quad w = -8$
The solutions are -8 and 1.

17. $v^2 + 4v + 1 = 0$
$v^2 + 4v = -1$
Complete the square.
$v^2 + 4v + 4 = -1 + 4$
$(v + 2)^2 = 3$
$\sqrt{(v + 2)^2} = \sqrt{3}$
$v + 2 = \pm\sqrt{3}$
$v + 2 = \sqrt{3}$ $\quad v + 2 = -\sqrt{3}$
$v = -2 + \sqrt{3}$ $\quad v = -2 - \sqrt{3}$
The solutions are $-2 - \sqrt{3}$ and $-2 + \sqrt{3}$.

19. $x^2 + 6x = 5$
Complete the square.
$x^2 + 6x + 9 = 5 + 9$
$(x + 3)^2 = 14$
$\sqrt{(x + 3)^2} = \sqrt{14}$
$x + 3 = \pm\sqrt{14}$
$x + 3 = \sqrt{14}$ $\quad x + 3 = -\sqrt{14}$
$x = -3 + \sqrt{14}$ $\quad x = -3 - \sqrt{14}$
The solutions are $-3 - \sqrt{14}$ and $-3 + \sqrt{14}$.

21. $2\left(\frac{z^2}{2}\right)^2 = 2\left(z + \frac{1}{2}\right)$
$z^2 = 2z + 1$
$z^2 - 2z = 1$
Complete the square.
$z^2 - 2z + 1 = 1 + 1$
$(z - 1)^2 = 2$
$\sqrt{(z - 1)^2} = \sqrt{2}$
$z - 1 = \pm\sqrt{2}$
$z - 1 = \sqrt{2}$ $\quad z - 1 = -\sqrt{2}$
$z = 1 + \sqrt{2}$ $\quad z = 1 - \sqrt{2}$
The solutions are $1 - \sqrt{2}$ and $1 + \sqrt{2}$.

23. $p^2 + 3p = 1$
Complete the square.
$p^2 + 3p + \frac{9}{4} = 1 + \frac{9}{4}$
$\left(p + \frac{3}{2}\right)^2 = \frac{13}{4}$
$\sqrt{\left(p + \frac{3}{2}\right)^2} = \sqrt{\frac{13}{4}}$
$p + \frac{3}{2} = \pm\frac{\sqrt{13}}{2}$
$p + \frac{3}{2} = \frac{\sqrt{13}}{2}$ $\quad p + \frac{3}{2} = -\frac{\sqrt{13}}{2}$
$p = -\frac{3}{2} + \frac{\sqrt{13}}{2}$ $\quad p = -\frac{3}{2} - \frac{\sqrt{13}}{2}$
$p = \frac{-3 + \sqrt{13}}{2}$ $\quad p = \frac{-3 - \sqrt{13}}{2}$
The solutions are $\frac{-3 - \sqrt{13}}{2}$ and $\frac{-3 + \sqrt{13}}{2}$.

25. $t^2 - 3t = -2$
Complete the square.
$t^2 - 3t + \frac{9}{4} = -2 + \frac{9}{4}$
$\left(t - \frac{3}{2}\right)^2 = \frac{1}{4}$
$\sqrt{\left(t - \frac{3}{2}\right)^2} = \sqrt{\frac{1}{4}}$
$t - \frac{3}{2} = \pm\frac{1}{2}$
$t - \frac{3}{2} = \frac{1}{2}$ $\quad t - \frac{3}{2} = -\frac{1}{2}$
$t = 2$ $\quad t = 1$
The solutions are 1 and 2.

27. $v^2 + v - 3 = 0$

$v^2 + v = 3$

Complete the square.

$v^2 + v + \frac{1}{4} = 3 + \frac{1}{4}$

$\left(v + \frac{1}{2}\right)^2 = \frac{13}{4}$

$\sqrt{\left(v + \frac{1}{2}\right)^2} = \sqrt{\frac{13}{4}}$

$v + \frac{1}{2} = \pm\frac{\sqrt{13}}{2}$

$v + \frac{1}{2} = \frac{\sqrt{13}}{2}$ $\quad$ $v + \frac{1}{2} = -\frac{\sqrt{13}}{2}$

$v = -\frac{1}{2} + \frac{\sqrt{13}}{2}$ $\quad$ $v = -\frac{1}{2} - \frac{\sqrt{13}}{2}$

$v = \frac{-1 + \sqrt{13}}{2}$ $\quad$ $v = \frac{-1 - \sqrt{13}}{2}$

The solutions are $\frac{-1 - \sqrt{13}}{2}$ and $\frac{-1 + \sqrt{13}}{2}$.

29. $y^2 = 7 - 10y$

$y^2 + 10y = 7$

Complete the square.

$y^2 + 10y + 25 = 7 + 25$

$(y + 5)^2 = 32$

$\sqrt{(y + 5)^2} = \sqrt{32}$

$y + 5 = \pm\sqrt{32}$

$y + 5 = \pm 4\sqrt{2}$

$y + 5 = 4\sqrt{2}$ $\quad$ $y + 5 = -4\sqrt{2}$

$y = -5 + 4\sqrt{2}$ $\quad$ $y = -5 - 4\sqrt{2}$

The solutions are $-5 - 4\sqrt{2}$ and $-5 + 4\sqrt{2}$.

31. $r^2 - 3r = 5$

Complete the square.

$r^2 - 3r + \frac{9}{4} = 5 + \frac{9}{4}$

$\left(r - \frac{3}{2}\right)^2 = \frac{29}{4}$

$\sqrt{\left(r - \frac{3}{2}\right)^2} = \sqrt{\frac{29}{4}}$

$r - \frac{3}{2} = \pm\frac{\sqrt{29}}{2}$

$r - \frac{3}{2} = \frac{\sqrt{29}}{2}$ $\quad$ $r - \frac{3}{2} = -\frac{\sqrt{29}}{2}$

$r = \frac{3}{2} + \frac{\sqrt{29}}{2}$ $\quad$ $r = \frac{3}{2} - \frac{\sqrt{29}}{2}$

$r = \frac{3 + \sqrt{29}}{2}$ $\quad$ $r = \frac{3 - \sqrt{29}}{2}$

The solutions are $\frac{3 - \sqrt{29}}{2}$ and $\frac{3 + \sqrt{29}}{2}$.

33. $t^2 - t = 4$

Complete the square.

$t^2 - t + \frac{1}{4} = 4 + \frac{1}{4}$

$\left(t - \frac{1}{2}\right)^2 = \frac{17}{4}$

$\sqrt{\left(t - \frac{1}{2}\right)^2} = \sqrt{\frac{17}{4}}$

$t - \frac{1}{2} = \pm\frac{\sqrt{17}}{2}$

$t - \frac{1}{2} = \frac{\sqrt{17}}{2}$ $\quad$ $t - \frac{1}{2} = -\frac{\sqrt{17}}{2}$

$t = \frac{1}{2} + \frac{\sqrt{17}}{2}$ $\quad$ $t = \frac{1}{2} - \frac{\sqrt{17}}{2}$

$t = \frac{1 + \sqrt{17}}{2}$ $\quad$ $t = \frac{1 - \sqrt{17}}{2}$

The solutions are $\frac{1 - \sqrt{17}}{2}$ and $\frac{1 + \sqrt{17}}{2}$.

35. $x^2 - 3x + 5 = 0$

$x^2 - 3x = 0$

Complete the square.

$x^2 - 3x + \frac{9}{4} = -5 + \frac{9}{4}$

$\left(x - \frac{3}{2}\right)^2 = -\frac{11}{4}$

$\sqrt{\left(x - \frac{3}{2}\right)^2} = \sqrt{-\frac{11}{4}}$

$\sqrt{-11/4}$ is not a real number.
The quadratic equation has
no real number solution.

37. $2t^2 - 3t + 1 = 0$

$2t^2 - 3t = -1$

$\frac{1}{2}(2t^2 - 3t) = \frac{1}{2}(-1)$

$t^2 - \frac{3}{2}t = -\frac{1}{2}$

Complete the square.

$t^2 - \frac{3}{2}t + \frac{9}{16} = -\frac{1}{2} + \frac{9}{16}$

$\left(t - \frac{3}{4}\right)^2 = \frac{1}{16}$

$\sqrt{\left(t - \frac{3}{4}\right)^2} = \sqrt{\frac{1}{16}}$

$t - \frac{3}{4} = \pm\frac{1}{4}$

$t - \frac{3}{4} = \frac{1}{4}$ $\quad$ $t - \frac{3}{4} = -\frac{1}{4}$

$t = 1$ $\quad$ $t = \frac{1}{2}$

The solutions are $\frac{1}{2}$ and 1.

39.
$$2r^2 + 5r = 3$$
$$\frac{1}{2}(2r^2 + 5r) = \frac{1}{2}(3)$$
$$r^2 + \frac{5}{2}r = \frac{3}{2}$$
Complete the square.
$$r^2 + \frac{5}{2}r + \frac{25}{16} = \frac{3}{2} + \frac{25}{16}$$
$$\left(r + \frac{5}{4}\right)^2 = \frac{49}{16}$$
$$\sqrt{\left(r + \frac{5}{4}\right)^2} = \sqrt{\frac{49}{16}}$$
$$r + \frac{5}{4} = \pm\frac{7}{4}$$
$$r + \frac{5}{4} = \frac{7}{4} \qquad r + \frac{5}{4} = -\frac{7}{4}$$
$$r = \frac{1}{2} \qquad r = -3$$
The solutions are -3 and $\frac{1}{2}$.

41.
$$2s^2 = 7s - 6$$
$$2s^2 - 7s = -6$$
$$\frac{1}{2}(2s^2 - 7s) = \frac{1}{2}(-6)$$
$$s^2 - \frac{7}{2}s = -3$$
Complete the square.
$$s^2 - \frac{7}{2}s + \frac{49}{16} = -3 + \frac{49}{16}$$
$$\left(s - \frac{7}{4}\right)^2 = \frac{1}{16}$$
$$\sqrt{\left(s - \frac{7}{4}\right)^2} = \sqrt{\frac{1}{16}}$$
$$s - \frac{7}{4} = \pm\frac{1}{4}$$
$$s - \frac{7}{4} = \frac{1}{4} \qquad s - \frac{7}{4} = -\frac{1}{4}$$
$$s = 2 \qquad s = \frac{3}{2}$$
The solutions are $\frac{3}{2}$ and 2.

43.
$$2v^2 = v + 1$$
$$2v^2 - v = 1$$
$$\frac{1}{2}(2v^2 - v) = \frac{1}{2}(1)$$
$$v^2 - \frac{1}{2}v = \frac{1}{2}$$
Complete the square.
$$v^2 - \frac{1}{2}v + \frac{1}{16} = \frac{1}{2} + \frac{1}{16}$$
$$\left(v - \frac{1}{4}\right)^2 = \frac{9}{16}$$
$$\sqrt{\left(v - \frac{1}{4}\right)^2} = \sqrt{\frac{9}{16}}$$
$$v - \frac{1}{4} = \pm\frac{3}{4}$$
$$v - \frac{1}{4} = \frac{3}{4} \qquad v - \frac{1}{4} = -\frac{3}{4}$$
$$v = 1 \qquad v = -\frac{1}{2}$$
The solutions are $-\frac{1}{2}$ and 1.

45.
$$3r^2 + 5r = 2$$
$$\frac{1}{3}(3r + 5r) = \frac{1}{3}(2)$$
$$r^2 + \frac{5}{3}r = \frac{2}{3}$$
Complete the square.
$$r^2 + \frac{5}{3}r + \frac{25}{36} = \frac{2}{3} + \frac{25}{36}$$
$$\left(r + \frac{5}{6}\right)^2 = \frac{49}{36}$$
$$\sqrt{\left(r + \frac{5}{6}\right)^2} = \sqrt{\frac{49}{36}}$$
$$r + \frac{5}{6} = \pm\frac{7}{6}$$
$$r + \frac{5}{6} = \frac{7}{6} \qquad r + \frac{5}{6} = -\frac{7}{6}$$
$$r = \frac{1}{3} \qquad r = -2$$
The solutions are -2 and $\frac{1}{3}$.

47. $3y^2 + 8y + 4 = 0$

$$3y^2 + 8y = -4$$
$$\frac{1}{3}(3y^2 + 8y) = \frac{1}{3}(-4)$$
$$y^2 + \frac{8}{3}y = -\frac{4}{3}$$

Complete the square.

$$y^2 + \frac{8}{3}y + \frac{16}{9} = -\frac{4}{3} + \frac{16}{9}$$
$$\left(y + \frac{4}{3}\right)^2 = \frac{4}{9}$$
$$\sqrt{\left(y + \frac{4}{3}\right)^2} = \sqrt{\frac{4}{9}}$$
$$y + \frac{4}{3} = \pm\frac{2}{3}$$
$$y + \frac{4}{3} = \frac{2}{3} \qquad y + \frac{4}{3} = -\frac{2}{3}$$
$$y = -\frac{2}{3} \qquad y = -2$$

The solutions are -2 and $-\frac{2}{3}$.

49. $4x^2 + 4x - 3 = 0$

$$4x^2 + 4x = 3$$
$$\frac{1}{4}(4x^2 + 4x) = \frac{1}{4}(3)$$
$$x^2 + x = \frac{3}{4}$$

Complete the square.

$$x^2 + x + \frac{1}{4} = \frac{3}{4} + \frac{1}{4}$$
$$\left(x + \frac{1}{2}\right)^2 = 1$$
$$\sqrt{\left(x + \frac{1}{2}\right)^2} = \sqrt{1}$$
$$x + \frac{1}{2} = \pm 1$$
$$x + \frac{1}{2} = -1 \qquad x + \frac{1}{2} = 1$$
$$x = -\frac{3}{2} \qquad x = \frac{1}{2}$$

The solutions are $-\frac{3}{2}$ and $\frac{1}{2}$.

51. $6s^2 + 7s = 3$

$$\frac{1}{6}(6s^2 + 7s) = \frac{1}{6}(3)$$
$$s^2 + \frac{7}{6}s = \frac{1}{2}$$

Complete the square.

$$s^2 + \frac{7}{6}s + \frac{49}{144} = \frac{1}{2} + \frac{49}{144}$$
$$\left(s + \frac{7}{12}\right)^2 = \frac{121}{144}$$
$$\sqrt{\left(s + \frac{7}{12}\right)^2} = \sqrt{\frac{121}{144}}$$
$$s + \frac{7}{12} = \pm\frac{11}{12}$$
$$s + \frac{7}{12} = \frac{11}{12} \qquad s + \frac{7}{12} = -\frac{11}{12}$$
$$s = \frac{1}{3} \qquad s = -\frac{3}{2}$$

The solutions are $-\frac{3}{2}$ and $\frac{1}{3}$.

53. $6p^2 = 5p + 4$

$$6p^2 - 5p = 4$$
$$\frac{1}{6}(6p^2 - 5p) = \frac{1}{6}(4)$$
$$p^2 - \frac{5}{6}p = \frac{2}{3}$$

Complete the square.

$$p^2 - \frac{5}{6}p + \frac{25}{144} = \frac{2}{3} + \frac{25}{144}$$
$$\left(p - \frac{5}{12}\right)^2 = \frac{121}{144}$$
$$\sqrt{\left(p - \frac{5}{12}\right)^2} = \sqrt{\frac{121}{144}}$$
$$p - \frac{5}{12} = \pm\frac{11}{12}$$
$$p - \frac{5}{12} = \frac{11}{12} \qquad p - \frac{5}{12} = -\frac{11}{12}$$
$$p = \frac{4}{3} \qquad p = -\frac{1}{2}$$

The solutions are $-\frac{1}{2}$ and $\frac{4}{3}$.

55. $w^2 + 5w = 2$

Complete the square.

$$w^2 + 5w + \frac{25}{4} = 2 + \frac{25}{4}$$
$$\left(w + \frac{5}{2}\right)^2 = \frac{33}{4}$$
$$\sqrt{\left(w + \frac{5}{2}\right)^2} = \sqrt{\frac{33}{4}}$$
$$w + \frac{5}{2} = \pm\frac{\sqrt{33}}{2}$$

$$w + \frac{5}{2} = \frac{\sqrt{33}}{2} \qquad w + \frac{5}{2} = -\frac{\sqrt{33}}{2}$$
$$w = -\frac{5}{2} + \frac{\sqrt{33}}{2} \qquad w = -\frac{5}{2} - \frac{\sqrt{33}}{2}$$
$$w = \frac{-5 + \sqrt{33}}{2} \qquad w = \frac{-5 - \sqrt{33}}{2}$$
$$w = 0.372 \qquad w = -5.372$$

The solutions are approximately -5.372 and 0.372.

57. $2x^2 + 3x = 11$

$$\frac{1}{2}(2x^2 + 3x) = \frac{1}{2}(11)$$
$$x^2 + \frac{3}{2}x = \frac{11}{2}$$

Complete the square.

$$x^2 + \frac{3}{2}x + \frac{9}{16} = \frac{11}{2} + \frac{9}{16}$$
$$\left(x + \frac{3}{4}\right)^2 = \frac{97}{16}$$
$$\sqrt{\left(x + \frac{3}{4}\right)^2} = \sqrt{\frac{97}{16}}$$
$$x + \frac{3}{4} = \pm\frac{\sqrt{97}}{4}$$

$$x + \frac{3}{4} = \frac{\sqrt{97}}{4} \qquad x + \frac{3}{4} = -\frac{\sqrt{97}}{4}$$
$$x = -\frac{3}{4} + \frac{\sqrt{97}}{4} \qquad x = -\frac{3}{4} - \frac{\sqrt{97}}{4}$$
$$x = \frac{-3 + \sqrt{97}}{4} \qquad x = \frac{-3 - \sqrt{97}}{4}$$
$$x = 1.712 \qquad x = -3.212$$

The solutions are approximately -3.212 and 1.712.

59. $4x^2 + 2x = 3$

$$\frac{1}{4}(4x^2 + 2x) = \frac{1}{4}(3)$$
$$x^2 + \frac{1}{2}x = \frac{3}{4}$$

Complete the square.

$$x^2 + \frac{1}{2}x + \frac{1}{16} = \frac{3}{4} + \frac{1}{16}$$
$$\left(x + \frac{1}{4}\right)^2 = \frac{13}{16}$$
$$\sqrt{\left(x + \frac{1}{4}\right)^2} = \sqrt{\frac{13}{16}}$$
$$x + \frac{1}{4} = \pm\frac{\sqrt{13}}{4}$$

$$x + \frac{1}{4} = \frac{\sqrt{13}}{4} \qquad x + \frac{1}{4} = -\frac{\sqrt{13}}{4}$$
$$x = -\frac{1}{4} + \frac{\sqrt{13}}{4} \qquad x = -\frac{1}{4} - \frac{\sqrt{13}}{4}$$
$$x = \frac{-1 + \sqrt{13}}{4} \qquad x = \frac{-1 - \sqrt{13}}{4}$$
$$x = 0.651 \qquad x = -1.151$$

The solutions are approximately -1.151 and 0.651.

61. $6 \cdot \left(\frac{x^2}{6} - \frac{x}{3}\right) = 6 \cdot (1)$

$$x^2 - 2x = 6$$

Complete the square.

$$x^2 - 2x + 1 = 6 + 1$$
$$(x - 1)^2 = 7$$
$$\sqrt{(x - 1)^2} = \sqrt{7}$$
$$x - 1 = \pm\sqrt{7}$$
$$x - 1 = \sqrt{7} \qquad x - 1 = -\sqrt{7}$$
$$x = 1 + \sqrt{7} \qquad x = 1 - \sqrt{7}$$

The solutions are $1 - \sqrt{7}$ and $1 + \sqrt{7}$.

63. $\sqrt{3x + 4} - x = 2$

$$\sqrt{3x + 4} = x + 2$$
$$\sqrt{(3x + 4)^2} = (x + 2)^2$$
$$3x + 4 = x^2 + 4x + 4$$
$$x^2 + 4x + 4 = 3x + 4$$
$$x^2 + x = 0$$
$$x(x + 1) = 0$$

Check: $\sqrt{3(0) + 4} - 0 = 2$

$$\sqrt{4} = 2$$
$$2 = 2$$
$$\sqrt{3(-1) + 4} - (-1) = 2$$
$$\sqrt{-3 + 4} + 1 = 2$$
$$\sqrt{1} + 1 = 2$$
$$1 + 1 = 2$$
$$2 = 2$$

$$x = 0 \qquad x + 1 = 0$$
$$x = -1$$

The solutions are -1 and 0.

65. $2(x-1)\left[\dfrac{x+1}{2}+\dfrac{3}{x-1}\right]=2(x-1)\cdot 4$

$x^2-1+6=8x-8$

$x^2-8x=-13$

Complete the square.

$x^2-8x+16=-13+16$

$(x-4)^2=3$

$\sqrt{(x-4)^2}=\sqrt{3}$

$x-4=\pm\sqrt{3}$

$x-4=\sqrt{3}$ $\qquad x-4=-\sqrt{3}$

$x=4+\sqrt{3}$ $\qquad x=4-\sqrt{3}$

The solutions are $4-\sqrt{3}$ and $4+\sqrt{3}$.

67. $4\sqrt{x+1}-x=4$

$4\sqrt{x+1}=x+4$

$(4\sqrt{x+1})^2=(x+4)^2$

$16(x+1)=x^2+8x+16$

$16x+16=x^2+8x+16$

$0=x^2-8x$

$0=x(x-8)$

Check: $4\sqrt{0+1}-0=4$

$4\sqrt{1}=4$

$4=4$

$4\sqrt{8+1}-8=4$

$4\sqrt{9}-8=4$

$4\cdot 3-8=4$

$12-8=4$

$4=4$

$x=0$ $\qquad x-8=0$

$x=8$

The solutions are 0 and 8.

69. $3\sqrt{x-1}+3=x$

$3\sqrt{x-1}=x-3$

$(3\sqrt{x+1})^2=(x-3)^2$

$9(x-1)=x^2-6x+9$

$9x-9=x^2-6x+9$

$x^2-6x+9=9x-9$

$x^2-15x=-18$

Complete the square.

$x^2-15x+\dfrac{225}{4}=-18+\dfrac{225}{4}$

$\left(x-\dfrac{15}{2}\right)^2=\dfrac{153}{4}$

$\sqrt{\left(x-\dfrac{15}{2}\right)^2}=\sqrt{\dfrac{153}{4}}$

$x-\dfrac{15}{2}=\pm\sqrt{\dfrac{153}{4}}$

$x-\dfrac{15}{2}=\sqrt{\dfrac{153}{4}}$ $\qquad x-\dfrac{15}{2}=-\sqrt{\dfrac{153}{4}}$

$x=\dfrac{15}{2}+\dfrac{3\sqrt{17}}{2}$ $\qquad x=\dfrac{15}{2}-\dfrac{3\sqrt{17}}{2}$

$x=\dfrac{15+3\sqrt{17}}{2}$ $\qquad x=\dfrac{15-3\sqrt{17}}{2}$

Check: $\dfrac{15}{2}+\dfrac{\sqrt{153}}{2}\approx 13.68$

$3\sqrt{13.68-1}+3=13.68$

$10.68+3=13.68$

$13.68=13.68$

$\dfrac{15}{2}-\dfrac{\sqrt{153}}{2}\approx 1.32$

$3\sqrt{1.32-1}+3\neq 1.32$

$1.68+3\neq 1.32$

$4.68\neq 1.32$

The solution is $\dfrac{15+3\sqrt{17}}{2}$.

71. $h=-16t^2+76t+5$

$0=-16t^2+76t+5$

$16t^2-76t=5$

$t^2-\dfrac{19}{4}t=\dfrac{5}{16}$

$t^2-\dfrac{19}{4}t+\dfrac{361}{64}=\dfrac{361}{64}+\dfrac{5}{16}$

$\left(t-\dfrac{19}{8}\right)^2=\dfrac{381}{64}$

$t-\dfrac{19}{8}=\sqrt{\dfrac{381}{64}}=\pm\dfrac{\sqrt{381}}{8}$

$t=\dfrac{19}{8}+\dfrac{\sqrt{381}}{8}$ $\qquad t=\dfrac{19}{8}-\dfrac{\sqrt{381}}{8}$

$t\approx 4.81$ $\qquad t=-0.065$

The ball hits the ground approximately 4.81 s after it is hit.

Section 11.3

Objective A Exercises

1. $x^2 - 4x - 5 = 0$
$a = 1, b = -4, c = -5$

$$x = \frac{-(-4) \pm \sqrt{(-4)^2 - 4(1)(-5)}}{2 \cdot 1} = \frac{4 \pm \sqrt{16 + 20}}{2} = \frac{4 \pm \sqrt{36}}{2} = \frac{4 \pm 6}{2}$$

$$x = \frac{4+6}{2} = \frac{10}{2} = 5 \qquad x = \frac{4-6}{2} = \frac{-2}{2} = -1$$

The solutions are -1 and 5.

3. $z^2 - 2z - 15 = 0$
$a = 1, b = -2, c = -15$

$$z = \frac{-(-2) \pm \sqrt{(-2)^2 - 4(1)(-15)}}{2 \cdot 1} = \frac{2 \pm \sqrt{4 + 60}}{2} = \frac{2 \pm \sqrt{64}}{2} = \frac{2 \pm 8}{2}$$

$$z = \frac{2+8}{2} = \frac{10}{2} = 5 \qquad z = \frac{2-8}{2} = \frac{-6}{2} = -3$$

The solutions are -3 and 5.

5. $y^2 = 2y + 3$
$y^2 - 2y - 3 = 0$
$a = 1, b = -2, c = -3$

$$y = \frac{-(-2) \pm \sqrt{(-2)^2 - 4(1)(-3)}}{2 \cdot 1} = \frac{2\sqrt{4 + 12}}{2} = \frac{2 \pm \sqrt{16}}{2} = \frac{2 \pm 4}{2}$$

$$y = \frac{2+4}{2} = \frac{6}{2} = 3 \qquad y = \frac{2-4}{2} = \frac{-2}{2} = -1$$

The solutions are -1 and 3.

7. $r^2 = 5 - 4r$
$r^2 + 4r - 5 = 0$
$a = 1, b = 4, c = -5$

$$r = \frac{-4 \pm \sqrt{(4)^2 - 4(1)(-5)}}{2 \cdot 1} = \frac{-4 \pm \sqrt{16 + 20}}{2} = \frac{-4 \pm \sqrt{36}}{2} = \frac{-4 \pm 6}{2}$$

$$r = \frac{-4+6}{2} = \frac{2}{2} = 1 \qquad r = \frac{-4-6}{2} = \frac{-10}{2} = -5$$

The solutions are -5 and 1.

9. $2y^2 - y - 1 = 0$
$a = 2, b = -1, c = -1$

$$y = \frac{-(-1) \pm \sqrt{(-1)^2 - 4(2)(-1)}}{2 \cdot 2} = \frac{1 \pm \sqrt{1 + 8}}{4} = \frac{1 \pm \sqrt{9}}{4} = \frac{1 \pm 3}{4}$$

$$y = \frac{1+3}{4} = \frac{4}{4} = 1 \qquad y = \frac{1-3}{4} = \frac{-2}{4} = -\frac{1}{2}$$

The solutions are $-\frac{1}{2}$ and 1.

11. $w^2 + 3w + 5 = 0$
$a = 1, b = 3, c = 5$

$$w = \frac{-3 \pm \sqrt{(3)^2 - 4(1)(5)}}{2 \cdot 1} = \frac{-3 \pm \sqrt{9 - 20}}{2} = \frac{-3 \pm \sqrt{-11}}{2}$$

$\sqrt{-11}$ is not a real number.
The quadratic equation has no real number solution.

13. $p^2 - p = 0$
$a = 1, b = -1, c = 0$

$$p = \frac{-(-1) \pm \sqrt{(-1)^2 - 4(1)(0)}}{2 \cdot 1} = \frac{1 \pm \sqrt{1}}{2} = \frac{1 \pm 1}{2}$$

$$p = \frac{1+1}{2} = \frac{2}{2} = 1 \qquad p = \frac{1-1}{2} = \frac{0}{2} = 0$$

The solutions are 0 and 1.

15. $4t^2 - 9 = 0$

$a = 4,\ b = 0,\ c = -9$

$$t = \frac{-0 \pm \sqrt{(0)^2 - 4(4)(-9)}}{2 \cdot 4} = \frac{\pm\sqrt{144}}{8} = \frac{\pm 12}{8}$$

$$t = \frac{12}{8} = \frac{3}{2} \quad t = \frac{-12}{8} = -\frac{3}{2}$$

The solutions are $-\frac{3}{2}$ and $\frac{3}{2}$.

17. $4y^2 + 4y = 15$

$4y^2 + 4y - 15 = 0$

$a = 4,\ b = 4,\ c = -15$

$$y = \frac{-4 \pm \sqrt{(4)^2 - 4(4)(-15)}}{2 \cdot 4} = \frac{-4 \pm \sqrt{16 + 240}}{8} = \frac{-4 \pm \sqrt{256}}{8} = \frac{-4 \pm 16}{8}$$

$$y = \frac{-4 + 16}{8} = \frac{12}{8} = \frac{3}{2} \quad y = \frac{-4 - 16}{8} = \frac{-20}{8} = -\frac{5}{2}$$

The solutions are $-\frac{5}{2}$ and $\frac{3}{2}$.

19. $2x^2 + x + 1 = 0$

$a = 2,\ b = 1,\ c = 1$

$$x = \frac{-1 \pm \sqrt{(1)^2 - 4(2)(1)}}{2 \cdot 2} = \frac{-1 \pm \sqrt{1 - 8}}{4} = \frac{-1 \pm \sqrt{-7}}{4}$$

$\sqrt{-7}$ is not a real number.

The quadratic equation has no real number solution.

21. $2\left(\frac{1}{2}t^2 - t\right) = 2\left(\frac{5}{2}\right)$

$t^2 - 2t = 5$

$t^2 - 2t - 5 = 0$

$a = 1,\ b = -2,\ c = -5$

$$t = \frac{-(-2) \pm \sqrt{(-2)^2 - 4(1)(-5)}}{2 \cdot 1} = \frac{2 \pm \sqrt{4 + 20}}{2} = \frac{2 \pm \sqrt{24}}{2} = \frac{2 \pm 2\sqrt{6}}{2} = 1 \pm \sqrt{6}$$

The solutions are $1 - \sqrt{6}$ and $1 + \sqrt{6}$.

23. $3\left(\frac{1}{3}t^2 + 2t - \frac{1}{3}\right) = 3(0)$

$t^2 + 6t - 1 = 0$

$a = 1,\ b = 6,\ c = -1$

$$t = \frac{-6 \pm \sqrt{(6)^2 - 4(1)(-1)}}{2 \cdot 1} = \frac{-6 \pm \sqrt{36 + 4}}{2} = \frac{-6 \pm \sqrt{40}}{2} = \frac{-6 \pm 2\sqrt{10}}{2} = -3 \pm \sqrt{10}$$

The solutions are $-3 - \sqrt{10}$ and $-3 + \sqrt{10}$.

25. $w^2 = 4w + 9$

$w^2 - 4w - 9 = 0$

$a = 1,\ b = -4,\ c = -9$

$$w = \frac{-(-4) \pm \sqrt{(-4)^2 - 4(1)(-9)}}{2 \cdot 1} = \frac{4 \pm \sqrt{16 + 36}}{2} = \frac{4 \pm \sqrt{52}}{2} = \frac{4 \pm 2\sqrt{13}}{2} = 2 \pm \sqrt{13}$$

The solutions are $2 - \sqrt{13}$ and $2 + \sqrt{13}$.

27. $9y^2 + 6y - 1 = 0$

$a = 9,\ b = 6,\ c = -1$

$$y = \frac{-6 \pm \sqrt{(6)^2 - 4(9)(-1)}}{2 \cdot 9} = \frac{-6 \pm \sqrt{36 + 36}}{18} = \frac{-6 \pm \sqrt{72}}{18} = \frac{-6 \pm 6\sqrt{2}}{18} = \frac{-1 \pm \sqrt{2}}{3}$$

The solutions are $\frac{-1 - \sqrt{2}}{3}$ and $\frac{-1 + \sqrt{2}}{3}$.

29. $4p^2 + 4p + 1 = 0$

$a = 4,\ b = 4,\ c = 1$

$$p = \frac{-4 \pm \sqrt{(4)^2 - 4(4)(1)}}{2 \cdot 4} = \frac{-4 \pm \sqrt{16 - 16}}{8} = \frac{-4 \pm \sqrt{0}}{8} = \frac{-4}{8} = -\frac{1}{2}$$

$-\frac{1}{2}$ is a double root of the equation.

The solution is $-\frac{1}{2}$.

31.
$$4\left(\frac{x^2}{2}\right) = 4\left(x - \frac{5}{4}\right)$$
$$2x^2 = 4x - 5$$
$$2x^2 - 4x + 5 = 0$$
$a = 2,\ b = -4,\ c = 5$
$$x = \frac{-(-4) \pm \sqrt{(-4)^2 - 4(2)(5)}}{2 \cdot 2}$$
$$= \frac{4 \pm \sqrt{16 - 40}}{4} = \frac{4 \pm \sqrt{-24}}{4}$$
$\sqrt{-24}$ is not a real number. The quadratic equation has no real number solution.

33.
$$4p^2 + 16p = -11$$
$$4p^2 + 16p + 11 = 0$$
$a = 4,\ b = 16,\ c = 11$
$$p = \frac{-16 \pm \sqrt{(16)^2 - 4(4)(11)}}{2 \cdot 4}$$
$$= \frac{-16 \pm \sqrt{256 - 176}}{8} = \frac{-16 \pm \sqrt{80}}{8}$$
$$= \frac{-16 \pm 4\sqrt{5}}{8} = \frac{-4 \pm \sqrt{5}}{2}$$
The solutions are $\frac{-4 - \sqrt{5}}{2}$ and $\frac{-4 + \sqrt{5}}{2}$.

35.
$$4x^2 = 4x + 11$$
$$4x^2 - 4x - 11 = 0$$
$a = 4,\ b = -4,\ c = -11$
$$x = \frac{-(-4) \pm \sqrt{(-4)^2 - 4(4)(-11)}}{2 \cdot 4}$$
$$= \frac{4 \pm \sqrt{16 + 176}}{8} = \frac{4 \pm \sqrt{192}}{8}$$
$$= \frac{4 \pm 8\sqrt{3}}{8} = \frac{1 \pm 2\sqrt{3}}{2}$$
The solutions are $\frac{1 - 2\sqrt{3}}{2}$ and $\frac{1 + 2\sqrt{3}}{2}$.

37.
$$9v^2 = -30v - 23$$
$$9v^2 + 30v + 23 = 0$$
$a = 9,\ b = 30,\ c = 23$
$$v = \frac{-30 \pm \sqrt{(30)^2 - 4(9)(23)}}{2 \cdot 9}$$
$$= \frac{-30 \pm \sqrt{900 - 828}}{18} = \frac{-30 \pm \sqrt{72}}{18}$$
$$= \frac{-30 \pm 6\sqrt{2}}{18} = \frac{-5 \pm \sqrt{2}}{3}$$
The solutions are $\frac{-5 - \sqrt{2}}{3}$ and $\frac{-5 + \sqrt{2}}{3}$.

39. $x^2 - 2x - 21 = 0$
$a = 1,\ b = -2,\ c = -21$
$$x = \frac{-(-2) \pm \sqrt{(-2)^2 - 4(1)(-21)}}{2 \cdot 1}$$
$$= \frac{2 \pm \sqrt{4 + 84}}{2} = \frac{2 \pm \sqrt{88}}{2}$$
$$= \frac{2 \pm 2\sqrt{22}}{2} = 1 \pm \sqrt{22}$$
$x = 1 + \sqrt{22} \approx 5.690$ $x = 1 - \sqrt{22} \approx -3.690$

The solutions are approximately -3.690 and 5.690.

41. $s^2 - 6s - 13 = 0$
$a = 1,\ b = -6,\ c = -13$
$$s = \frac{-(-6) \pm \sqrt{(-6)^2 - 4(1)(-13)}}{2 \cdot 1}$$
$$= \frac{6 \pm \sqrt{36 + 52}}{2} = \frac{6 \pm \sqrt{88}}{2}$$
$$= \frac{6 \pm 2\sqrt{22}}{2} = 3 \pm \sqrt{22}$$
$s = 3 + \sqrt{22} \approx 7.690$ $s = 3 - \sqrt{22} \approx -1.690$

The solutions are approximately -1.690 and 7.690.

43. $2p^2 - 7p - 10 = 0$
$a = 2,\ b = -7,\ c = -10$
$$p = \frac{-(-7) \pm \sqrt{(-7)^2 - 4(2)(-10)}}{2 \cdot 2}$$
$$= \frac{7 \pm \sqrt{49 + 80}}{4} = \frac{7 \pm \sqrt{129}}{4}$$
$p = \frac{7 + \sqrt{129}}{4} \approx 4.589$ $p = \frac{7 - \sqrt{129}}{4} \approx -1.089$

The solutions are approximately -1.089 and 4.589.

45. $4z^2 + 8z - 1 = 0$
$a = 4,\ b = 8,\ c = -1$
$$z = \frac{-8 \pm \sqrt{(8)^2 - 4(4)(-1)}}{2 \cdot 4}$$
$$= \frac{-8 \pm \sqrt{64 + 16}}{8} = \frac{-8 \pm \sqrt{80}}{8}$$
$$= \frac{-8 \pm 4\sqrt{5}}{8} = \frac{-2 \pm \sqrt{5}}{2}$$
$z = \frac{-2 + \sqrt{5}}{2} \approx \frac{-2 + 2.236}{2} = 0.118$ $z = \frac{-2 - \sqrt{5}}{2} \approx \frac{-2 - 2.236}{2} = -2.118$

The solutions are approximately -2.118 and 0.118.

47. $5v^2 - v - 5 = 0$

$a = 5, b = -1, c = -5$

$$v = \frac{-(-1) \pm \sqrt{(-1)^2 - 4(5)(-5)}}{2 \cdot 5}$$

$$= \frac{1 \pm \sqrt{1 + 100}}{10} = \frac{1 \pm \sqrt{101}}{10}$$

$$v = \frac{1 + \sqrt{101}}{10} \qquad v = \frac{1 - \sqrt{101}}{10}$$

$$\approx \frac{1 + 10.050}{10} \qquad \approx \frac{1 - 10.050}{10}$$

$$= 1.105 \qquad = -0.905$$

The solutions are approximately -0.905 and 1.105.

Applying the Concepts

49. $0x^2 + 3x + 4 = 0$ cannot be solved by the quadratic formula because it is not a quadratic equation. A quadratic equation's form is $ax^2 + bx + c = 0$, where $a \neq 0$. In $0x^2 + 3x + 4 = 0$, $a = 0$.
Students may note that if $a = 0$, the denominator of the quadratic formula is 0. Because division by zero is undefined, the quadratic formula cannot be used when $a = 0$.

51. **a.** False

b. False

c. False

d. True

53.
$$\sqrt{x+4} = x + 4$$
$$(\sqrt{x+4})^2 = (x+4)^2$$
$$x + 4 = x^2 + 8x + 16$$
$$0 = x^2 + 7x + 12$$
$$0 = (x+3)(x+4)$$

Check: $\sqrt{-3+4} = -3 + 4$
$$\sqrt{1} = 1$$
$$1 = 1$$
$$\sqrt{-4+4} = -4 + 4$$
$$\sqrt{0} = 0$$
$$0 = 0$$

$$x + 3 = 0 \qquad x + 4 = 0$$
$$x = -3 \qquad x = -4$$

The solutions are -4 and -3.

55.
$$\sqrt{x^2 + 2x + 1} = x - 1$$
$$(\sqrt{x^2 + 2x + 1})^2 = (x-1)^2$$
$$x^2 + 2x + 1 = x^2 - 2x + 1$$
$$4x = 0$$
$$x = 0$$

Check: $\sqrt{0^2 + 2(0) + 1} = 0 - 1$
$$\sqrt{1} = -1$$
$$1 \neq -1$$

The equation has no solution.

57. $5(x-1)\left[\frac{x+1}{5} - \frac{4}{x-1}\right] = 5(x-1) \cdot 2$
$$x^2 - 1 - 20 = 10x - 10$$
$$x^2 - 10x - 11 = 0$$
$$(x - 11)(x + 1) = 0$$
$$x - 11 = 0 \qquad x + 1 = 0$$
$$x = 11 \qquad x = -1$$

The solutions are -1 and 11.

59. $d^2 = 800^2 + 600^2$
$$d^2 = 1000000$$
$$d = \pm 1000$$

The planes are 1000 mi apart after 2 h.

Section 11.4

Objective A Exercises

1. Because the coefficient of x^2 is negative, the graph opens down.

3. Because the coefficient of x^2 is positive, the graph opens up.

5. $f(x) = x^2 - 2x + 1$
$f(3) = 3^2 - 2(3) + 1 = 9 - 6 + 1 = 4$

7. $f(x) = 4 - x^2$
$f(-3) = 4 - (-3)^2 = 4 - 9 = -5$

9. $f(x) = -x^2 + 5x - 6$
$f(-4) = -(-4)^2 + 5(-4) - 6$
$= -16 - 20 - 6 = -42$

11.

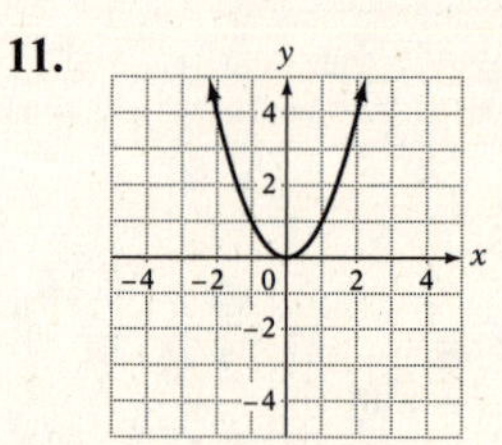

13.

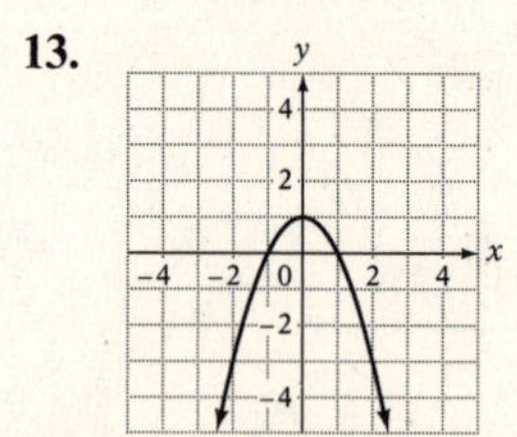

15.

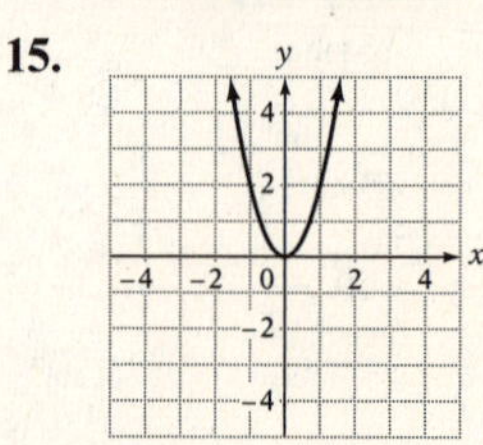

17.

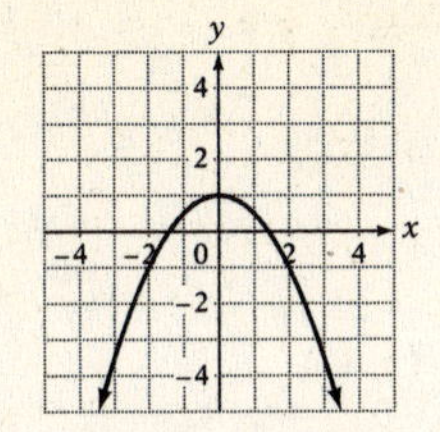

19.

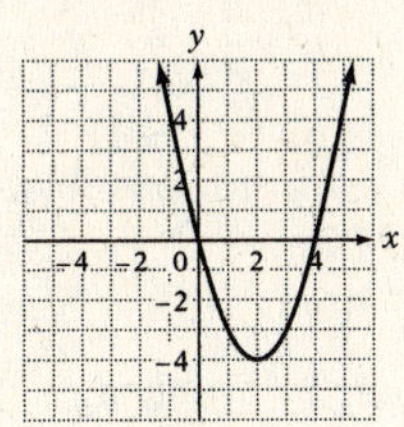

21.

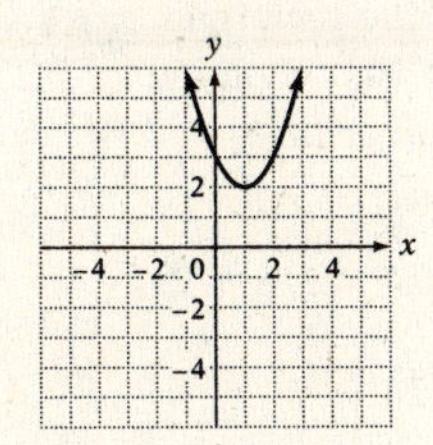

23.

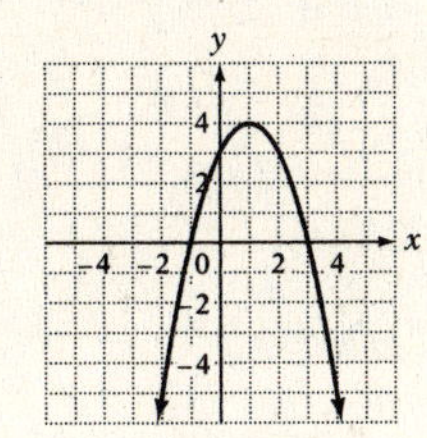

25.

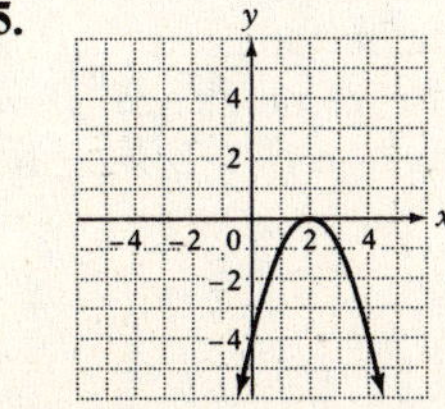

Applying the Concepts

27. x-intercepts: $y = 0$

$$x^2 + 5x - 6 = 0$$
$$(x+6)(x-1) = 0$$
$$x + 6 = 0 \qquad x - 1 = 0$$
$$x = -6 \qquad x = 1$$

$(-6, 0)$ and $(1, 0)$

y-intercept: $x = 0$

$y = 0^2 + 5(0) - 6 = -6$

$(0, -6)$

29. x-intercepts: $y = 0$

$$x^2 + 12x + 36 = 0$$
$$(x+6)^2 = 0$$
$$x + 6 = 0$$
$$x = -6$$

$(-6, 0)$

y-intercept: $x = 0$

$f(0) = 0^2 + 12(0) + 36 = 36$

$(0, 36)$

31. x-intercepts: $f(x) = 0$

$$x^2 + 4x - 2 = 0$$
$$x = \frac{-4 \pm \sqrt{4^2 - 4(1)(-2)}}{2(1)}$$
$$= \frac{-4 \pm \sqrt{16+8}}{2} = \frac{-4 \pm \sqrt{24}}{2}$$
$$= \frac{-4 \pm 2\sqrt{6}}{2} = -2 \pm \sqrt{6}$$

$(-2 - \sqrt{6}, 0)$ and $(-2 + \sqrt{6}, 0)$

y-intercept: $x = 0$

$f(0) = 0^2 + 4(0) - 2 = -2$

$(0, -2)$

33. x-intercepts: $y = 0$

$$x^2 - x + 1 = 0$$
$$x = \frac{1 \pm \sqrt{(-1)^2 - 4(1)(1)}}{2(1)}$$
$$= \frac{1 \pm \sqrt{1-4}}{2} = \frac{1 \pm \sqrt{-3}}{2}$$

no real solutions

no x-intercepts

y-intercept: $x = 0$

$y = 0^2 + (0) + 1 = 1$

$(0, 1)$

35. x-intercepts: $f(x) = 0$

$$2x^2 - 13x + 15 = 0$$
$$(2x-3)(x-5) = 0$$
$$2x - 3 = 0 \qquad x - 5 = 0$$
$$2x = 3 \qquad x = 5$$
$$x = \frac{3}{2}$$

$\left(\frac{3}{2}, 0\right)$ and $(5, 0)$

y-intercept: $x = 0$

$f(0) = 2(0)^2 - 13(0) + 15 = 15$

$(0, 15)$

37. x-intercepts: $y = 0$

$2 - 3x - 3x^2 = 0$

$$x = \frac{3 \pm \sqrt{(-3)^2 - 4(-3)(2)}}{2(-3)}$$

$$= \frac{3 \pm \sqrt{9+24}}{-6} = \frac{3 \pm \sqrt{33}}{-6}$$

$$\left(\frac{-3-\sqrt{33}}{6}, 0\right) \text{ and } \left(\frac{-3+\sqrt{33}}{6}, 0\right)$$

y-intercept: $x = 0$

$y = 2 - 3(0) - 3(0)^2 = 2$

$(0, 2)$

39.
$$y - 2 = 3(x+1)^2$$
$$y - 2 = 3(x^2 + 2x + 1)$$
$$y - 2 = 3x^2 + 6x + 3$$
$$y = 3x^2 + 6x + 5$$

41.
$$y + 3 = 3(x-1)^2$$
$$y + 3 = 3(x^2 - 2x + 1)$$
$$y + 3 = 3x^2 - 6x + 3$$
$$y = 3x^2 - 6x$$

43.
$$x^3 - 4x^2 - 5x = 0$$
$$x(x^2 - 4x - 5) = 0$$
$$x(x-5)(x+1) = 0$$

$x = 0 \quad x - 5 = 0 \quad x + 1 = 0$

$x = 5 \quad x = -1$

The x-intercepts are $(-1, 0)$, $(0, 0)$, and $(5, 0)$.

45.
$$x^3 + 3x^2 - x - 3 = 0$$
$$(x^3 + 3x^2) + (-x - 3) = 0$$
$$x^2(x+3) - (x+3) = 0$$
$$(x+3)(x^2 - 1) = 0$$
$$(x+3)(x+1)(x-1) = 0$$

$x + 3 = 0 \quad x + 1 = 0 \quad x - 1 = 0$

$x = -3 \quad x = -1 \quad x = 1$

The x-intercepts are $(-3, 0)$, $(-1, 0)$, and $(1, 0)$.

Section 11.5

Objective A Exercises

1. Strategy This is a geometry problem.

- Length of the triangle base: x
  Height of triangle: $2x + 2$
- Use the equation: $A = \frac{1}{2} bh$.

Solution
$$A = \frac{1}{2} bh$$
$$20 = \frac{1}{2}x(2x+2)$$
$$20 = x^2 + x$$
$$0 = x^2 + x - 20$$
$$0 = (x+5)(x-4)$$

$x + 5 = 0 \quad x - 4 = 0$

$x = -5 \quad x = 4$

$2x + 2 = 10$

The solution -5 is not possible.
The length is 4 m.
The height is 10 m.

3. Strategy This is a geometry problem.

- Width of the rectangle: x
  Length of the rectangle: $x + 2$
- Use the equation $A = LW$.

Solution
$$A = LW$$
$$24 = x(x+2)$$
$$24 = x^2 + 2x$$
$$x^2 + 2x - 24 = 0$$
$$(x+6)(x-4) = 0$$
$$x - 4 = 0$$
$$x = 4$$

$$x + 2 = 6$$

The width is 4 ft.
The length is 6 ft.

5. Strategy This is a geometry problem.

- Width of the rectangle: x
  Length of the rectangle: $2x$
- Use the equation $A = LW$.

Solution
$$A = LW$$
$$5000 = x(2x)$$
$$5000 = 2x^2$$
$$2500 = x^2$$
$$50 = x$$

$$100 = 2x$$

The width is 50 ft.
The length is 100 ft.

7.
$$s = -16t^2 + 88t + 1$$
$$0 = -16t^2 + 88t + 1$$
$$16t^2 - 88t - 1 = 0$$
$$t = \frac{-(88) \pm \sqrt{(-88)^2 - 4(16)(-1)}}{2(16)}$$
$$t = \frac{88 \pm \sqrt{7808}}{32}$$

$t \approx 5.5 \quad t \approx -0.01$

The hang time of the football is approximately 5.5 s.
The solution -0.01 s is not possible.

9.

$$s = 0.0344v^2 - 0.758v$$
$$150 = 0.0344v^2 - 0.758v$$
$$0 = 0.0344v^2 - 0.758v - 150$$
$$v = \frac{-(-0.758) \pm \sqrt{(-0.758)^2 - 4(0.0344)(-150)}}{2(0.0344)}$$
$$v = \frac{0.758 \pm \sqrt{21.214564}}{0.0688}$$

$v \approx 78 \quad v \approx -56$

The maximum velocity is 78 ft/s.

The solution -56 is not possible.

11. Strategy This is a geometry problem.

- Radius of small pizza: x
  Radius of large pizza: $2x - 1$
- The difference between the areas of the two pizzas is 33π in^2.

Solution

$$\pi(2x - 1)^2 - \pi x^2 = 33\pi$$
$$\pi(2x - 1)^2 - \pi x^2 = \pi[33]$$
$$(2x - 1)^2 - x^2 = 33$$
$$4x^2 - 4x + 1 - x^2 = 33$$
$$3x^2 - 4x - 32 = 0$$
$$x = \frac{4 \pm \sqrt{(4)^2 - 4(3)(-32)}}{2(3)}$$
$$x = \frac{4 \pm \sqrt{400}}{6}$$

$x = 4 \quad x \approx -2.67$

The solution -2.67 is not reasonable.

$2x - 1 = 2(4) - 1 = 7$

The large pizza has a radius of 7 in.

13. Strategy This is a work problem.

- Time for second computer to solve the equation: t
  Time for first computer to solve the equation: $t + 21$

	Rate	Time	Part
Second Computer	$\frac{1}{t}$	10	$\frac{10}{t}$
First computer	$\frac{1}{t+21}$	10	$\frac{10}{t+21}$

- The sum of the parts of the task completed by each computer must equal 1.

Solution

$$\frac{10}{t} + \frac{10}{t + 21} = 1$$
$$t(t + 21)\left(\frac{10}{t} + \frac{10}{t + 21}\right) = t(t + 21)(1)$$
$$10(t + 21) + 10t = t(t + 21)$$
$$10t + 210 + 10t = t^2 + 21t$$
$$t^2 + t - 210 = 0$$
$$(t + 15)(t - 14) = 0$$

$t + 15 = 0 \qquad t - 14 = 0$

$t = -15 \qquad t = 14$

$t + 21 = 35$

The solution -15 is not possible.

The first computer can solve the equation in 35 min.

The second computer can solve the equation in 14 min.

15. Strategy This is a work problem.

- Time for using the second engine: t
  Time for using the first engine: $t + 6$

	Rate	Time	Part
Second engine	$\frac{1}{t}$	4	$\frac{4}{t}$
First engine	$\frac{1}{t+6}$	4	$\frac{4}{t+6}$

- The sum of the parts of the task completed by each engine must equal 1.

Solution

$$\frac{4}{t} + \frac{4}{t + 6} = 1$$
$$t(t + 6)\left(\frac{4}{t} + \frac{4}{t + 6}\right) = t(t + 6)(1)$$
$$4(t + 6) + 4t = t(t + 6)$$
$$4t + 24 + 4t = t^2 + 6t$$
$$t^2 - 2t - 24 = 0$$
$$(t - 6)(t + 4) = 0$$

$t - 6 = 0 \qquad t + 4 = 0$

$t = 6 \qquad t = -4$

$t + 6 = 12$

The solution -4 is not possible.

Using the first engine it would take 12 h.

Using the second engine it would take 6 h.

17. Strategy This is a distance-rate problem.

- Rate of the plane in calm air: r

	Distance	Rate	Part
With wind	375	$r+25$	$\frac{375}{r+25}$
Against wind	375	$r-25$	$\frac{375}{r-25}$

- The time spent traveling against the wind is 2 h more than the time spent traveling with the wind.

Solution

$$\frac{375}{r-25} = 2 + \frac{375}{r+25}$$

$$(r-25)(r+25)\left(\frac{375}{r-25}\right) = (r-25)(r+25)\left(2+\frac{375}{r+25}\right)$$

$$375(r+25) = 2(r-25)(r+25) + 375(r-25)$$

$$375r + 9375 = 2(r^2 - 625) + 375r - 9375$$

$$375r + 9375 = 2r^2 - 1250 + 375r - 9375$$

$$2r^2 - 20{,}000 = 0$$

$$2(r^2 - 10{,}000) = 0$$

$$(r+100)(r-100) = 0$$

$r + 100 = 0 \qquad r - 100 = 0$

$r = -100 \qquad r = 100$

The solution -100 is not possible.

The rate of the plane in calm air is 100 mph.

Applying the Concepts

19. Strategy

- Find the cost per square inch using the information given for the 8-inch pizza.
- Multiply the cost per square inch by the area of a 16-inch pizza.

Solution Area of 8-inch pizza:

$A = \pi r^2$

$= \pi(4)^2 = 16\pi$ sq. in.

Cost per square inch of pizza:

$C = \frac{\$6}{16\pi \text{ sq. in.}} = \$\frac{3}{8\pi}/\text{sq. in.}$

Area of 16-inch pizza:

$A = \pi r^2$

$= \pi(8)^2 = 64\pi$ sq. in.

Cost for 16-inch pizza:

$C = 64\pi \cdot \frac{3}{8\pi} = 8 \cdot 3 = 24$

The 16-inch pizza should cost \$24.

Chapter 11 Review Exercises

1. $6x^2 + 13x - 28 = 0$

$(2x+7)(3x-4) = 0$

$2x + 7 = 0 \qquad 3x - 4 = 0$

$2x = -7 \qquad 3x = 4$

$x = -\frac{7}{2} \qquad x = \frac{4}{3}$

The solutions are $-\frac{7}{2}$ and $\frac{4}{3}$.

2. $49x^2 = 25$

$7x = \pm 5$

$x = \frac{\pm 5}{7}$

$x = \frac{5}{7} \quad x = -\frac{5}{7}$

The solutions are $-\frac{5}{7}$ and $\frac{5}{7}$.

3. $x^2 + 2x - 24 = 0$

$x^2 + 2x = 24$

$x^2 + 2x + 1 = 24 + 1$

$(x+1)^2 = 25$

$x + 1 = \pm 5$

$x = -1 \pm 5$

$x = -1 + 5 = 4$

$x = -1 - 5 = -6$

The solutions are -6 and 4.

4. $x^2 + 5x - 6 = 0$

$x = \frac{-5 \pm \sqrt{25+24}}{2}$

$x = \frac{-5 \pm \sqrt{49}}{2}$

$x = \frac{-5 \pm 7}{2}$

$x = \frac{-5+7}{2} = \frac{2}{2} = 1$

$x = \frac{-5-7}{2} = -\frac{12}{2} = -6$

The solutions are -6 and 1.

5.
$$2x^2 + 5x = 12$$
$$x^2 + \frac{5}{2}x = 6$$
$$x^2 + \frac{5}{2}x + \frac{25}{16} = 6 + \frac{25}{16}$$
$$\left(x + \frac{5}{4}\right)^2 = \frac{96}{16} + \frac{25}{16}$$
$$\left(x + \frac{5}{4}\right)^2 = \frac{121}{16}$$
$$x + \frac{5}{4} = \pm\frac{11}{4}$$
$$x = -\frac{5}{4} \pm \frac{11}{4}$$
$$x = -\frac{5}{4} + \frac{11}{4} \qquad x = -\frac{5}{4} - \frac{11}{4}$$
$$x = \frac{6}{4} = \frac{3}{2} \qquad x = -\frac{16}{4} = -4$$
The solutions are -4 and $\frac{3}{2}$.

6.
$$12x^2 + 10 = 29x$$
$$12x^2 - 29x + 10 = 0$$
$$(12x - 5)(x - 2) = 0$$
$$12x - 5 = 0 \qquad x - 2 = 0$$
$$12x = 5 \qquad x = 2$$
$$x = \frac{5}{12}$$
The solutions are $\frac{5}{12}$ and 2.

7.
$$(x + 2)^2 - 24 = 0$$
$$(x + 2)^2 = 24$$
$$x + 2 = \pm\sqrt{24}$$
$$x = -2 \pm \sqrt{24}$$
$$x = -2 \pm 2\sqrt{6}$$
$$x = -2 + 2\sqrt{6} \qquad x = -2 - 2\sqrt{6}$$
The solutions are $-2 - 2\sqrt{6}$ and $-2 + 2\sqrt{6}$.

8.
$$2x^2 + 3 = 5x$$
$$2x^2 - 5x + 3 = 0$$
$$x = \frac{5 \pm \sqrt{25 - 24}}{4}$$
$$x = \frac{5 \pm \sqrt{1}}{4}$$
$$x = \frac{5 \pm 1}{4}$$
$$x = \frac{5 + 1}{4} = \frac{6}{4} = \frac{3}{2}$$
$$x = \frac{5 - 1}{4} = \frac{4}{4} = 1$$
The solutions are 1 and $\frac{3}{2}$.

9.
$$6x(x + 1) = x - 1$$
$$6x^2 + 6x = x - 1$$
$$6x^2 + 5x + 1 = 0$$
$$(3x + 1)(2x + 1) = 0$$
$$3x + 1 = 0 \qquad 2x + 1 = 0$$
$$3x = -1 \qquad 2x = -1$$
$$x = -\frac{1}{3} \qquad x = -\frac{1}{2}$$
The solutions are $-\frac{1}{2}$ and $-\frac{1}{3}$.

10. $4y^2 + 9 = 0$
The equation has no real number solution.

11.
$$x^2 - 4x + 1 = 0$$
$$x^2 - 4x + 4 = -1 + 4$$
$$(x - 2)^2 = 3$$
$$x - 2 = \pm\sqrt{3}$$
$$x = 2 \pm \sqrt{3}$$
$$x = 2 + \sqrt{3} \qquad x = 2 - \sqrt{3}$$
The solutions are $2 - \sqrt{3}$ and $2 + \sqrt{3}$.

12.
$$x^2 - 3x - 5 = 0$$
$$x = \frac{3 \pm \sqrt{9 + 20}}{2}$$
$$x = \frac{3 \pm \sqrt{29}}{2}$$
$$x = \frac{3 + \sqrt{29}}{2}$$
$$x = \frac{3 - \sqrt{29}}{2}$$
The solutions are $\frac{3 - \sqrt{29}}{2}$ and $\frac{3 + \sqrt{29}}{2}$.

13.
$$x^2 + 6x + 12 = 0$$
$$x^2 + 6x = -12$$
$$x^2 + 6x + 9 = -12 + 9$$
$$(x + 3)^2 = -3$$
The equation has no real number solution.

14.
$$(x + 9)^2 = x + 11$$
$$x^2 + 18x + 81 = x + 11$$
$$x^2 + 17x + 70 = 0$$
$$(x + 7)(x + 10) = 0$$
$$x + 7 = 0 \qquad x + 10 = 0$$
$$x = -7 \qquad x = -10$$
The solutions are -10 and -7.

15.
$$\left(x - \frac{1}{2}\right)^2 = \frac{9}{4}$$
$$x - \frac{1}{2} = \pm\frac{3}{2}$$
$$x = \frac{1}{2} \pm \frac{3}{2}$$
$$x = \frac{1}{2} + \frac{3}{2} = \frac{4}{2} = 2$$
$$x = \frac{1}{2} - \frac{3}{2} = -\frac{2}{2} = -1$$
The solutions are -1 and 2.

16. $4x^2 + 16x = 7$

$$x^2 + 4x = \frac{7}{4}$$

$$x^2 + 4x + 4 = \frac{7}{4} + 4$$

$$x^2 + 4x + 4 = \frac{7}{4} + \frac{16}{4}$$

$$x^2 + 4x + 4 = \frac{23}{4}$$

$$(x+2)^2 = \frac{23}{4}$$

$$x + 2 = \pm\frac{\sqrt{23}}{2}$$

$$x = -2 \pm \frac{\sqrt{23}}{2}$$

$$x = \frac{-4 \pm \sqrt{23}}{2}$$

$$x = \frac{-4 + \sqrt{23}}{2} \quad x = \frac{-4 - \sqrt{23}}{2}$$

The solutions are $\frac{-4 - \sqrt{23}}{2}$ and $\frac{-4 + \sqrt{23}}{2}$.

17. $x^2 - 4x + 8 = 0$

$$x = \frac{4 \pm \sqrt{16 - 32}}{2}$$

$$x = \frac{4 \pm \sqrt{-16}}{2}$$

The equation has no real number solution.

18. $2x^2 + 5x + 2 = 0$

$$x = \frac{-5 \pm \sqrt{25 - 16}}{4}$$

$$x = \frac{-5 \pm \sqrt{9}}{4}$$

$$x = \frac{-5 \pm 3}{4}$$

$$x = \frac{-5 + 3}{4} = -\frac{2}{4} = -\frac{1}{2}$$

$$x = \frac{-5 - 3}{4} = -\frac{8}{4} = -2$$

The solutions are -2 and $-\frac{1}{2}$.

19.

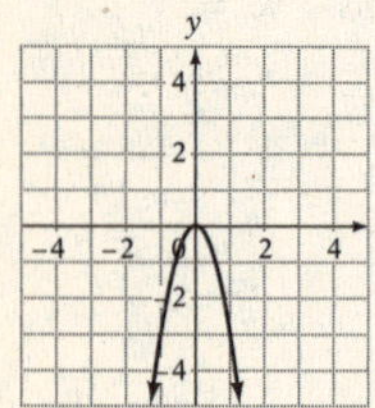

20.

21.

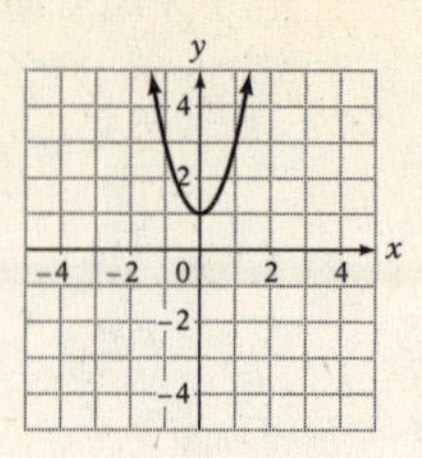

22.

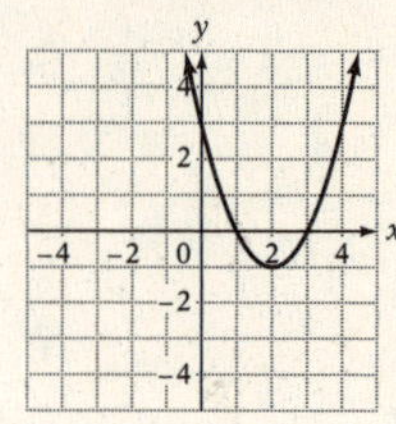

23.

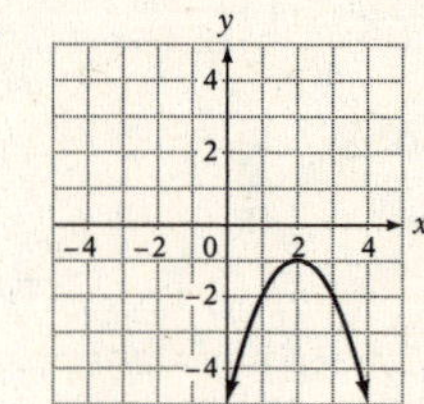

24. x-intercepts: $y = 0$

$$x^2 - 2x - 15 = 0$$

$$(x+3)(x-5) = 0$$

$$x + 3 = 0 \qquad x - 5 = 0$$

$$x = -3 \qquad x = 5$$

$(-3, 0)$ and $(5, 0)$

y-intercept: $x = 0$

$y = 0^2 - 2(0) - 15 = -15$

$(0, -15)$

25. Strategy Rate of the hawk in calm air: r

	Distance	Rate	Part
Against the wind	70	$r-5$	$\frac{70}{r-5}$
With wind	40	$r+5$	$\frac{40}{r+5}$

The time traveling against the wind is half an hour more than the time traveling with the wind.

Solution

$$\frac{70}{r-5} = \frac{1}{2} + \frac{40}{r+5}$$
$$2(r-5)(r+5)\left(\frac{70}{r-5}\right) = 2(r-5)(r+5)\left(\frac{1}{2} + \frac{40}{r+5}\right)$$
$$70(2)(r+5) = (r-5)(r+5) + 40(2)(r-5)$$
$$140(r+5) = r^2 - 25 + 80(r-5)$$
$$140r + 700 = r^2 - 25 + 80r - 400$$
$$r^2 - 60r - 1125 = 0$$
$$(r-75)(r+15) = 0$$
$$(r-75) = 0 \qquad r+15 = 0$$
$$r = 75 \qquad r = -15$$

The solution -15 is not possible.
The rate of the hawk in calm air is 75 mph.

Chapter 11 Test

1.
$$x^2 - 5x - 6 = 0$$
$$(x-6)(x+1) = 0$$
$$x - 6 = 0 \qquad x + 1 = 0$$
$$x = 6 \qquad x = -1$$
The solutions are -1 and 6.

2.
$$3x^2 + 7x = 20$$
$$3x^2 + 7x - 20 = 0$$
$$(3x-5)(x+4) = 0$$
$$3x - 5 = 0 \qquad x + 4 = 0$$
$$3x = 5 \qquad x = -4$$
$$x = \frac{5}{3}$$
The solutions are -4 and $\frac{5}{3}$.

3.
$$2(x-5)^2 - 50 = 0$$
$$2(x-5)^2 = 50$$
$$(x-5)^2 = 25$$
$$x - 5 = \pm 5$$
$$x = 5 \pm 5$$
$$x = 5 + 5 = 10 \qquad x = 5 - 5 = 0$$
The solutions are 0 and 10.

4.
$$3(x+4)^2 - 60 = 0$$
$$3(x+4)^2 = 60$$
$$(x+4)^2 = 20$$
$$x + 4 = \pm\sqrt{20}$$
$$x = -4 \pm 2\sqrt{5}$$
$$x = -4 + 2\sqrt{5} \qquad x = -4 - 2\sqrt{5}$$
The solutions are $-4 - 2\sqrt{5}$ and $-4 + 2\sqrt{5}$.

5.
$$x^2 + 4x - 16 = 0$$
$$x^2 + 4x + 4 = 16 + 4$$
$$(x+2)^2 = 20$$
$$x + 2 = \pm\sqrt{20}$$
$$x = -2 \pm 2\sqrt{5}$$
$$x = -2 + 2\sqrt{5} \qquad x = -2 - 2\sqrt{5}$$
The solutions are $-2 - 2\sqrt{5}$ and $-2 + 2\sqrt{5}$.

6.
$$x^2 + 3x = 8$$
$$x^2 + 3x + \frac{9}{4} = 8 + \frac{9}{4}$$
$$\left(x + \frac{3}{2}\right)^2 = \frac{32}{4} + \frac{9}{4}$$
$$\left(x + \frac{3}{2}\right)^2 = \frac{41}{4}$$
$$x + \frac{3}{2} = \pm\frac{\sqrt{41}}{2}$$
$$x = -\frac{3}{2} \pm \frac{\sqrt{41}}{2}$$
$$x = \frac{-3 + \sqrt{41}}{2} \qquad x = \frac{-3 - \sqrt{41}}{2}$$
The solutions are $\frac{-3 - \sqrt{41}}{2}$ and $\frac{-3 + \sqrt{41}}{2}$.

7. $2x^2 - 6x + 1 = 0$

$$2x^2 - 6x = -1$$
$$x^2 - 3x = -\frac{1}{2}$$
$$x^2 - 3x + \frac{9}{4} = -\frac{1}{2} + \frac{9}{4}$$
$$\left(x - \frac{3}{2}\right)^2 = \frac{7}{4}$$
$$x - \frac{3}{2} = \pm\frac{\sqrt{7}}{2}$$
$$x = \frac{3}{2} \pm \frac{\sqrt{7}}{2}$$
$$x = \frac{3 + \sqrt{7}}{2} \quad x = \frac{3 - \sqrt{7}}{2}$$

The solutions are $\frac{3 - \sqrt{7}}{2}$ and $\frac{3 + \sqrt{7}}{2}$.

8. $2x^2 + 8x = 3$

$$x^2 + 4x = \frac{3}{2}$$
$$x^2 + 4x + 4 = \frac{3}{2} + 4$$
$$(x + 2)^2 = \frac{3}{2} + \frac{8}{2}$$
$$(x + 2)^2 = \frac{11}{2}$$
$$(x + 2)^2 = \frac{22}{4}$$
$$x + 2 = \pm\frac{\sqrt{22}}{2}$$
$$x = -2 \pm \frac{\sqrt{22}}{2}$$
$$x = \frac{-4 \pm \sqrt{22}}{2}$$
$$x = \frac{-4 + \sqrt{22}}{2} \quad x = \frac{-4 - \sqrt{22}}{2}$$

The solutions are $\frac{-4 - \sqrt{22}}{2}$ and $\frac{-4 + \sqrt{22}}{2}$.

9. $x^2 + 4x + 2 = 0$

$$x = \frac{-4 \pm \sqrt{16 - 8}}{2}$$
$$x = \frac{-4 \pm \sqrt{8}}{2}$$
$$x = \frac{-4 \pm 2\sqrt{2}}{2}$$
$$x = -2 \pm \sqrt{2}$$
$$x = -2 + \sqrt{2} \quad x = -2 - \sqrt{2}$$

The solutions are $-2 - \sqrt{2}$ and $-2 + \sqrt{2}$.

10. $x^2 - 3x = 6$

$$x^2 - 3x - 6 = 0$$
$$x = \frac{3 \pm \sqrt{9 + 24}}{2}$$
$$x = \frac{3 \pm \sqrt{33}}{2}$$
$$x = \frac{3 + \sqrt{33}}{2} \quad x = \frac{3 - \sqrt{33}}{2}$$

The solutions are $\frac{3 - \sqrt{33}}{2}$ and $\frac{3 + \sqrt{33}}{2}$.

11. $2x^2 - 5x - 3 = 0$

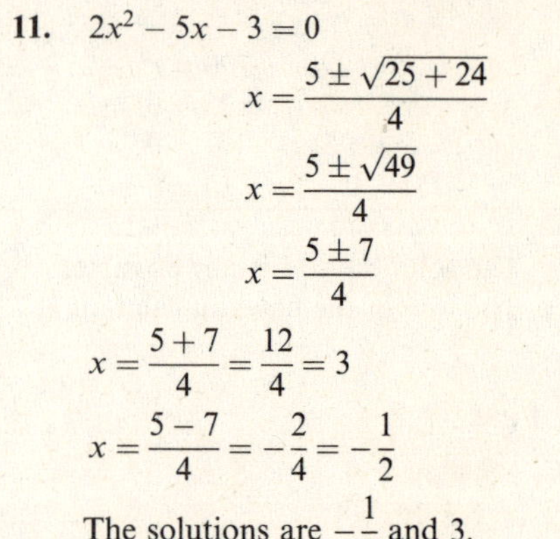

$$x = \frac{5 \pm \sqrt{25 + 24}}{4}$$
$$x = \frac{5 \pm \sqrt{49}}{4}$$
$$x = \frac{5 \pm 7}{4}$$
$$x = \frac{5 + 7}{4} = \frac{12}{4} = 3$$
$$x = \frac{5 - 7}{4} = -\frac{2}{4} = -\frac{1}{2}$$

The solutions are $-\frac{1}{2}$ and 3.

12. $3x^2 - x = 1$

$$3x^2 - x - 1 = 0$$
$$x = \frac{1 \pm \sqrt{1 + 12}}{6}$$
$$x = \frac{1 \pm \sqrt{13}}{6}$$
$$x = \frac{1 + \sqrt{13}}{6} \quad x = \frac{1 - \sqrt{13}}{6}$$

The solutions are $\frac{1 - \sqrt{13}}{6}$ and $\frac{1 + \sqrt{13}}{6}$.

13.

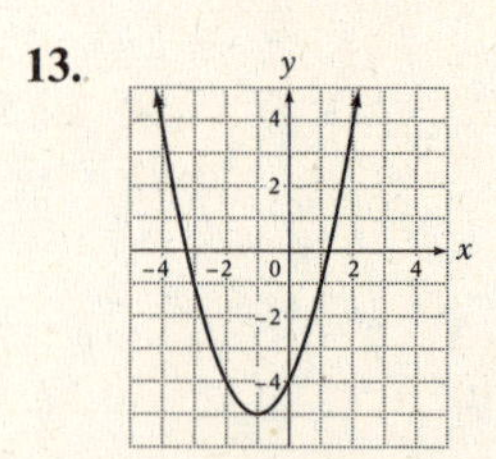

14. x-intercepts: $f(x) = 0$

$$x^2 + x - 12 = 0$$
$$(x + 4)(x - 3) = 0$$
$$x + 4 = 0 \qquad x - 3 = 0$$
$$x = -4 \qquad x = 3$$

$(-4, 0)$ and $(3, 0)$

y-intercept: $x = 0$

$$f(0) = 0^2 + 0 - 12 = -12$$

$(0, -12)$

15. Strategy • Width: x
Length: $2x - 2$
• Use the equation for the area of a rectangle.

Solution

$$\begin{aligned} LW &= A \\ x(2x-2) &= 40 \\ 2x^2 - 2x &= 40 \\ 2x^2 - 2x - 40 &= 0 \\ 2(x^2 - x - 20) &= 0 \\ (x-5)(x+4) &= 0 \end{aligned}$$

$x - 5 = 0 \qquad x + 4 = 0$

$x = 5 \qquad x = -4$

$2x - 2 = 8$

The solution -4 is not possible.
The width is 5 ft.
The length is 8 ft.

16. Strategy • Rate of the boat in calm water: r

	Distance	Rate	Part
Against current	60	$r-1$	$\frac{60}{r-1}$
With current	60	$r+1$	$\frac{60}{r+1}$

• The time spent traveling against the current is 1 h more than the time spent traveling with the current.

Solution

$$\frac{60}{r-1} = 1 + \frac{60}{r+1}$$

$$(r-1)(r+1)\left(\frac{60}{r-1}\right) = (r-1)(r+1)\left(1 + \frac{60}{r+1}\right)$$

$$\begin{aligned} 60(r+1) &= (r-1)(r+1) + 60(r-1) \\ 60r + 60 &= r^2 - 1 + 60r - 60 \\ r^2 - 121 &= 0 \\ (r+11)(r-11) &= 0 \end{aligned}$$

$r + 11 = 0 \qquad r - 11 = 0$

$r = -11 \qquad r = 11$

The solution -11 is not possible.
The rate of the boat in calm water is 11 mph.

Cumulative Review Exercises

1.
$$\begin{aligned} &2x - 3[2x - 4(3 - 2x) + 2] - 3 \\ &= 2x - 3[2x - 12 + 8x + 2] - 3 \\ &= 2x - 3[10x - 10] - 3 \\ &= 2x - 30x + 30 - 3 \\ &= -28x + 27 \end{aligned}$$

2.
$$-\frac{3}{5}x = -\frac{9}{10}$$

$$-\frac{5}{3} \cdot -\frac{3}{5}x = -\frac{\cancel{9}^{3}}{\cancel{10}_{2}} \cdot -\frac{\cancel{5}}{\cancel{3}}$$

$$x = \frac{3}{2}$$

3.
$$\begin{aligned} 2x - 3(4x - 5) &= -3x - 6 \\ 2x - 12x + 15 &= -3x - 6 \\ -10x + 15 &= -3x - 6 \\ -7x &= -21 \\ x &= 3 \end{aligned}$$

4.
$$\begin{aligned} &(2a^2b)^2(-3a^4b^2) \\ &= (4a^4b^2)(-3a^4b^2) \\ &= -12a^8b^4 \end{aligned}$$

5.
$$\begin{array}{r} x + 2 \\ x - 2\overline{)x^2 + 0x - 8} \\ \underline{x^2 - 2x} \\ 2x - 8 \\ \underline{2x - 4} \\ -4 \end{array}$$

$$(x^2 - 8) \div (x - 2) = x + 2 - \frac{4}{x - 2}$$

6. $3x^3 + 2x^2 - 8x$
$= x(3x^2 + 2x - 8)$
$= x(3x - 4)(x + 2)$

7.
$$\frac{3x^2 - 6x}{4x - 6} \div \frac{2x^2 + x - 6}{6x^3 - 24x}$$
$$= \frac{3x^2 - 6x}{4x - 6} \cdot \frac{6x^3 - 24x}{2x^2 + x - 6}$$
$$= \frac{3x(x-2)}{2(2x-3)} \cdot \frac{6x(x^2 - 4)}{(2x-3)(x+2)}$$
$$= \frac{3x(x-2)}{\cancel{2}(2x-3)} \cdot \frac{\overset{3}{\cancel{6}}x\cancel{(x+2)}(x-2)}{(2x-3)\cancel{(x+2)}}$$
$$= \frac{9x^2(x-2)^2}{(2x-3)^2}$$

8.
$$\frac{x}{2(x-1)} - \frac{1}{(x-1)(x+1)}$$
LCM $= 2(x-1)(x+1)$
$$\frac{x(x+1)}{2(x-1)(x+1)} - \frac{2}{2(x-1)(x+1)}$$
$$= \frac{x^2 + x - 2}{2(x-1)(x+1)}$$
$$= \frac{(x+2)\cancel{(x-1)}}{2\cancel{(x-1)}(x+1)}$$
$$= \frac{x+2}{2(x+1)}$$

9.
$$\frac{1 - \frac{7}{x} + \frac{12}{x^2}}{2 - \frac{1}{x} - \frac{15}{x^2}}$$
$$\frac{x^2\left(1 - \frac{7}{x} + \frac{12}{x^2}\right)}{x^2\left(2 - \frac{1}{x} - \frac{15}{x^2}\right)} = \frac{x^2 - 7x + 12}{2x^2 - x - 15}$$
$$= \frac{(x-4)\cancel{(x-3)}}{(2x+5)\cancel{(x-3)}}$$
$$= \frac{x-4}{2x+5}$$

10. $4x - 3y = 12$

x-intercept	y-intercept
$4x = 12$	$-3y = 12$
$x = 3$	$y = -4$
$(3, 0)$	$(0, -4)$

11. $m = -\frac{4}{3}$, $(x_1, y_1) = (-3, 2)$
$y - y_1 = m(x - x_1)$
$y - 2 = -\frac{4}{3}(x + 3)$
$y - 2 = -\frac{4}{3}x - 4$
The equation of the line is $y = -\frac{4}{3}x - 2$.

12. (1) $3x - y = 5$
(2) $y = 2x - 3$
Substitute in Equation (1).
$3x - (2x - 3) = 5$
$3x - 2x + 3 = 5$
$x + 3 = 5$
$x = 2$
Substitute in Equation (2).
$y = 2x - 3$
$y = 4 - 3$
$y = 1$
The solution is (2, 1).

13. Add the equations.

(1) $3x + 2y = 2$
(2) $5x - 2y = 14$
$8x = 16$
$x = 2$
Replace x in Equation (1).
$3x + 2y = 2$
$6 + 2y = 2$
$2y = -4$
$y = -2$
The solution is (2, −2).

14. $2x - 3(2 - 3x) > 2x - 5$
$2x - 6 + 9x > 2x - 5$
$11x - 6 > 2x - 5$
$9x > 1$
$x > \frac{1}{9}$

15. $(\sqrt{a} - \sqrt{2})(\sqrt{a} + \sqrt{2}) = \sqrt{a^2} - \sqrt{4} = a - 2$

16.
$$\frac{\sqrt{108a^7b^3}}{\sqrt{3a^4b}} = \sqrt{36a^3b^2}$$
$$= \sqrt{36a^2b^2}\sqrt{a}$$
$$= 6ab\sqrt{a}$$

17.
$$\frac{\sqrt{3}}{5 + 2\sqrt{3}} = \frac{\sqrt{3}}{5 + 2\sqrt{3}} \cdot \frac{5 - 2\sqrt{3}}{5 - 2\sqrt{3}}$$
$$= \frac{\sqrt{3}(5 - 2\sqrt{3})}{25 - 12}$$
$$= \frac{5\sqrt{3} - 2\sqrt{9}}{13}$$
$$= \frac{-6 + 5\sqrt{3}}{13}$$

18. $3 = 8 - \sqrt{5x}$
$-5 = \sqrt{5x}$
$5 = \sqrt{5x}$
$(5)^2 = (\sqrt{5x})^2$
$25 = 5x$
$5 = x$

19.
$$6x^2 - 17x = -5$$
$$6x^2 - 17x + 5 = 0$$
$$(3x - 1)(2x - 5) = 0$$
$$3x - 1 = 0 \quad 2x - 5 = 0$$
$$3x = 1 \quad 2x = 5$$
$$x = \frac{1}{3} \quad x = \frac{5}{2}$$

The solutions are $\frac{1}{3}$ and $\frac{5}{2}$.

20.
$$2(x - 5)^2 = 36$$
$$(x - 5)^2 = 18$$
$$x - 5 = \pm\sqrt{18}$$
$$x = 5 \pm 3\sqrt{2}$$
$$x = 5 + 3\sqrt{2} \quad x = 5 - 3\sqrt{2}$$

The solutions are $5 - 3\sqrt{2}$ and $5 + 3\sqrt{2}$.

21.
$$3x^2 + 7x = -3$$
$$x^2 + \frac{7}{3}x = -1$$
$$x^2 + \frac{7}{3}x + \frac{49}{36} = -1 + \frac{49}{36}$$
$$\left(x + \frac{7}{6}\right)^2 = -\frac{36}{36} + \frac{49}{36}$$
$$\left(x + \frac{7}{6}\right)^2 = \frac{13}{36}$$
$$x + \frac{7}{6} = \pm\frac{\sqrt{13}}{6}$$
$$x = -\frac{7}{6} \pm \frac{\sqrt{13}}{6}$$
$$x = \frac{-7 \pm \sqrt{13}}{6}$$
$$x = \frac{-7 + \sqrt{13}}{6} \quad x = \frac{-7 - \sqrt{13}}{6}$$

The solutions are $\frac{-7 - \sqrt{13}}{6}$ and $\frac{-7 + \sqrt{13}}{6}$.

22.
$$2x^2 - 3x - 2 = 0$$
$$x = \frac{3 \pm \sqrt{9 + 16}}{4}$$
$$x = \frac{3 \pm \sqrt{25}}{4}$$
$$x = \frac{3 \pm 5}{4}$$
$$x = \frac{3 + 5}{4} = \frac{8}{4} = 2$$
$$x = \frac{3 - 5}{4} = -\frac{2}{4} = -\frac{1}{2}$$

The solutions are $-\frac{1}{2}$ and 2.

23. Strategy
- Selling price of mixture: x

	Amount	Cost	Value
Cashews	20	3.50	20(3.50)
Peanuts	50	1.75	50(1.75)
Mixture	70	x	$70x$

- The sum of the values before mixing equals the value after mixing.

Solution
$$20(3.50) + 50(1.75) = 70x$$
$$70.00 + 87.50 = 70x$$
$$157.50 = 70x$$
$$2.25 = x$$

The selling price of the mixture is \$2.25 per pound.

24. Strategy
- Write and solve a proportion using x to represent the additional shares.

Solution
$$\frac{100}{215} = \frac{x + 100}{752.50}$$
$$215(x + 100) = 100(752.50)$$
$$215x + 21{,}500 = 75{,}250$$
$$215x = 53{,}750$$
$$x = 250$$

250 additional shares are required.

25. Strategy
- Rate of the plane in still air: p
  Rate of the wind: w

	Rate	Time	Distance
With wind	$p + w$	3	$3(p + w)$
Against wind	$p - w$	4.5	$4.5(p - w)$

- The distance traveled with the wind is equal to the distance traveled against the wind.

Solution
$$3(p + w) = 720$$
$$4.5(p - w) = 720$$
$$\frac{1}{3} \cdot 3(p + w) = 720 \cdot \frac{1}{3}$$
$$\frac{1}{4.5} \cdot 4.5(p - w) = 720 \cdot \frac{1}{4.5}$$
$$p + w = 240$$
$$p - w = 160$$
$$2p = 400$$
$$p = 200$$
$$p + w = 240$$
$$200 + w = 240$$
$$w = 40$$

The rate of the plane in still air is 200 mph.
The rate of the wind is 40 mph.

26. Strategy
- Score on last test: x
- The sum of all the scores must give a total minimum of 400 points.

Solution

$$70 + 91 + 85 + 77 + x \geq 400$$
$$323 + x \geq 400$$
$$x \geq 77$$

The score on the last test must be 77 or better.

27. Strategy
- First odd integer: n
  Second odd integer: $n + 2$
  Third odd integer: $n + 4$
- The sum of the squares of the three consecutive odd integers is 83.

Solution

$$n^2 + (n+2)^2 + (n+4)^2 = 83$$
$$n^2 + n^2 + 4n + 4 + n^2 + 8n + 16 = 83$$
$$3n^2 + 12n + 20 = 83$$
$$3n^2 + 12n - 63 = 0$$
$$3(n^2 + 4n - 21) = 0$$
$$(n+7)(n-3) = 0$$

$n + 7 = 0 \qquad n - 3 = 0$
$n = -7 \qquad n = 3$
$n + 2 = -5 \qquad n + 2 = 5$
$n + 4 = -3 \qquad n + 4 = 7$

The middle integer can be −5 or 5.

28. Strategy
- Constant rate: r
  Reduced rate: $r - 3$

	Distance	Rate	Time
First 7 miles	7	r	$\frac{7}{r}$
Second 8 miles	8	$r - 3$	$\frac{8}{r-3}$

- The total time for the trip was 3 h.

Solution

$$\frac{7}{r} + \frac{8}{r-3} = 3$$
$$r(r-3)\left(\frac{7}{r} + \frac{8}{r-3}\right) = r(r-3)(3)$$
$$7(r-3) + 8r = 3r(r-3)$$
$$7r - 21 + 8r = 3r^2 - 9r$$
$$15r - 21 = 3r^2 - 9r$$
$$3r^2 - 24 + 21 = 0$$
$$3(r^2 - 8r + 7) = 0$$
$$3(r-7)(r-1) = 0$$

$r - 7 = 0 \qquad r - 1 = 0$
$r = 7 \qquad r = 1$
$r - 3 = 4$

The solution 1 is not possible.
The rate for the last 8 mi is 4 mph.

29.

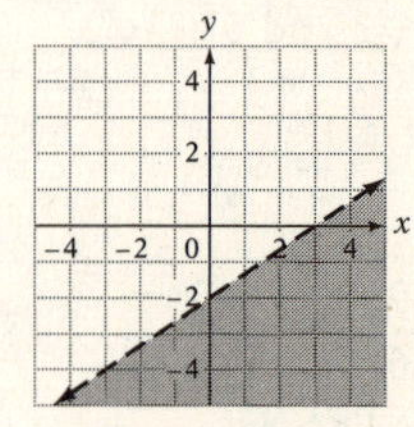

30.

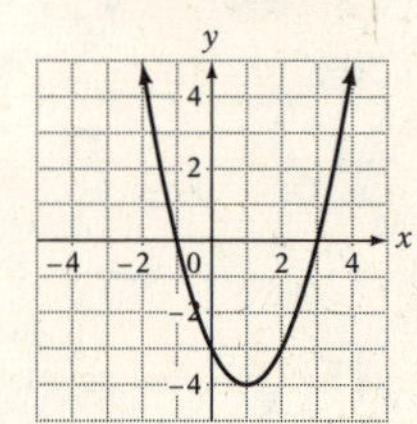

Final Exam

1. $-|-3| = -3$

2. $$\begin{aligned}&-15-(-12)-3\\&=-15+12-3\\&=-3-3\\&=-6\end{aligned}$$

3. $$\begin{aligned}&-2^4(-2)^4\\&=-16(16)\\&=-256\end{aligned}$$

4. $$\begin{aligned}&-7-\frac{12-15}{2-(-1)}\cdot(-4)\\&=-7-\frac{-3}{3}\cdot(-4)\\&=-7-(-1)\cdot(-4)\\&=-7-(4)\\&=-11\end{aligned}$$

5. $$\frac{a^2-3b}{2a-2b^2}=\frac{(3)^2-3(-2)}{2(3)-2(-2)^2}=\frac{9+6}{6-8}=\frac{15}{-2}=-\frac{15}{2}$$

6. $$\begin{aligned}&6x-(-4y)-(-3x)+2y\\&=6x+4y+3x+2y\\&=9x+6y\end{aligned}$$

7. $$(-\cancel{15}^{3}z)\left(-\frac{2}{\cancel{5}}\right)=6z$$

8. $$\begin{aligned}&-2[5-3(2x-7)-2x]\\&=-2[5-6x+21-2x]\\&=-2[-8x+26]\\&=16x-52\end{aligned}$$

9. $$\begin{aligned}20&=-\frac{2}{5}x\\-\frac{5}{2}\cdot\cancel{20}^{10}&=-\frac{2}{5}x\cdot-\frac{5}{2}\\-50&=x\end{aligned}$$

10. $$\begin{aligned}4-2(3x+1)&=3(2-x)+5\\4-6x-2&=6-3x+5\\-6x+2&=11-3x\\-3x&=9\\x&=-3\end{aligned}$$

11. $\frac{1}{8}=0.125=12.5\%$

12. $$\begin{aligned}P\times B&=A\\19\%\times 80&=A\\0.19\times 80&=A\\15.2&=A\end{aligned}$$

13. $$\begin{aligned}&(2x^2-5x+1)-(5x^2-2x-7)\\&\quad 2x^2-5x+1\\&\underline{-5x^2+2x+7}\\&-3x^2-3x+8\end{aligned}$$

14. $(-3xy^3)^4=81x^4y^{12}$

15. $$\begin{aligned}&(3x^2-x-2)(2x+3)\\&=(2x+3)(3x^2-x-2)\\&=2x(3x^2-x-2)+3(3x^2-x-2)\\&=6x^3-2x^2-4x+9x^2-3x-6\\&=6x^3+7x^2-7x-6\end{aligned}$$

16. $$\frac{(-2x^2y^3)^3}{(-4xy^4)^2}=\frac{-8x^6y^9}{16x^2y^8}=-\frac{x^4y}{2}$$

17. $$\begin{aligned}&\frac{12x^2y-16x^3y^2-20y^2}{4xy^2}\\&=\frac{12x^2y}{4xy^2}-\frac{16x^3y^2}{4xy^2}-\frac{20y^2}{4xy^2}\\&=\frac{3x}{y}-4x^2-\frac{5}{x}\end{aligned}$$

18. $$\begin{array}{r}5x-12\\x+2\overline{)5x^2-\ \ 2x-1}\\\underline{5x^2+10x}\\-12x-1\\\underline{-12x-24}\\23\end{array}$$
$$(5x^2-2x-1)\div(x+2)=5x-12+\frac{23}{x+2}$$

19. $$\begin{aligned}&(4x^{-2}y)^2(2xy^{-2})^{-2}\\&=(16x^{-4}y^2)(2^{-2}x^{-2}y^4)\\&=\left(\frac{\cancel{16}^{4}y^2}{x^4}\right)\left(\frac{y^4}{\cancel{4}x^2}\right)=\frac{4y^6}{x^6}\end{aligned}$$

20. $$\begin{aligned}f(t)&=\frac{t}{t+1}\\f(3)&=\frac{3}{3+1}=\frac{3}{4}\end{aligned}$$

21. $x^2-5x-6=(x-6)(x+1)$

22. $6x^2-5x-6=(3x+2)(2x-3)$

23. $$\begin{aligned}8x^3-28x^2+12x&=4x(2x^2-7x+3)\\&=4x(2x-1)(x-3)\end{aligned}$$

24. $$\begin{aligned}25x^2-16&=(5x)^2-(4)^2\\&=(5x+4)(5x-4)\end{aligned}$$

25. $$\begin{aligned}2a(4-x)-6(x-4)&=2a(4-x)+6(4-x)\\&=(4-x)(2a+6)\\&=2(a+3)(4-x)\end{aligned}$$

26. $$\begin{aligned}75y-12x^2y&=3y(25-4x^2)\\&=3y[(5)^2-(2x)^2]\\&=3y(5+2x)(5-2x)\end{aligned}$$

27.
$$
\begin{aligned}
2x^2 &= 7x - 3 \\
2x^2 - 7x + 3 &= 0 \\
(2x-1)(x-3) &= 0 \\
2x - 1 = 0 \quad & \quad x - 3 = 0 \\
2x = 1 \quad & \quad x = 3 \\
x = \frac{1}{2} \quad &
\end{aligned}
$$

The solutions are $\frac{1}{2}$ and 3.

28.
$$
\begin{aligned}
&\frac{2x^2 - 3x + 1}{4x^2 - 2x} \cdot \frac{4x^2 + 4x}{x^2 - 2x + 1} \\
&= \frac{(\cancel{2x - 1})(\cancel{x - 1})}{\cancel{2x}(\cancel{2x - 1})} \cdot \frac{\overset{2}{\cancel{4x}}(x+1)}{(\cancel{x - 1})(x-1)} \\
&= \frac{2(x+1)}{x-1}
\end{aligned}
$$

29.
$$\frac{5}{x+3} - \frac{3x}{2x-5}$$

$$
\begin{aligned}
&\text{LCM} = (x+3)(2x-5) \\
&\frac{5(2x-5)}{(x+3)(2x-5)} - \frac{3x(x+3)}{(x+3)(2x-5)} \\
&= \frac{10x - 25 - 3x^2 - 9x}{(x+3)(2x-5)} \\
&= \frac{-3x^2 + x - 25}{(x+3)(2x-5)}
\end{aligned}
$$

30.
$$
\begin{aligned}
x - \frac{1}{1 - \frac{1}{x}} &= x - \frac{1}{\frac{x-1}{x}} \\
&= x - \frac{x}{x-1} = \frac{x(x-1) - x}{x-1} \\
&= \frac{x^2 - x - x}{x-1} = \frac{x^2 - 2x}{x-1}
\end{aligned}
$$

31.
$$
\begin{aligned}
\frac{5x}{3x-5} - 3 &= \frac{7}{3x-5} \\
(3x-5)\left(\frac{5x}{3x-5} - 3\right) &= (3x-5)\left(\frac{7}{3x-5}\right) \\
5x - 3(3x-5) &= 7 \\
5x - 9x + 15 &= 7 \\
-4x &= -8 \\
x &= 2
\end{aligned}
$$

32.
$$
\begin{aligned}
a &= 3a - 2b \\
a - 3a &= -2b \\
-2a &= -2b \\
a &= b
\end{aligned}
$$

33.
$$
\begin{aligned}
m &= \frac{y_2 - y_1}{x_2 - x_1} \\
&= \frac{-1 - (-3)}{2 - (-1)} = \frac{-1 + 3}{2 + 1} = \frac{2}{3}
\end{aligned}
$$

34.
$$
\begin{aligned}
m &= -\frac{2}{3}, (x_1, y_1) = (3, -4) \\
y + 4 &= -\frac{2}{3}(x - 3) \\
y + 4 &= -\frac{2}{3}x + 2 \\
y &= -\frac{2}{3}x - 2
\end{aligned}
$$

35. $y = 4x - 7$
$y = 2x + 5$
Substitute in Equation (2).
$$
\begin{aligned}
4x - 7 &= 2x + 5 \\
2x &= 12 \\
x &= 6
\end{aligned}
$$
Substitute in Equation (1).
$$
\begin{aligned}
y &= 4x - 7 \\
y &= 24 - 7 \\
y &= 17
\end{aligned}
$$
The solution is (6, 17).

36. (1) $4x - 3y = 11$
(2) $2x + 5y = -1$
Eliminate x and add the equations.
$$
\begin{aligned}
4x - 3y &= 11 & \qquad 4x - 3y &= 11 \\
-2(2x + 5y) &= -1(-2) & \qquad -4x - 10y &= 2 \\
& & \qquad -13y &= 13 \\
& & \qquad y &= -1
\end{aligned}
$$

Replace the y in Equation (2).
$$
\begin{aligned}
2x + 5y &= -1 \\
2x - 5 &= -1 \\
2x &= 4 \\
x &= 2
\end{aligned}
$$
The solution is (2, –1).

37.
$$
\begin{aligned}
4 - x &\geq 7 \\
-x &\geq 3 \\
x &\leq -3
\end{aligned}
$$

38.
$$
\begin{aligned}
2 - 2(y-1) &\leq 2y - 6 \\
2 - 2y + 2 &\leq 2y - 6 \\
-2y + 4 &\leq 2y - 6 \\
-4y &\leq -10 \\
y &\geq \frac{10}{4} \\
y &\geq \frac{5}{2}
\end{aligned}
$$

39. $\sqrt{49x^6} = \sqrt{7^2x^6} = 7x^3$

40.
$$
\begin{aligned}
&2\sqrt{27a} + 8\sqrt{48a} \\
&= 2\sqrt{9}\sqrt{3a} + 8\sqrt{16}\sqrt{3a} \\
&= 2 \cdot 3\sqrt{3a} + 8 \cdot 4\sqrt{3a} \\
&= 6\sqrt{3a} + 32\sqrt{3a} \\
&= 38\sqrt{3a}
\end{aligned}
$$

41. $\dfrac{\sqrt{3}}{\sqrt{5}-2}=\dfrac{\sqrt{3}}{\sqrt{5}-2}\cdot\dfrac{\sqrt{5}+2}{\sqrt{5}+2}$

$=\dfrac{\sqrt{3}(\sqrt{5}+2)}{5-4}=\dfrac{\sqrt{15}+2\sqrt{3}}{1}$

$=\sqrt{15}+2\sqrt{3}$

42. $\sqrt{2x-3}+4=5$

$\sqrt{2x-3}=1$

$(\sqrt{2x-3})^2=(1)^2$

$2x-3=1$

$2x=4$

$x=2$

Check: $\sqrt{2(2)-3}+4=5$

$\sqrt{4-3}+4=5$

$\sqrt{1}+4=5$

$1+4=5$

$5=5$

The solution is 2.

43. $3x^2-x=4$

$3x^2-x-4=0$

$(3x-4)(x+1)=0$

$3x-4=0 \qquad x+1=0$

$3x=4 \qquad x=-1$

$x=\dfrac{4}{3}$

The solutions are -1 and $\dfrac{4}{3}$.

44. $4x^2-2x-1=0$

$x=\dfrac{2\pm\sqrt{4+16}}{8}$

$x=\dfrac{2\pm\sqrt{20}}{8}=\dfrac{2\pm2\sqrt{5}}{8}$

$=\dfrac{\cancel{2}(1\pm\sqrt{5})}{\underset{4}{\cancel{8}}}$

$x=\dfrac{1+\sqrt{5}}{4} \qquad x=\dfrac{1-\sqrt{5}}{4}$

The solutions are $\dfrac{1-\sqrt{5}}{4}$ and $\dfrac{1+\sqrt{5}}{4}$.

45. **Strategy** • Unknown number: x

Solution $2x+3(x-2)=2x+3x-6$
$=5x-6$

46. $P\times B=A$

$80\%\times x=2400$

$0.80x=2400$

$x=3000$

The original value is $3000.

47. $S=C+rC$

$1485=900+900r$

$585=900r$

$0.65=r$

The markup rate is 65%.

48. **Strategy** • Amount of money to be invested at 11%: x

	Principal	Rate	Interest
Amount at 8%	3000	0.08	0.08(3000)
Additional amount	x	0.11	$0.11(x)$
Total	$x+3000$	0.10	$0.10(x+3000)$

• The total interest earned is 10% of the total investment.

Solution $0.08(3000)+0.11x=0.10(x+3000)$

$240+0.11x=0.10x+300$

$0.01x=60$

$x=6000$

$6000 must be invested at 11%.

49. **Strategy** • Cost per pound for the mixture: x

	Amount	Cost	Value
Peanuts	4	2	2(4)
Walnuts	2	5	5(2)
Mixture	6	x	$6x$

• The sum of the values before mixing equals the value after mixing.

Solution $2(4)+5(2)=6x$

$8+10=6x$

$18=6x$

$3=x$

The cost for the mixture is $3 per pound.

50. **Strategy** • Percent concentration of acid in the mixture: x

	Amount	Percent	Quantity
60% acid	20	0.60	0.60(20)
20% acid	30	0.20	0.20(30)
Mixture	50	x	$50x$

• The sum of the quantities before mixing is equal to the quantity after mixing.

Solution $0.60(20)+0.20(30)=50x$

$12+6=50x$

$18=50x$

$0.36=x$

The percent concentration of acid in the mixture is 36%.

51. Strategy
- Rate for first part of trip: x
  Rate for second part of trip: $2x$

	Rate	Time	Distance
First part	x	1	$1x$
Second part	$2x$	1.5	$1.5(2x)$

- The total distance is 860 km.

Solution
$$1x + 1.5x(2x) = 860$$
$$1x + 3x = 860$$
$$4x = 860$$
$$x = 215$$
The distance for the first part of the trip is 215 km.

52. Strategy
- First angle: x
  Second angle: $x + 10$
  Third angle: $x + 20$
- The sum of the angles of a triangle is 180°.

Solution
$$x + x + 10 + x + 20 = 180$$
$$3x + 30 = 180$$
$$3x = 150$$
$$x = 50$$
$$x + 10 = 60$$
$$x + 20 = 70$$
The angles are 50°, 60°, and 70°.

53. Strategy
- First consecutive integer: x
  Second consecutive integer: $x + 1$
  Third consecutive integer: $x + 2$
- The sum of the squares of the three consecutive integers is 50.

Solution
$$x^2 + (x+1)^2 + (x+2)^2 = 50$$
$$x^2 + x^2 + 2x + 1 + x^2 + 4x + 4 = 50$$
$$3x^2 + 6x - 45 = 0$$
$$3(x^2 + 2x - 15) = 0$$
$$3(x+5)(x-3) = 0$$
$$x + 5 = 0 \qquad x - 3 = 0$$
$$x = -5 \qquad x = 3$$
$$x + 1 = -4 \qquad x + 1 = 4$$
$$x + 2 = -3 \qquad x + 2 = 6$$
The middle integer can be −4 or 4.

54. Strategy
- Width: x
  Length: $x + 5$
- Use the equation for the area of a rectangle.

Solution
$$A = LW$$
$$50 = x(x+5)$$
$$50 = x^2 + 5x$$
$$0 = x^2 + 5x - 50$$
$$0 = (x+10)(x-5)$$
$$x + 10 = 0 \qquad x - 5 = 0$$
$$x = -10 \qquad x = 5$$
The solution −10 is not possible.
The width is 5 m.
The length is 10 m.

55. Strategy
- Write and solve a proportion.

Solution
$$\frac{2}{15} = \frac{x}{120}$$
$$2(120) = 15 \cdot x$$
$$240 = 15x$$
$$x = 16$$
16 oz of dye are required.

56. Strategy
- This is a work problem.
- Time for chef and apprentice working together: t

	Rate	Time	Part
Chef	$\frac{1}{60}$	t	$\frac{t}{60}$
Apprentice	$\frac{1}{90}$	t	$\frac{t}{90}$

- The sum of the parts of the task completed must equal 1.

Solution
$$\frac{t}{60} + \frac{t}{90} = 1$$
$$60 \cdot 90\left(\frac{t}{60} + \frac{t}{90}\right) = 1 \cdot 60 \cdot 90$$
$$90t + 60t = 5400$$
$$150t = 5400$$
$$t = 36$$
Working together, it would take them 36 min or 0.6 h.

57. **Strategy**

- Rate of the boat in calm water: x
  Rate of the current: c

	Rate	Time	Distance
With current	$x + c$	2.5	$2.5(x + c)$
Against current	$x - c$	5	$5(x - c)$

- The distance traveled with the current is equal to the distance traveled against the current.

Solution

$2.5(x + c) = 50 \quad \frac{2}{5} \cdot \frac{5}{2}(x + c) = 50 \cdot \frac{2}{5}$

$5(x - c) = 50 \quad \frac{1}{5} \cdot 5(x - c) = 50 \cdot \frac{1}{5}$

$$x + c = 20$$
$$x - c = 10$$
$$2x = 30$$
$$x = 15$$
$$x + c = 20$$
$$15 + c = 20$$
$$c = 5$$

The rate of the boat in calm water is 15 mph. The rate of the current is 5 mph.

58. **Strategy**

- Rate of the wind: r

	Distance	Rate	Time
With wind	500	$225 + r$	$\frac{500}{225+r}$
Against wind	500	$225 - r$	$\frac{500}{225-r}$

- The time spent traveling against the wind is $\frac{1}{2}$ h more than the time spent traveling with the wind.

Solution

$$\frac{500}{225 + r} + \frac{1}{2} = \frac{500}{225 - r}$$

$$2(225 + r)(225 - r)\left(\frac{500}{225 + r} + \frac{1}{2}\right) = 2(225 + r)(225 - r)\left(\frac{500}{225 - r}\right)$$

$$2(225 - r)(500) + (225 - r)(225 + r) = 2(225 + r)(500)$$

$$225{,}000 - 1000r + 50{,}625 - r^2 = 225{,}000 + 1000r$$

$$r^2 + 2000r - 50{,}625 = 0$$

$$(r + 2025)(r - 25) = 0$$

$r + 2025 = 0 \qquad r - 25 = 0$

$r = -2025 \qquad r = 25$

The solution -2025 is not possible.
The rate of the wind is 25 mph.

59.

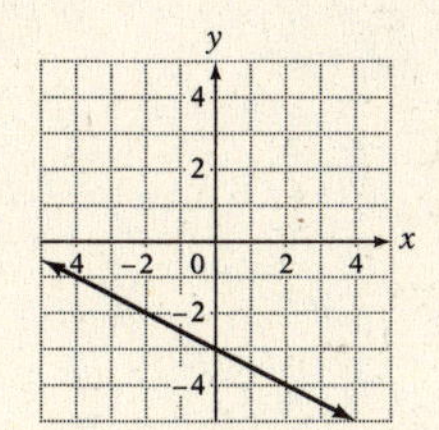

60.

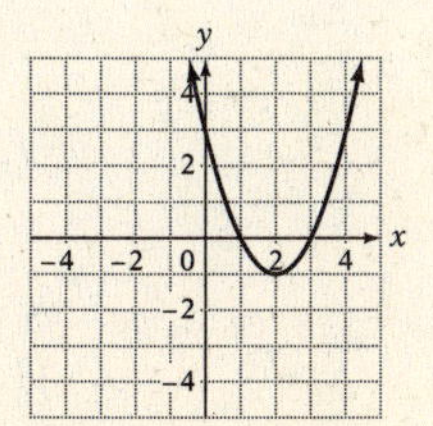